Time for Science Education

How Teaching the History and Philosophy of Pendulum Motion can Contribute to Science Literacy

INNOVATIONS IN SCIENCE EDUCATION AND TECHNOLOGY

Series Editor:

Karen C. Cohen, Harvard University, Cambridge, Massachusetts

A Continuation Order Plan is available for this series. A continuation order will bring delivery of each new volume immediately upon publication. Volumes are billed only upon actual shipment. For further information please contact the publisher.

Time for Science Education

How Teaching the History and Philosophy of Pendulum Motion can Contribute to Science Literacy

Michael R. Matthews

University of New South Wales
Sydney, Australia

Kluwer Academic / Plenum Publishers

New York, Boston, Dordrecht, London, Moscow

Library of Congress Cataloging-in-Publication Data

Matthews, Michael R.
　　Time for science education: how teaching the history and philosophy of pendulum motion can contribute to science literacy/Michael R. Matthews.
　　　　p.　cm. — (Innovations in science education and technology)
　　Includes bibliographical references and index.
　　ISBN 0-306-45880-2
　　　1. Pendulum.　2. Time measurements.　3. Science—Study and teaching.　I. Title.　II. Series.

QA862.P4 M37 2000
507.1—dc21

00-028578

ISBN: 0-306-45880-2

©2000 Kluwer Academic/Plenum Publishers, New York
233 Spring Street, New York, New York 10013

http://www.wkap.nl

10　9　8　7　6　5　4　3　2　1

A C.I.P. record for this book is available from the Library of Congress

for

Amelia Kathleen

Preface to the Series

The mandate to expand and improve science education is an educational imperative and an enormous challenge. Implementing change, however, is complicated given that science as well as science education is dynamic, continually incorporating new ideas, practices, and procedures. Science and science education take place in varying contexts and must deal with amazingly rapid technological advances. Lacking clear paths for improvement, we can and should learn from the results of all types of science education, traditional as well as experimental. Successful reform of science education requires careful orchestration of a number of factors which take into account technological developments, cognitive development, societal impacts and relationships, organizational issues, impacts of standards and assessment, teacher preparation and enhancement, as well as advances in the scientific disciplines themselves. Understanding and dealing with such a complex mission is the focus of this book series. Each book in this series deals in depth with one or more of potential factors for understanding, creating and sustaining effective science education reform.

In 1992, a multidisciplinary forum was launched for sharing the perspectives and research findings of the widest possible community of people involved in addressing the challenge of science education reform. Those who had something to share regarding impacts on science education were invited to contribute. This forum was the *Journal of Science Education and Technology*. Since the inception of the journal, many articles have highlighted relevant themes and topics and expanded the context of understanding to include historical, current, and future perspectives in an increasingly global context. Recurring topics and themes have emerged as foci requiring expanded treatment and presentation. This book series, "Innovations in Science Education and Technology" is the result.

It is a privilege to be able to continue to elucidate and effect improvement and reform in science education by providing this in-depth forum for the work of others. The series brings focus and understanding to efforts worldwide, helping readers to understand, to incorporate, and to utilize what we know, what we are learning, and what we are inventing technologically to advance the mission of science education reform worldwide.

Karen C. Cohen
Cambridge, Massachusetts

Acknowledgments

This book has been a long time in gestation, its birth being delayed by a number of other writing, editorial, administrative, and teaching projects and commitments. The seed was planted when I was introduced to Galileo's texts by Abner Shimony during a happy and productive sabbatical leave at the Boston University's Centre for History and Philosophy of Science in 1978. As with many science and philosophy graduates, I had read a lot about Galileo but precious little by him. Professor Shimony's semester course was based entirely on Galileo's *Dialogues*, and it introduced me to the rich field of Galilean scholarship that has remained a continuing interest. Through Robert Cohen, Marx Wartofsky, Michael Martin, and others I was introduced to the "Boston University Style" of history and philosophy of science. The "Style," exemplified in the *Boston Studies in Philosophy of Science* series, was characterised by its diversity, its attention to the history and cultural dimensions of science, as well as more standard philosophical analysis. If this book were seen as a minor, but worthy, exemplar of that Boston style, I would feel that my labours were justified.

Some years later I worked on Galileo's pendulum discoveries during another productive sabbatical leave in the Philosophy Department of Florida State University in the first half of 1987. David Gruender was a hospitable host, and a careful commentator on a paper dealing with the pendulum that I presented to a departmental seminar. That year I presented a paper on Galileo's pendulum discoveries at a meeting of the United States Philosophy of Education Society in Boston (Matthews 1987a), and read a revised version of the paper at the Second International Conference on Misconceptions in Science and Mathematics (Matthews, 1987b). Epistemological considerations were, at the time, my chief interest in the pendulum story, and on these matters I benefitted from discussion, as many others have also done, with the late Wallis Suchting, a friend, teacher, and illustrious member of the University of Sydney Philosophy Department.

During that 1987 leave, by serendipitous circumstance, Jaakko Hintikka, then at Florida State University, invited me to guest edit a special issue of the journal *Synthese* to be devoted to "History, Philosophy and Science Teaching" (*Synthese*,

x   Time for Science Education

1989, vol. 80, no. 1). This invitation lead, two years later, to the formation of the International History, Philosophy and Science Teaching Group, a group whose members, conferences and publications have been a continuing source of inspiration for my own work.[1] I began, then, to think of the pendulum as a case study for the contribution of the history and philosophy of science to science teaching, and presented a paper on that subject to the 1989 meeting of the Australian Science Education Research Association (Matthews, 1989a). The paper's concern was, again, largely philosophical. It dealt with Galileo's argument with del Monte over the properties of pendula, and how this argument might be used in the classroom to increase students' understanding of the methodological changes enacted in the scientific revolution.

Further reading and research widened my understanding of the pendulum, and its place in the foundation of modern science. This research was advanced during a sabbatical leave in early 1991, and papers on the subject were given at a number of conferences and universities in the United States and Europe. The work was written up during my period as Foundation Professor of Science Education at the University of Auckland and formed a central chapter in my book, *Science Teaching: The Role of History and Philosophy of Science* (Matthews 1994a, chap. 6). The chapter was certainly richer than my earlier published pieces, with epistemological matters spelled out, and some pedagogical considerations introduced. Nevertheless the chapter can be read without learning anything about the cultural or horological impact of the pendulum, matters that are addressed in this book.

On return to the University of New South Wales in 1994 I was awarded an Australian Research Council Small Grant to assist my work on the the role of the pendulum in Galileo's physics. Over the next couple of years I was able to make use of the enormous holdings of the University of New South Wales library. This book contains approximately twelve hundred references, nearly all of which are in the University of New South Wales library which has enormous holdings in the fields covered by this work—science education, history and philosophy of science and horology. No matter how obscure the book, or how minor the journal, it is in the library. Each trip to the library with a "wish list" of references was an occasion of joy and relief, because they were mostly all there. Scholars who have worked in humanities research will appreciate the importance of such a library for maintaining continuity of work and minimising frustration. I am deeply indebted to former university staff who conscientiously ordered books and journals, and to the librarians, who in difficult times, have so wonderfully maintained the Univer-

---

1. The Group's first conference was held at Florida State University in 1989 (Herget, 1989, 1990), the second at Queen's University Ontario in 1992 (Hills, 1992), the third at the University of Minnesota in 1995 (Finley et al., 1995), the fourth at the University of Calgary (1997) and the fifth at the University of Pavia (1999). The group is associated with the journal *Science & Education* which commenced publication in 1992. Membership inquiries can be made to the author.

sity of New South Wales history and philosophy of science, science education, and horology collections.

During 1995 and 1996, I focused more on the role of the pendulum in 17th- and 18th-century culture and horology. This included the pendulum's role in solving one of the period's most pressing problems: the problem of longitude determination. I also began to address the implications for science education of the historical and philosophical research I had been conducting. I presented papers on these themes at conferences of the European Physical Society's Interdivisional Group on the History of Physics and Physics Teaching (Bratislava, 1996), and the United States National Association for Research in Science Teaching (Chicago, 1997). I benefited from discussion and comments on those occasions.

During another sabbatical leave in the first half of 1997, I began to write the material up in book form. Papers on aspects of this were given at the University of Delaware, University of Montana, Texas A&M University, University of Texas El Paso, Federal University of Florianopolis, the Federal Rural University at Recife, and the Fourth Conference of the International History, Philosophy and Science Teaching Group held in Calgary. I am indebted to, respectively, Nancy Brickhouse, Robert Carson, Cathleen Loving, Douglas Allchin, Arden Zylbersztaijn, Alexandre Medeiros, and Ian Winchester, for these opportunities. The work benefited from innumerable comments and discussions.

It was my misfortune not to have known about the Longitude Conference held at Harvard University in November 1993. Along with countless thousands of others, I first became aware of the conference in 1995 when I read Dava Sobel's masterful *Longtude* (Sobel, 1995)—a book she acknowledged as being based upon papers presented at the conference. Unfortunately I did not read the Conference Proceedings (Andrewes, 1998a) until January 2000, some year or more after this book had been completed and sent to New York for copyediting. Thankfully Robert Maged, Kluwer's Production Editor, listened to my plaintive call to "Hold the Press!" while minimal account of the rich contributions to the 1993 conferences could be incorporated into the text. The Conference Proceedings, and the references therein cited, are an invaluable lode for those interested in the horological and cartographical dimensions of the pendulum story.

Having a manuscript and having a publisher are, as is well known, two different things. I am deeply thankful to Karen Cohen, Kluwer Academic/Plenum Pubishers' editor of *Innovations in Science Education and Technology* series, for her offer of a contract that generously allowed a fairly wide-ranging, and cross-disciplinary examination of pendulum matters.

My wife, Julie House, painstakingly read and copyedited the first draft of the book manuscript. Her concern for clear and straightforward writing resulted in most manuscript pages being covered in red ink. Readers will have benefit considerably from her efforts. The penultimate draft was read by Ron Good at Louisiana State University whose encouragement and suggestions were valuable. Alan

Chalmers at the University of Sydney read two chapters and made some important corrections and suggestions. Colin Gauld, formerly at the University of New South Wales, also read the penultimate draft. His painstaking attention to bibliographical and other detail, and corrections of some of my gross errors concerning 17th century physics, considerably improved the manuscript. Harshi Gunawardena at University of New South Wales read Chapter 3 and provided helpful comments and advice on literature, especially concerning Chinese horology and early mechanical timekeepers. During a most productive conference of the International History, Philosophy and Science Teaching Group which was organised at Lake Como by Fabio Bevilacqua and Enrico Giannetto in late 1999, I had the good fortune to meet John Heilbron who read and made valuable comments on the pages in Chapter 6 dealing with Huygens' proposal for an international unit of length. At the same conference, I meet Jürgen Renn who read Chapter 5, which deals with del Monte's important exchange with Galileo. His comments and corrections were also invaluable. Finally, my friend and University of New South Wales colleague, Peter Slezak, laboured over the draft, making innumerable detailed and thoughtful corrections, comments, and suggestions on every page. I am deeply indebted to these readers who have been patient, informed, careful, and critical.

    The book contains illustrations and diagrams that have come from various sources. These are acknowledged in the Credits section. I am grateful to the relevant copyright holders for permission to reproduce this material.

    Finally, Amelia McNamara, Sheryl Levart, Eve Tsirigotakis, and Robert Wheeler at Kluwer's New York editorial office, and Robert Maged in the Production Department, have been prompt and professional in all matters as the manuscript has travelled its rocky road from submission to production.

Contents

Introduction

The pendulum has played a significant role in the development of Western science, culture, and society. Its initial Galileo-inspired, utilization in clockwork provided the world's first accurate measure of time. The accuracy of mechanical clocks went, in the space of a couple of decades, from plus or minus 30 minutes a day to a few seconds per day. This quantum increase in accuracy of timing enabled hitherto unimagined degrees of precision measurement in mechanics and astronomy. It ushered in the world of precision so characteristic of the scientific revolution. Time could, for the first time, be expressed as an independent variable in the investigation of nature. The pendulum was studied by Galileo, Huygens, Newton, Hooke, and all the leading figures of 17th-century science. The pendulum was crucial for, among other things, establishing the collision laws, the conservation laws, the value of the acceleration due to gravity g, and ascertaining the variation in g from equatorial to polar regions and hence discovering the oblate shape of the earth. Perhaps most important, it provided the crucial evidence for Newton's synthesis of terrestrial and celestial mechanics. Also tied up with the scientific revolution of the mid-17th century was a horological revolution, and the pendulum played a central role in this as well.

The pendulum played more than a scientific and technical role in the formation of the modern world. Accurate time measurement was long seen as the solution to the problem of longitude determination which had vexed European maritime nations in their efforts to sail beyond Europe's shores. If an accurate and reliable clock were carried on voyages from London, Lisbon, Genoa, or any other port, then by comparing its time with local noon (as determined by the sun's shadow), the longitude of any place in the journey could be ascertained. As latitude could already be determined, this enabled the world to be mapped. In turn, this provided a firm base on which European exploitation, colonization, and commerce could proceed.

The clock also transformed social life and customs: Labor could be regulated by clockwork and, because time duration could be accurately measured, there could be debate and struggle about the length of the working day. Timetables for transport could be enacted. The starting time for religious and cultural events could be specified. Punctuality could become a virtue, and so on. The clock also did duty in philosophy. It was a metaphor for the new mechanical worldview that was challenging the entrenched Aristotelian, organismic and teleological view of the world that had sustained so much of European intellectual and religious life. In theology, the clock was appealed to in the very influential argument from design for God's existence—if the world functions regularly like a clock, then there must be, as Newton insisted, a cosmic clockmaker.

This book outlines something of the pendulum story and its scientific, philosophical, and cultural ramifications. It also tries to indicate how understanding that story can assist teachers to improve science education by suggesting pendulum-related curricular content, experimental and project work, and points of connection with other parts of science and other school subjects. Although the pendulum is a minor topic in most curricula, it is argued that a richer approach to its treatment can result in enhanced science literacy and enhanced appreciation of the part played by science in the development of society and culture.

I have argued elsewhere that some knowledge of the history and philosophy of science is essential for proper understanding and resolution of many of the theoretical debates that occur in contemporary science education: debates about constructivism, multiculturalism, feminism, religion, and so on (Matthews, 1994a). I have also argued that the history and philosophy of science can contribute to more practical, pedagogical matters, such as curriculum development and classroom teaching. For the most part this book is a contribution to this second area, although here and there the arguments bear upon some theoretical disputes in contemporary science education.

The book argues first, that education is best defined in traditional liberal terms as the initiation of students into a range of intellectual disciplines in such a way that, to some appropriate degree, they understand the discipline and its relationship to other disciplines and culture; second, that science education needs to be understood as contributing to the overall education of students, and that considerations about aims and purposes of education constrain decisions about science education; third, current broad interpretations of science literacy that stress the importance of students understanding and appreciating the nature of science, including its history and interrelations with culture, mesh very well with liberal educational goals; and fourth, teaching the history and philosophy of pendulum motion is an ideal vehicle for realising some of these more ambitious aspirations for scientific literacy.

At least five strands run through the book: history of science, philosophy of science, horology, science education, and philosophy of education. When writing a book that straddles different disciplines the dangers of trivilization and misrepresentation are well known. Nevertheless, I hope that enough justice has been rendered to each strand that the book's central argument can be sustained.

The book's argument depends, as do most proposals in education, upon certain positions in the philosophy of education. I believe that education should be primarily concerned with developing understanding, with initiation into worthwhile traditions of intellectual achievement, and with developing capacities for clear, analytic and critical thought. These have been the long-accepted goals of liberal education. In a liberal education, students should come to know and appreciate a variety of disciplines, know them at an appropriate depth, see the interconnectedness of the disciplines, or the modes of thought, and finally have some critical disposition toward what is being learned, to be genuinely open-minded about intellectual things. These liberal goals are contrasted with goals such as professional training, job preparation, promotion of self-esteem, social engineering, entertainment, or countless other putative purposes of schooling that are enunciated by politicians, administrators, and educators.

The book's argument might be consistent with other views of education—especially ones about the training of specialists (sometimes called a professional view of education)—but the argument fits best with a liberal view of education. The liberal hope has always been that if education is done well, then other personal and social goods will follow. The development of informed, critical, and moral capacities is the cornerstone for personal and social achievements. Conversely, educational concentration on the latter achievements—avoiding AIDS, saving the whale, minimizing greenhouse gas emission, eradicating racism and sexism—can easily result in indoctrination being substituted for education, and a failure to develop the mental and critical capacities that are are needed to face new challenges intelligently. Although philosophy of education is academically unfashionable, it is clear that most curricular, lesson planning, class management, and assessment decisions depend on views about the purposes of education. If we do not know what we want to achieve in schools, then knowing how to proceed becomes truly problematic, and educational decisions are more likely to be made on faddish, expedient, or merely political grounds. If teachers and administrators have no clear philosophy of education, then it is more likely that education will merely take the shape of the last political foot that trod upon it.

In its science education strand, the book embraces the frequently made, but often ignored, claims for widening the definition of science literacy to incorporate some appreciation of the history of science, the nature of science, and the interrelationships of science with culture and society. In the United States, the *National Standards for Science Education*, and the AAAS' *Project 2061* both endorse this wider, liberal idea of scientific literacy. So also does the national curriculum in the United Kingdom, a number of provincial science curricula in Canada, the Norwegian science curriculum, the Danish science curriculum, and the New South Wales state syllabus in Australia. Most science programs aspire to having students know more than just a certain amount of science content, and having a certain level of competence in scientific method and scientific thinking.

Most programs want students to have some sense of the "big picture" of science: Its history, philosophy, and relationship to social ideologies, institutions, and practices. In most countries, science education has dual goals: promoting learning *of* science, and also learning *about* science. Or, as it has been stated, science education has both *disciplinary* and *cultural* goals (Gauld, 1977). The argument of this book will be that the history and philosophy of science contribute to *both* these goals of science education—it helps students learn science and learn about science. And knowledge of the history and philosophy of science enables teachers to better understand their increasingly important cultural role in the continuation of the scientific tradition: Without teachers there would be no science.

The book argues that more than lip-service should be paid to these wider, or liberal, definitions of science literacy. And if serious attention is to be given to realizing these wider definitions of literacy, then the historical, philosophical, and cultural dimensions of curricular topics must be addressed. Or at least addressed for those topics where these dimensions of the topic clearly contribute to an understanding of the "big picture" of science. This is certainly the case for the pendulum.

In its science education strand, the book takes up some theoretical and practical issues raised by constructivism. Although constructivism as a philosopical basis for science education, and as a theory of pedagogy, has been very influential over the past few decades, it is suggested that constructivism has only a limited amount to offer when teaching about the pendulum, and has deleterious implications if taken too literally.

Theories in science and social science all designate certain elements of a domain as the basic explanatory units—natural motions in Aristotelianism, impacts in Cartesian mechanics, forces in Newtonian mechanics, selection and adaptation in Darwinian biology, changes in productive forces for Marxist political economy, reinforcement schedules for Skinnerian learning theory, and so on. At the center of constructivism's theoretical stage is a knowing subject (student) who experiences phenomena and tries to make sense of them either individually or in cooperation with a group of fellow learners, perhaps with some guidance from a teacher-facilitator. These elements constitute the explanatory core for constructivist accounts of education and learning. Removed from center stage is the notion of education as the transmission of a body of concepts, intellectual techniques, and practical skills all of which predate the individual's appearance, and whose worth is independent of what sense the student may or may not make of them. The importance of telling the historical, philosophical, or cultural story of the pendulum is simply not suggested by the central core of constructivist theory. Consequently there is no constructivist reason why teachers should learn anything about these aspects of the pendulum, or base class discussion, projects, and laboratory work on the historical, philosophical, and cultural stories. This analysis is supported by the fact that among the thousands of constructivist-inspired research papers published in the past three decades, there are no more than two or three that deal with the pendulum, and these leave aside all of the pendulum's histori-cal, philosophical and cultural dimensions (Pfundt and Duit, 1994). For constructivism, these dimensions basically do not count; they are invisible.

That constructivism leaves out, or does not encourage, the teaching of cer-tain interesting aspects of pendulum motion, is one problem, but there is a larger, more theoretically interesting, problem for constructivism. Students are not go-ing to learn the fundamentals of Newtonianism by looking at or merely playing around with things. No matter how challenging the experience and how good the facilitation of it, the law of isochronic motion does not emerge from student's experience or laboratory work. On the contrary, all pendulums when set swinging come to a halt, some sooner, some later, but all stop swinging. This fact is incon-sistent with isochrony. As one philosopher has provocatively stated the matter, "The laws of physics lie" (Cartwright, 1983). Isochronic motion means that every swing takes the same time. Thus if the first swing takes some time, all the others take the same time. This means that an isochronic pendulum should never come to rest. Isochrony is consistent with the amplitude diminishing, but not with the pendulum ceasing to swing.

This is simply another case where scientific laws outrun experience. Inas-much as the theoretical center of constructivism is occupied by a thinking subject confronting the world and trying to construct knowledge of it, and inasmuch as constructivism maintains that knowledge and meaning cannot be transmitted, then the student must invariably get intellectually stuck where Galileo's opponents got stuck. Natural pendula are not strictly isochronic, even laboratory pendula fall short of this ideal. Experience must somehow be overcome, or transcended, to "see" or make sense of, the pendulum laws. If transmission of meaning and knowl-edge is allowed, this is still a problem, as all teachers can attest, but not an insur-mountable one. If transmission is not allowed, and if students can only construct their own knowledge from their own experience, then the laws of pendulum mo-tion simply cannot be constructed. In brief, the book is critical of the Robinson Crusoe epistemological model so commonly endorsed by constructivists, and criti-cal even when that model is expanded to Robinson Crusoe and his friends. I claim that constructivism is basically a form of empiricism, and that the pendulum story is inconsistent with empiricist epistemology. Further, constructivism has in many areas functioned more as an ideology than a learning theory, and although fitting comfortably with Western individualism and relativism, it has had unhappy edu-cational and cultural ramifications (Matthews, 1995a, 2000).

In its horology strand, the book makes no claim for originality. There are scores of excellent books and hundreds of research articles available on the tech-nical, social, and comparative history of timekeeping. Many of these are men-tioned in the text. Where this horological research bears upon the analysis and utilization of pendulum motion, I have tried to convey a little of its results. The book shows the importance that the longitude problem presented to European maritime powers in the 16th, 17th, and 18th centuries, and how the development of clockwork was associated with solving this problem. All the major figures of 17th century science—Galileo, Huygens, Hooke, Newton, and, in a small way, Leibniz—contributed to this development. Unfortunately, their involvement in this practical and technical endeavor is usually ignored in both science and philosophy programs.

If science texts routinely ignore the cultural and horological context of Galileo's pendulum discoveries, then it is equally true that cultural histories frequently ignore the crucial philosophical, or methodological, context of his discoveries. Perhaps the one novel contribution of this book to horology, is to point out how Galileo's pendulum discoveries were tied to, and dependent upon, his new epistemology, or theory of scientific method. Galileo's law of isochronic motion, and hence his suggestions for using the pendulum in timekeeping, could not be accepted until he threw off the strait-jacket placed on science by the epistemological primacy given to experience and the evidence of the senses by Aristotelian science. As long as scientific claims were judged by what could be seen, and as long as mathematics and physics were kept separate, then Galileo's pendular claims could not be substantiated. Their justification required not just a new science, but a new way of doing science, a new methodology of science. Galileo provided this. It was the beginning of the Galilean–Newtonian Paradigm (GNP) which quickly came to characterize the scientific revolution, and the subsequent centuries of modern science.

In its philosophy of science strand, the book adopts the view that philosophy of science should genuinely be informed by history of science. This is one of the important contributions of the Kuhnian legacy. The idea of proposing a methodology of science independently of examining how science has been conducted is decidedly odd. There are, admittedly, problems with this Kuhnian position. One is that history can be mined merely to find support for antecedently arrived at epistemological positions. History is "reconstructed" to suit whatever philosophical position is being advocated (Lakatos 1971). Albert Schweitzer, in his monumental 1910 work on *The Quest of the Historical Jesus* that traced the history of Christian interpretation of Jesus, remarked that "each successive epoch of theology found its own thoughts in Jesus . . . But it was not only each epoch that found its reflection in Jesus; each individual created Him in accordance with his own character" (Schweitzer, 1910, p. 4). Schweitzer could equally have been talking of Galileo. It is notorious that Galileo has been made out to be a shining example of the full range of epistemological positions: from rationalist through empiricist and experimentalist, to positivist, and to methodological anarchist. The common thread is that the epistemology attributed to Galileo is usually the one favored by the biographer or interpreter.

There is a chicken-and-egg problem with the Kuhnian stance. If philosophy of science emerges from history of science, how is the history first demarcated?

Independent of a philosophical, normative, position what will count as the subject matter of history from which our methodological lesson is to be drawn? Do we draw lessons equally from Christian Science, National Socialist Science, Lysenkoism, Astrological Science, Islamic Science, Hindu Science, New Age Science, as well as classical mechanics, thermodynamics, and quantum mechanics?

The two standard ways around these problems are essentialist approaches on the one hand, and nominalist approaches on the other. For essentialists, history is ignored and science is characterized on *a priori* grounds—usually philosophical, political, or sometimes religious. For nominalists, philosophy is ignored, and science is taken to be whatever people claiming to do science actually do. This option is popular among cultural historians of science and sociologists of science. The method of this book is to try to steer a path between these two alternatives by focusing on an episode that all can agree upon as being good science and then teasing out some methodological lessons from that. If the achievements of Galileo and Newton are not considered good, or at least, representative, science, then the very question of the epistemology of science loses its cogency. This is a version of the common "paradigm case" argument in philosophy: To understand something, find an exemplary instance of it and examine its features and ramifications.

I claim that the 17th century's analysis of pendulum motion is a particularly apt window through which to view the methodological heart of the scientific revolution. More particularly, the debate between the Aristotelian Guidobaldo del Monte and Galileo over the latter's pendular claims, represents, in microcosm, the larger methodological struggle between Aristotelianism and the new science. I see this struggle as being over the legitimacy of idealization in science, and the utilization of mathematics in the construction and interpretation of experiments. del Monte was a prominent mathematician, engineer, and patron of Galileo. He kept indicating how the behavior of pendulums contradicted Galileo's claims about them. Galileo kept maintaining that refined and ideal pendulums would behave according to his theory. Del Monte said Galileo was a great mathematician, but a hopeless physicist. This is the methodological kernel of the scientific revolution. The development of pendular analyses by Huygens, and then Newton, beautifully illustrates the interplay between mathematics and experiment so characteristic of the emerging Galilean–Newtonian Paradigm in modern science.

It goes, almost without saying, that there are other windows, and the view through those may give a different picture of the scientific revolution. I do draw my lessons from physics, and in particular from mechanics. Astronomy, chemistry, and medical sciences are ignored. This selectivity might very well result in a skewed understanding of what was methodologically important to the scientific revolution. It might be more modest to say that my characterisation of the scientific revolution applies just to the revolution in mechanics occasioned by Galileo and perfected by Newton. However, given the importance of mechanics in the history of physics, this more modest claim is still interesting. And if students can be made familiar with it, through pendular investigations, then that is an achievement which contributes something important to their understanding of science.

Even here, some may say that science has moved on, and that understanding 17th century debates about the pendulum is simply irrelevant to understanding modern techno-industrial science and its methodology. This is a complex issue. In brief, understanding origins is important for understanding and judging the present. This is true in just about all spheres—political, religious, social, and personal. Further, I do not believe that modern science has so outgrown its methodological roots as to make irrelevant an examination of central seventeenth century epistemological debates. Even if, contrary to my assumption, it could be shown that modern science is methodologically different from its origins, nevertheless understanding where modern science has come from and, consequently, what occasioned the change, is still important. Finally, in education it is sensible to begin with simple or idealized cases. Presenting students with the full story—the truth, the whole truth, and nothing but the truth—is rarely a good idea. Concentrating on just some key aspects of a topic, be it in history, economics, biology, or what ever, makes pedagogical sense. The Galileo–del Monte debate does capture in comprehensible form some of the central issues in epistemology, and this gives an educational justification for its presentation. Provided students are made aware that the complete picture, or the modern picture, might be more complex, and provided they are encouraged to examine how science may have changed, then dealing with the 17th century is educationally and philosophically justified.

In its history of science strand, the book paints a big picture about the interrelatedness of timekeeping, the pendulum, the foundation of modern science, and the role of philosophy in all of this. It deals with both *internal* matters concerning the development and refinement of scientific concepts, and *external* matters concerning the social and cultural context in which the development of science occurs. There are good grounds for being suspicious of these "big pictures," especially when they are tied to "grand narratives" purporting to convey some philosophical message. The fine detail, which is so important for historians and on which the integrity of the tale so often depends, is frequently missing in the big picture. Further, the big picture can easily be anachronistic in its concepts and judgments. It is notorious that philosophers frequently deal with past thinkers as if they were contributors to the latest issue of *Mind*. Scientists also are prone to interpret the words of earlier writers as having the meaning that the same words have today. Some potential mistakes are fairly obvious and easy to avoid. For example, *forza* in Galileo's work, clearly does not mean *force* in contemporary, or Newtonian, terms. Other cases are not so easily avoided. For Newton and moderns, the term *momentum* is a vector quantity (it is direction dependent), however for Descartes, *motion (mv)* was a scalar quanity. Major problems arise if Descartes' quantity is interpreted as a vector. These, and other such pitfalls, are a problem in a book like this, but I hope that they have been avoided.

An appreciation that the meaning of words and concepts depend upon their intellectual context, and that this changes over time, might be one of the contributions that history of science can make to students' education. Identity of words, or mathematical definitions, does not mean identity of concepts and meaning. This is an important point to grasp in historical study. Grasping it is an antidote to crude fundamentalisms in politics and religion where reading contemporary meaning back into canonical texts is a standard error.

This possible historical function for science education is especially important at a time when history is fast disappearing from school programs. It might well be that science is the only place in a student's schooling where history can be encountered. Science might be the only place where students can learn that there was life before the 20th century and that people did not always think as we think and see things as we see them; that there were different priorities in people's dealings with nature; that there was a Middle Age, a Renaissance, a scientific revolution, an industrial revolution; and that our thinking is indebted to individuals such as Aristotle, Galileo, Newton, Darwin, and Einstein. As students, and school curricula, desert history, the science program might be the only place where students have the chance to understand how thinking occurs within a tradition, and that the healthy continuation of that tradition depends upon learning key lessons from its past.

Although I have said that the book deals in part with the *internal* history of pendulum analysis, and in part with the *external* history of the analysis, the book's story does give grounds for saying that this distinction is, in reality, blurred. I stress that a change in epistemology was fundamental to Galileo's achievements in understanding pendulum motion. Is then epistemology internal or external to science? Huygens' achievement rested upon the new geometrical analysis of the cycloid curve. Is then mathematics internal or external to science? Neither Galileo's or Huygens' proposals for utilizing the pendulum in timekeeping could be experimentally tested until technological advances in gear-cutting and escapement mechanisms were made. Is then technology internal or external to science? Once science is seen and treated as part of the intellectual culture of a society, and a historical period, then the separation of "internal" and "external" elements borders on being artificial and arbitrary.

That the distinction is blurred, does not mean that it cannot be made in some form. The book indicates the overwhelming influence that the longitude problem played in the development of clockwork. Solving longitude was one of the major preoccupations of European nations from the 15th to the 18th centuries. Kings' ransoms were offerred for its solution. Despite all the external financial and political pressure, a solution had to wait on scientific, methodological and mathematical progress. The world was the judge of putative solutions, not political or ideological interests. This is an important point to be appreciated at a time when many maintain that science simply dances to the tune of the last patron who payed the fiddler. In science, paying the fiddler and getting a good dance, are two different things.

The recently adopted United States *National Science Education Standards* (NRC, 1996) are briefly discussed in the final chapter. In the *Standards*, two pages are devoted to the pendulum, and they well illustrate many of the claims made in this book. Sadly there is no mention of the history, philosophy, or cultural impact of pendulum motion, no mention of the pendulum's connection with timekeeping, and no mention of the longitude problem. In the suggested assessment exercise, the obvious opportunity to connect standards of length with standards of time is not taken. The *Standards* document was reviewed by tens of thousands of teachers and educators, and putatively represents current best practice in science education. This book argues that a little historical and philosophical knowledge about the pendulum could have transformed the treatment of the subject in the *Standards*, and consequently could have resulted in a much richer and more meaningful science education for American students. That this historical and philosophical knowledge is not manifest in the *Standards*, indicates the amount of work that needs to be done in having science educators become more familiar with the history and philosophy of the subject they teach. And, of course, this problem is not confined to the United States. It is a worldwide problem. It is hoped this book will demonstrate the educational worth, for classroom teaching, curriculum development, and teacher education, of such interaction between science education and the history and philosophy of science.

CHAPTER 1

Learning about the Pendulum and Improving Science Education

The scientific revolution is clearly one of the great episodes in human history. Its scientific, philosophical, and cultural impact, initially on Europe, and subsequently on the rest of the world, has been without parallel. The core of the scientific revolution occurred in the half-century between the publication of Galileo's *Dialogue Concerning the Two Chief World Systems* (1633) and Newton's *Principia* (1687).[1] In this period Newton brought to fruition what Galileo had begun.[2] A mathematical–experimental way of interrogating nature displaced the varied common-sensical–observational–philosophical ways that had hitherto dominated attempts to understand and control the world. From the perspective of world history it was a singular and contingent event. Great civilizations had existed in Europe, Africa, Australasia, and the Americas for thousands of years, yet none of them came close to producing that way of understanding the world that was characteristic of the new science of 17th-century Europe. The Galilean–Newtonian method (or Galilean–Newtonian Paradigm, GNP, as it has been called) originated in physics and quickly flowed to other scientific and to nonscientific fields. The new science, frequently in conjunction with colonial and commercial interests, spread rapidly from its origins in western Europe to the far reaches of the globe.[3] The science of western Europe, became Western science, or "universal science." Of its cultural significance for Europe, the historian Herbert Butterfield has written:

> Since the Scientific Revolution overturned the authority in science not only of the Middle Ages but of the ancient world—since it ended not only in the eclipse of scholastic philosophy but in the destruction of Aristotelian physics—it outshines everything since the rise of Christianity and reduces the Renaissance and Reformation to the rank of mere episodes, mere internal displacements, within the system of medieval Christendom. . . . it changed the character of men's habitual mental operations even in the conduct of non-material sciences, while transforming the whole diagram of the physical universe and the very texture of human life itself, it looms so large as

1

the real origin both of the modern world and of the modern mentality . . . (Butterfield, 1949, p. viii)

THE PENDULUM IN WESTERN SCIENCE, TECHNOLOGY, AND CULTURE

The pendulum played a pivotal role in the scientific revolution. Among other things, the pendulum provided the first effective measure of time, without which modern quantitative mechanics (as distinct from statics that depended just on length and weight measures) would be impossible. Stillman Drake identifies Galileo's discovery of the pendulum laws as "marking the commencement of the early modern era in physics" (Drake, 1990, p. 6).

Galileo used pendulum motion to establish his law of free fall, his law of conservation of energy, and to undermine the crucial Aristotelian conceptual distinction between violent and natural motions. In a 1632 letter, Galileo surveyed his achievements in physics and recorded his debt to the pendulum for enabling him to measure the time of free-fall, which, he said, "we shall obtain from the marvellous property of the pendulum, which is that it makes all its vibrations, large or small, in equal times" (Drake, 1978, p. 399).

The pendulum played a comparable role in Newton's work. He used the pendulum in the following ways: to determine the gravitational constant **g**; to improve timekeeping; to disprove the existence of the mechanical philosophers' aether presumption; to show the proportionality of mass to weight; to determine the coefficient of elasticity of bodies; to investigate the laws of impact; and to determine the speed of sound.

The importance of the pendulum for the scientific revolution has not been as widely recognized as it deserves. However René Descartes (1596–1650), who was notorious for his low opinion of Galileo's contribution to physics,[4] did recognize the importance of the pendulum for Galileo. He commented that Galileo "seems to have written all his three dialogues for no other purpose than to demonstrate that the descents and ascents of a pendulum are equal" (*Works*, letter 146). Richard Westfall, among contemporary commentators on the scientific revolution, has perhaps given greatest recognition to the pendulum's role, saying that, "the pendulum became the most important instrument of 17th-century science . . . Without it, the 17th century could not have begot the world of precision" (Westfall, 1990, p. 67). Concerning the pendulum's role in Newton's science, Westfall said that "It is not too much to assert that without the pendulum there would have been no *Principia*" (Westfall, 1990, p. 82). No small praise. Another historian, Bert Hall, wrote:

> In the history of physics the pendulum plays a role of singular importance. From the early years of the 17th century, when Galileo announced his formulation of the laws governing pendular motion, to the early years of this century, when it was displaced by devices of superior accuracy, the pendulum was either an object of study or a

means to study questions in astronomy, gravitation, and mechanics. (Hall, 1978, p. 441)

The pendulum was not just of scientific importance. When it was incorporated into the pendulum clock, it had an enormous influence on navigation, commerce, the development of the industrial revolution, and on the transformation of public life and culture. Lewis Mumford commented that:

the clock, not the steam-engine, is the key-machine of the modern industrial age. . . . by its essential nature it dissociated time from human events and helped create the belief in an independent world of mathematically measurable sequences: the special world of science. (Mumford, 1934, pp. 14–15)

This special world of science is captured by Newton in his opening definitions of the *Principia*: "Absolute, true and mathematical time, of itself, and from its own nature, flows equably without relation to anything external." A prominent historian of the cultural impact of timekeeping observed:

Work time has been one of the great themes of social conflict since the beginning of industrialization. . . . the new concept of the "economy of time" that arose along with workshops and factories . . . has been widely discussed in the last decades under the catchphrases "social disciplining" and "the loss of individual control of time." (Rossum, 1996, p. 289)

The clock played a part in philosophy. When, in the 17th and 18th centuries, the clock was used metaphorically, it contributed to the spread of the new "mechanical world view." The idea that changes in phenomena are the result of deterministic pushes and pulls was made tangible and plausible by appeal to the world being "like clockwork" (Laudan, 1981a, Price, 1964).

The scientific, technical, commercial and cultural aspects of the pendulum are brought together in its contribution to solving the longstanding problem of longitude determination. The solution of the longitude problem was of major historical significance because it opened the world to European exploration, exploitation, commerce, and colonization.

The humble pendulum thus played a large role in the formation of the modern world. The thesis of this book is that the pendulum can play a larger role in the modern science classroom. A proper education in science should encourage children to see and understand something of the "big picture" of science; that is, the interrelatedness of science with culture. A liberal or contextual approach to teaching about the pendulum contributes to this understanding.

THE PENDULUM AND SCIENCE TEXTBOOKS

There is a striking imbalance between the importance of the pendulum in the history of science, and the meager attention it commands in science curricula. Through its utilization in timekeeping, the pendulum had an enormous cultural and economic impact. Yet most students can complete a school science program, or get a university physics degree without learning anything of the scientific, cultural, or economic impact of the pendulum. This contrast between the pendulum's scientific and social importance and its educational neglect highlights an increasingly recognized deficiency in science education. There is little sense of students being introduced to a tradition of thought, and almost no attention to the cultural context of science. So much effort has been put into having students understand the trees, that they often do not see the forest.

Knowledge of science (science content and method), and knowledge about science (its history, philosophy and sociology) are both important components of scientific literacy. Both components should have a place in school curricula. New curricular proposals in many countries are attempting to make the science program more engaging, reflective, informative, and contextual. The scientific, historical, philosophical, and social study of the pendulum is an ideal way to promote the broader understanding of science to which many new science programs aspire. Because the pendulum is so tangible, and its history so multifaceted, it can be discussed intelligently and informatively with all ability levels, from elementary school, through high school, and on to college.

However science textbooks pay little attention to the historical, methodological, and cultural dimension of pendulum motion. It is sometimes given a cameo appearance in the story of Galileo who supposedly during a church sermon observed a swaying chandelier and timed its swings with his pulse, and presto! there was the law of isochronic motion. This account is found in Fredrick Wolf's text:

> When he [Galileo] was barely seventeen years old, he made a passive observation of a chandelier swinging like a pendulum in the church at Pisa where he grew up. He noticed that it swung in the gentle breeze coming through the half-opened church door. Bored with the sermon, he watched the chandelier carefully, then placed his fingertips on his wrist, and felt his pulse. He noticed an amazing thing. . . . Sometimes the chandelier swings widely and sometimes it hardly swings at all . . . [yet] it made the same number of swings every sixty pulse beats. (Wolf, 1981, p. 33)

Wolf's story, sans boredom, appeared in the opening pages of the most widely used high school physics text in the world: the Physical Science Study Committee's *Physics* (PSSC, 1960).

Whatever the problems with Wolf and the PSSC text might be, the Galileo story is at least presented. However the pendulum more often appears in physics texts without any historical context; it is standardly introduced merely as an instance of simple harmonic motion. The extent to which the pendulum has been plucked from its historical and cultural roots can be seen in the Harvard Project Physics text, an excellent and most contextual of texts, where the equation

$$T = 2\pi\sqrt{l/\mathbf{g}}$$

is abruptly introduced for the period of the pendulum, and students are told "you may learn in a later physics course how to derive the formula."

The influential contemporary curriculum proposal of the U.S. National Science Teachers Association (NSTA), *Scope, Sequence and Coordination* (Aldridge, 1992), highlights the pendulum to illustrate its claims for sequencing and coordination in science instruction. Yet nowhere in its discussion of the pendulum is history, philosophy, or technology mentioned.[5]

The utilization, in the 17th century, of the pendulum's isochronic properties to produce the first reliable clock is hardly mentioned in science texts. Also not discussed is the clock's role in solving the problem of longitude or even the length of the seconds-pendulum as the putative original universal standard of length.

Even the best and most detailed horological histories—which are full of technical details about clock construction, and rich with the cultural context of timekeeping—leave out the epistemological change, or revolution, inherent in Galileo's discovery of isochronic motion. For example, J. Drummond Robertson in his classic and definitive work, *The Evolution of Clockwork* (1931), said this of Galileo:

> Already at the age of seventeen he had made his discovery [of isochronism] in 1581 from observing the swing of a lamp suspended in the Cathedral of Pisa, and had thus laid the foundation of the third step in the achievement of exact time measurement. (Robertson, 1931, p. 76)

Origins of the Cathedral Story

The story of Galileo and the cathedral lamp is perhaps rivaled in popular imagination only by that of Newton and the falling apple and Archimedes in his bath. These are the three episodes from the history of science that have entered into Western folk culture. A hint of the lamp-watching story first appeared in Galileo's *Two New Sciences*, where he had the bystander, Sagredo, say:

> You give me frequent occasion to admire the wealth and profusion of nature when, from such common and even trivial phenomena, you derive facts which are not only striking and new but which are often far removed from what we would have imagined. Thousands of times I have observed vibrations especially in churches where lamps, suspended by long cords, have been inadvertently set into motion; but the most which I could infer from these observations was that the view of those who think that such vibrations are maintained by the medium is highly improbable...But I never dreamed of learning that one and the same body, when suspended from a string a hundred cubits long and pulled aside through an arc of 90° or even 1° or ½°, would employ the same time in passing through the least as through the largest of these arcs; and, indeed, it still strikes me as somewhat unlikely. (Galileo, 1638/1954, p. 97)

Vincent Viviani's brief biography of Galileo (1654), written 12 years after Galileo's death, had on p. 603 the story of Galileo watching the lamp in the Cathedral of Pisa. This biography is included in Antonio Favaro's edited *Le Opere di Galileo Galilei, Edizione Nazionale,* (20 volumes, Florence, Barbéra, 1890–1909, vol. 19, pp. 597–632). Viviani's story is repeated and embellished by subsequent authors. Antonio Favaro praised Galileo's ability:

A first example of his very sharp intelligence is given us in the original and famous observation concerning the isochronism of the oscillations of a lamp which was suspended in the vault of the cathedral of Pisa. (Favaro, 1883, vol. 1, p. 13; in Kutschmann, 1986, p. 116)

The story appears in Cajori's *History of Physics* (Cajori, 1929, p. 42), from which perhaps it gets its wide currency in popular histories and science texts.

Knowledge and Perception: A Problem with the Textbook Story

The science textbook account fails to explain why it was that the supposed isochronism of the pendulum was only seen in the 16th century, when countless people of genius, who had acute powers of observation, had for thousands of years been pushing children on swings, looking at swinging lamps and swinging weights, and using suspended bobs in tuning musical instruments, without seeing their isochronism. For centuries people had been concerned to find a reliable measure of time, both for scientific purposes and for everyday life, to determine the duration of activities and events and to determine longitude. As the isochronic pendulum was the answer to all these questions, the widespread failure to recognize something so apparently obvious is perplexing. It suggests that there is not just a problem of perception, but a deeper problem, a problem of epistemology or cultural presuppositions, involved. Seemingly no one in the rich Chinese, Indian, and Islamic scientific traditions noted the isochrony of the pendulum.[6]

Thomas Kuhn made a passing reference to this problem in his *Structure of Scientific Revolutions*, saying that Galileo's "genius does not here manifest itself in more accurate or objective observation of the swinging body . . . rather, what seems to have been involved was the exploitation by genius of perceptual possibilities made available by a medieval paradigm shift [impetus theory]" (Kuhn, 1970, p. 119). Kuhn intimated that the medieval Aristotelians could not, on Aristotelian terms, analyze the pendulum's motion until attempts to analyze that other longstanding embarrassment to Aristotelian theory—projectile motion—had produced the face-saving concept of *impetus* to account for the continued motion of the projectile after leaving the thrower's hand. This same concept then explained the continued motion of the pulled-aside bob at the end of the length of rope, and of the continued peal of bells. However Kuhn's cognitive and epistemological explanation for a failure of perception does not apply to Platonists, Stoics, Chinese thinkers, and other non-Aristotelians who also failed to "see" the pendulum's isochronic motion.

Knowledge of how Galileo came to recognize and prove the laws of pendulum motion can contribute to the teaching of the topic. Teachers want students to recognize, understand, and prove the properties of pendulum motion—period being independent of mass and amplitude, and varying inversely as the square root of length. How these properties were initially discovered can throw light on current attempts to teach and learn the topic. And this historical knowledge can illuminate some important methodological features of the scientific revolution, it can cast light on the "new way of knowing" initiated in western Europe in the 16th and 17th centuries.

PROBLEMS WITH SCIENCE EDUCATION

It is apparent to everyone that something has gone wrong with science education. During the final decades of the 20th century there was, in the Western world, a flight from the science classroom and, more generally, from scientific understanding. The *Australian* newspaper (9th September 1998) reported 500,000 people in Rome admitted they regularly contacted clairvoyants, and it reported that horoscopes are read out on Italian state-run radio before the morning news. Stories about an astrologer giving advice in the Oval Office well captures the gravity of the current situation. The predicament of science education in the United States has been well documented.[7] In 1990, only four states required the three years of basic science recommended by the sobering 1983 report *A Nation at Risk*, the rest allowed high school graduation after only two years science (Beardsley, 1992, p. 80). Irrespective of years required, 70% of all students drop science at the first available opportunity: this is one reason why in 1986 fewer than one in five high school graduates had studied any physics. The science-test performance of 17-year-olds in 1986 was substantially lower than it had been in 1969 (Rutherford and Ahlgren, 1990 p. vii). In 1991 the Carnegie Commission on Science, Technology, and Government warned that the failings of science education were so great that they posed a "chronic and serious threat to our nation's future" (Beardsley, 1992, p. 79). A recent National Science Board report indicated that of the few high school graduates who do major in natural science or engineering nearly 50% drop out, or change majors before completion (NSB, 1996, p. 83).

In the United Kingdom, recent reports of the National Commission on Education and the Royal Society have documented alarming levels of science illiteracy. One commentator said "wherever you look, students are turning away from science. . . . Those that do go to university are often of a frighteningly low calibre" (Brown, 1993, p. 12). The percentage of United Kingdom school leavers sitting for A-level science subjects has dropped from 35–16% in recent years.[8]

In New Zealand, a 1991 Ministry of Research, Science and Technology study[9] revealed how little citizens knew and cared about science. The study of 1,012 representative adults showed that:

- Fully 90 percent were scientifically illiterate, having less than a minimum understanding of the processes, terms, and social impact of science.
- Only 13 percent were even interested in science.

- Only 3 percent were both literate and interested; that is most of the 10 percent who were scientifically literate, were not interested in science!
- Overall there was a negative attitude to science. (MORST, 1991, p. 4).

In Australia in the mid-1990s there was a drop in enrollment in university science and engineering programs. The entry requirements had to be dramatically lowered to fill places, and numerous science faculty are being retrenched from universities. A number of physics departments have been, or are facing, closure. Between 1992 and 1997, the number of students in Australian universities increased by 16%, yet the numbers enrolled in science dropped by 12%. In 1990–1995, the number of high school students in the state of New South Wales doing *any* science subject for their Higher School Certificate dropped by 40%. In 1997, after 118 years of existence, the Australian and New Zealand Association for the Advancement of Science (ANZAAS) voted to dissolve itself. The act of near professional hari-kari was performed at a meeting of a dozen members during a poorly attended annual conference. The association barely managed to survive.

There are complex economic, social, and cultural causes for this flight from university and school science, and for the depressing rise of *anti*scientific views in the wider population.[10] Most of the causes for this flight, especially from university science, are beyond the scope of teachers and education systems to rectify: poor salaries, other prestigious career paths, and so on. This flight from university science programs makes more pressing the question of an appropriate high school science curriculum. If high school science is as much formal science as the overwhelming bulk of the population is exposed to, then schools should try to get it right. As one British commentator, and member of the United Kingdom's "Understanding Science Project," said: "What we need are alternative syllabuses, suitable for the nonspecialist and specialist alike" (Hughes, 1993, p. 47). Such syllabuses are being developed. Major educational reform programs have commenced in the United States,[11] Canada,[12] Denmark,[13] Norway,[14] the United Kingdom,[15] and other parts of the world to address the flight from science. The sponsors of education reform realize the educational response cannot be confined to the creation of new syllabuses. Other areas need to be overhauled, including teacher training, assessment, teaching practices, teachers' salaries, and textbooks. The problems of science education require a systematic response.

Moreover scientific literacy cannot be divorced from general literacy and numeracy. If children cannot read, cannot multiply or divide, cannot manipulate simple proportions and equations, cannot reason correctly or think logically, then they cannot achieve scientific literacy. If children have no patience for learning, if the bulk of their time is spent watching brain-dead television, then the idea of "science for all" is fanciful. The problems of science education cannot be divorced from the problem of education in general. If schools give up on education and devote themselves to social engineering, socialization, and entertainment, then the prospects for science education are dim.

Some New Zealand research indicates the gravity of the situation. Of the 338 students who entered the Primary Teacher Education program at the Auckland College of Education in 1991, 41% (138 students) could not determine the retail sales tax (at 12.5%) payable on a bill of $62, and 27% (90 students) of the same primary intake could not work out the length of a pencil when its end was placed on the 2 cm mark of a ruler (Buzeika, 1993, p. 36). The 353 students in the 1992 intake did slightly worse on this simple 12-item mathematics test. For instance, one question was:

> Gweneth bought 4.15 litres of unleaded petrol for her motor mower. She paid $3.69 for the petrol. Gweneth should use which one of the following calculations to check the price for one litre?
> a. 4.15 + 3.69
> b. 3.69 ÷ 4.15
> c. 4.15 − 3.69
> d. 3.69 − 4.15
> e. 4.15 ÷ 3.69
> f. 3.69 × 4.15

Despite completing 12 years of schooling, 41% (144 students) could not answer the question correctly. They did not have to give a numerical answer, just identify what *procedure* would be used. At best, these results suggest that such poor arithmetic skills place a low ceiling on possible science literacy attainments. They reinforce the view that, before attending to science literacy, basic arithmetic, literacy and reasoning have to be improved. In New Zealand, as elsewhere, there is no shortage of educational talk about "empowerment" and "critical literacy," but these aspirations are hollow, if not counterproductive, when basic skills are so lacking.[16] If a person cannot think straight, then they cannot hardly think critically.

The problem is exemplified in a small Australian study by Gordon Cochaud (1989). He gave a brief, ten-item, logic test to first-year science students at an Australian university. Among the items was this one where students had to fill in the conclusion:

> If one adds chloride ions to a silver solution then a white precipitate is produced.
> Addition of chloride ions to solution K produced a white precipitate.
> Therefore ____

Out of a group of 65 students, 48 concluded that solution *K* contained silver. Thus nearly 3/4 of a group of high-achieving high school graduates who had studied science for at least six years went along with fundamentally flawed reasoning. The majority of these students, despite 12 years of schooling, happily committed the logical fallacy of affirming the consequent; comparable results were obtained by Ehud Jungwirth (1987) in a larger, more international study of science students' reasoning skills. Little wonder that as citizens they are easily swayed by arguments such as:

> Communists support unionism.
> Fred supports unionism.
> Therefore Fred is a Communist.

Literacy in the United States has been studied and debated more than any-where else. What brought the issue to popular attention, and galvanized the government to action, was the publication in 1983 of *A Nation at Risk*. Its conclusion was stark: "the educational foundations of our society are presently being eroded by a rising tide of mediocrity that threatens our very future as a nation and as a people" (NCEE, 1983). It expressed a particular concern about the abysmal state of scientific and mathematical knowledge of high school graduates, and proposed the educational goal of "science for all." In the five years after the publication of *A Nation At Risk*, over 300 reports had documented the sorry state of American education. In 1983 20 bills, designed to offer solutions to the crisis,[17] were put before Congress. With much fanfare and hoopla, President George Bush in 1991 launched *America 2000: An Educational Strategy* (Department of Education, 1991). On the opening page, President Bush claimed that "by the year 2000 United States students will be the first in the world in science and mathematics achievement." For its monumental failure to grasp reality, this claim ranks with Neville Chamberlain's 1939 pronouncement of "peace in our time"—but of course it might be said that political rhetoric is not even meant to "grasp reality."

Five years later when the political landscape changed, the same department released a more sobering report titled *Adult Literacy in America* (Department of Education, 1995). This report was based on interviews with, and test scores of, 26,000 citizens above the age of 16. The report estimated that 90 million of the 191 million American adults (nearly 50%) had inadequate skills in maths and reading. It ranked the sample into five literacy levels, with the bottom two levels being judged inadequate. The test based its questions on everyday situations such as reading a bus schedule, making out a bank deposit slip, and understanding information contained in a newspaper. Given that in 1993 it was estimated that 86% of the adult population had received high school diplomas, the report's conclusions suggest President Bush's 1991 predictions are completely illusory. Scientific literacy is parasitic on general literacy. When every year hundreds of thousands of high school students who are functionally illiterate nevertheless graduate, then achieving simple science literacy—never mind being first in the world—is impossible.

Science and Liberal Education

There have been a variety of responses to the science literacy crisis. These constitute the agenda of Science Education Reform in a host of countries. Some curriculum developers and educators want to continue business as usual, but only for the select few going on to tertiary science-related studies; they want to abandon the 1970s goal of "science for all." Others want to substitute various vocational and applied science courses for the dominant professional curricula in schools. Still others want to promote forms of Science–Technology–Society (STS) programs. Some even wish to resurrect aspects of the minority liberal education tradition in science teaching, a tradition that gives more recognition to the history and philosophy of science in science programs.[18]

One advocate of the STS approach said: "Schooling in science ought to enhance the personal development of students and contribute to their lives as citizens. Achieving this purpose requires us to reinstate the personal and social goals that were eliminated in the curriculum reform movement of the 1960s and 1970s" (McFadden, 1989, p. 261).[19] A problem with STS education is that contemporary science–technology–society interactions are exceedingly complex and notoriously difficult to grasp. This problem is identified by Morris Shamos in his recent book on *The Myth of Science Literacy*:

> The main advantage of a historical approach is that the science it deals with, when viewed from a modern perspective, is generally much simpler and the reasoning involved is likely to be more transparent than in a contemporary example. (Shamos, 1995, p. 199)

Advocates of more liberal, or contextural, approaches to science education point out that, for instance, understanding the science and technology of a pendulum clock is within the reach of a school student, but understanding the science and technology of a cesium clock or a quartz watch is beyond their reach. Here, as with most contemporary technology, explanations have to be taken on faith. Even mechanics and technicians who work with modern technology rarely understand *why* it works. High tech is beyond the comprehension of all but the super elite. In the middle of the century, people with a modicum of scientific and technical education could fathom the principles of everyday technology: the telephone, steam engine, pressure cooker, carburetor, valve radio, record player, typewriter, movie projector, and so on. The transistor, microchip, semiconductors, and computer have taken the principles to another level of comprehension, a level beyond the cognitive reach of the overwhelming majority. To primitive people, the natural world was mysterious. To moderns, it is the technical, created, world which is mysterious. People are strangers in their own living room. This is a problem that advocates of STS education need to recognize, the more so as they typically advocate "Science Education for All."

In the same vein, 17th-century arguments about pendulum motion provide an accessible "window" for viewing the methodological, mathematical, experimental, technical, and commercial dimensions of the scientific revolution. Current debates about acid rain, greenhouse emissions or the social value of the CERN project, do not reveal the methodological dimension of science. Contextural teaching about the pendulum can be fitted into STS programs, but these programs need to be more accommodating of historical and philosophical questions.

The liberal tradition contends that science programs should promote the dual goals of learning *of* science (scientific facts, theories, and methods) and learning *about* science (scientific methodology, history, philosophy, and cultural interactions).[20] Sometimes this dual goal is expressed as the need for students to learn about the *nature of science* as well as learning science; or that students need to learn the *culture* of science as well as the *discipline* of science (Gauld, 1977). The American Association for the Advancement of Science, in its *The Liberal Art of Science*, proposed that:

> Science courses should place science in its historical perspective. Liberally educated students—the science major and the non-major alike—should complete their science courses with an appreciation of science as part of an intellectual, social, and cultural tradition. . . . Science courses must convey these aspects of science by stressing its ethical, social, economic, and political dimensions. (AAAS, 1990, p. 24)

After World War II, the Harvard Committee, under the direction of James Conant, produced an important report (the "Red Book") saying that:

> Science instruction in general education should be characterized mainly by broad integrative elements—the comparison of scientific with other modes of thought, the comparison and contrast of the individual sciences with one another, the relations of science with its own past and with general human history, and of science with problems of human society. These are areas in which science can make a lasting contribution to the general education of all students . . . Below the college level, virtually all science teaching should be devoted to general education. (Conant, 1945, pp. 155–156)

Harvard's General Education program had enormous influence.[21] Among other things, it gave rise to Harvard Project Physics (Rutherford, Holton, and Watson, 1970), and it was responsible for opening Thomas Kuhn's eyes to the history of science, which in turn led to the Kuhnian revolution in philosophy of science.

The key to Harvard's general education science courses, its science for nonscience majors, was that the "facts of science must be learned in another context, cultural, historical, and philosophical." A problem for all such courses is the tension between the dual goals of learning *of*, and the learning *about*, science. Professional programs (for example, PSSC Physics) stress the former often to the exclusion of the latter, while some generalist programs stress the latter often to the exclusion of the former. During Conant's period at Harvard, *nonscience* majors learned about the history and philosophy of science in their general education course, but *science* majors, on the other hand, had no such exposure to the history and philosophy of their discipline. Unfortunately what Conant advocated for school science programs, was not carried over to Harvard University science.

A similar approach to general education, under the guidance of Robert M. Hutchins, was instituted at the University of Chicago. Of this program the astronomer Carl Sagan wrote:

I also was lucky enough to go through a general education program devised by Robert M. Hutchins, where science was presented as an integral part of the gorgeous tapestry of human knowledge. It was considered unthinkable for an aspiring physicist not to know Plato, Aristotle, Bach, Shakespeare, Gibbon, Malinowski, and Freud—among many others. In an introductory science class, Ptolemy's view that the Sun revolved around the Earth was presented so compellingly that some students found themselves re-evaluating their commitment to Copernicus. . . . teachers were valued for their teaching, their ability to inform and inspire the next generation. (Sagan, 1997, p. 5)

The rationale for general education, particularly as it is practiced in the United States, basically amounts to a statement of liberal education principles. The core of both approaches is that education involves knowing, at some appropriate depth, a *range* of disciplines, or better still *forms of knowledge*, and how these forms relate to each other. The liberal education tradition stresses the importance of developing an educated mind, or developing intellectual virtues—an ability to reason, to be critical, to be open-minded, and to be objective in judgments. A liberal education aims not just at having students know a discipline, or a range of disciplines, in a spectator sense, but know them in a way that transforms their own behavior, understanding, attitude, and outlook on both the natural and social world. Knowledge is not to be inert.

The outcome of a liberal education should be a certain habit of mind. Students can be *trained* in a discipline or technique, but liberal education requires more than narrow specialization. Incorporating historical material into the science program can facilitate the integration of subjects in the school curriculum. Routinely incorporating philosophical questions (What does X mean? How do we know Y? Can Z be justified?) into the science program can facilitate the intellectual virtues or habits that liberal education aspires to develop. The content of the school day, or at least year, can be more of a tapestry, rather than a curtain of unconnected curricular beads. For instance, pendulum motion, if taught from a historical and philosophical perspective, allows connections to be made with topics in religion, history, mathematics, philosophy, music, and literature, as well as other topics in the science program. Such teaching promotes greater understanding of science, its methodology, and its contribution to society and culture. Figure 1, using the circles to represent curriculum topics, displays the integrative function of history and philosophy.[22]

SCIENCE LITERACY

Morris Shamos argued that our lack of success in developing scientific literacy for all (in the sense of competence, or knowledge *of* science), meant we should now abandon this goal, recognize that this competence is the preserve of an intelligent minority in the population, and instead redefine literacy in terms of knowledge *about* science. According to Shamos the bulk of school science programs

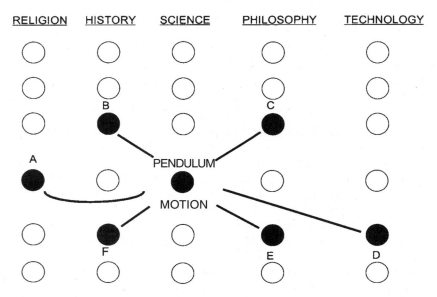

Figure 1. HPS-informed curriculum. a. The design argument; b. European voyages of discovery; c. Aristotelian physics and methodology; d. Pendulum clock; e. Idealization and theory testing; f. Industrial Revolution.

should not attempt to teach science, but should teach scientific and technological awareness. In his words: "The science and engineering communities, and our nation generally, would be better served by a society that, while perhaps illiterate in science in the formal sense, at least is aware of what science is, how it works, and its horizons and limitations" (Shamos, 1995, p. 198). The physicist James Trefil[23] also defended this idea:

> Too often, when scientists start to talk about scientific education, they skip the ninety-nine percent of the population that is not going to be trained in a technical career, and zero in on the one percent that are. . . . Scientific literacy consists of a potpouri of facts, concepts, history, philosophy, and ideas, all connected to each other by strands of logic. . . . scientific literacy does not require the ability to *do* science. (Trefil, 1996, p. 547)

Requirements for scientific literacy have been long debated.[24] Sometimes the issue is expressed as a choice between scientific competence and scientific literacy, or as a choice between knowledge *of* science and knowledge *about* science, or even sometimes as knowledge about the *discipline* of science, versus knowledge about the *culture* of science. Disciplinary knowledge *of* science has occasionally been elaborated as knowledge of scientific *products* and at other times as knowledge of scientific *processes*, or *inquiry* skills. Aside from these differences about areas of *cognitive* skills and competence, educators have been concerned with *affective* or *attitudinal* aspects of scientific literacy. As well as knowing something of or about science, how much does a scientifically literate person need to *appreciate*, or *value* science?

Nearly a century ago John Dewey, in his influential article "Science as Subject-Matter and as Method" (Dewey 1910), expressed dismay at the failure of schools to attract students to science: "Considering the opportunities, students have not flocked to the study of science in the numbers predicted, nor has science modified the spirit and purport of all education in a degree commensurate with the claims made for it" (Dewey, 1910, p. 122). For Dewey, a large part of the problem was that "science has been taught too much as an accumulation of ready-made material with which students are to be made familiar, not enough as a method of thinking, an attitude of mind, after the pattern of which mental habits are to be transformed" (Dewey, 1910, p. 122).

The affective, or valuing, dimension is often taken for granted, but it need not be. We might well wish to consider a person literate in say Maoism, Leninism, Islam, feminism, or Catholicism if they have an agreed knowledge *of* Maoism, Leninism, Islam, feminism, or Catholicism and an agreed knowledge *about* these doctrines or systems of ideas. Literacy here entails absolutely nothing about *affective* aspects. A person literate in each of these fields might not value any of them, indeed this person could think that the doctrines are mistaken and misguided. We might call this *neutral* literacy, the sort of literacy that anthropologists develop about foreign systems of ideas, whereby literacy has no implications for positive or negative appraisals of the systems studied. Some may want to consider science literacy in the same way. Indeed some fundamentalists might reasonably claim to be scientifically literate in the sense of knowledge *of* and *about* science, yet also be hostile to science.

Figure 2 illustrates the connections between the above two *cognitive* domains (knowledge *of* and knowledge *about* science), and the *affective* domain (appreciation and valuing of science). The liberal tradition in education has generally taken a broad view of science literacy and has identified it with the *intersection* of these three domains (A). We could call this *positive* literacy. The technical, or professional, tradition has generally taken a narrow view, and has identified scientific literacy with knowledge *of*, or competence *in*, science. Informed, but hostile, critics occupy domain (B).

In passing it is worth mentioning a subsidiary issue concerning the *scope* of the *affective* domain. Dewey well expressed this matter in his 1910 article: "One of the only two articles that remain in my creed of life is that the future of our civilization depends upon the widening spread of the scientific habit of mind; and that the problem of problems in our education is therefore to discover how to mature and make effective this scientific habit" (Dewey, 1910, p. 127). He went on to say:

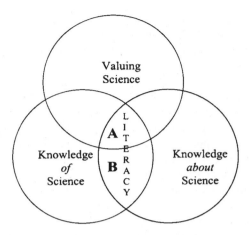

Figure 2. Liberal and technical accounts of scientific literacy. a. Positive literacy; b. Neutral literacy

> Scientific method is not just a method which it has been found profitable to pursue in
> this or that abstruse subject for purely technical reasons. It represents the only method
> of thinking that has proved fruitful in any subject. (Dewey, 1910, p. 127)

Dewey is arguing for, what we might call, the *universal* scope of scientific
thinking. His attitude to, and appreciation of, science is *comprehensive,* or some
might say *imperialistic.* He even regretted that science "has as yet had next to
nothing to do with forming the social and moral ideals for the sake of which she is
used" (Dewey, 1910, p. 127). He wanted the Western scientific tradition to permeate all of culture.

Others argue for a *compartmental* view of scientific thought, saying that it is
preeminent in its own sphere, but only in its own sphere of natural events and processes. Politics, art, religion, human affairs, etc., are beyond the reach of scientific
thinking and analysis. They have their own forms of intelligibility. Paul Hirst
provided an influential statement of this position in his much anthologized "Liberal Education and the Nature of Knowledge." He said that education "lays the
foundation for the various modes of understanding, scientific, historical, religious,
moral and so on" (Hirst, 1974, p. 51). He calls these modes of understanding
"forms of knowledge" and says that they are identified by distinct central concepts; logical structures; tests against experience; and techniques and skills for
their development (Hirst, 1974, p. 44). At a surface level, Hirst's characterization
is reasonable, and it can shape and give guidance to curricular deliberations. Problems arise when his view is subject to closer scrutiny: Are the forms so distinct?
Are there unique truth tests in the different forms? Do not ethics, aesthetics, logic,
metaphysics, even religion, play a role in scientific decision making? And so on.[25]

At the other end of the spectrum from Dewey, some science educators have recently been highlighting not the achievements, but the failures of science; not the wisdom, but the conceits of a scientific attitude. This position, which we might call, at best a *utilitarian*, and at worst, a *hostile* view of scientific thought, is common in feminist and multiculturalist writings on science education.

For example, in contradiction to Dewey, the feminist scholar Sandra Harding is concerned with the gains that scientific rationality has made in Western society. She asked whether "we can locate anything morally and politically worth redeeming or reforming in the scientific world view, its underlying epistemology, or the practices these legitimate?" (Harding, 1986, p. 29). Glen Aikenhead, in writing on Native American education, says that Western science is "a repository to be raided for what it can contribute to the achievement of practical ends" (Aikenhead, 1997, p. 218). He referred to recent social studies of science that characterized science as mechanistic, materialistic, reductionist, decontextualized, ideological, masculine, elitist, competitive, exploitive, impersonal, and violent (Aikenhead, 1997, p. 220). In other words, not very nice.

Where Dewey urged the spread of scientific rationality and saw its flourishing in society as one of the chief benefits of good education, Aikenhead warned that "because the subculture of science tends to permeate the culture of those who engage it, curriculum specialists and teachers need to develop a science curriculum that explicitly eschews assimilation and vigilantly circumvents unwanted acculturation" (Aikenhead, 1997, p. 228). Seemingly, if a culture is laden with mythological and erroneous—to say nothing of dangerous, racist, sexist and casteist—views about the natural and social world, then science teachers should be careful not to disturb the milieu. Good teachers, from Socrates to Scopes to Sagan, have not followed this advice. The practice of science has clearly had its faults, but nevertheless it has been scientific practice, research, and thinking that has undermined most of the mythological, erroneous, racist, sexist, and casteist views of the societies in which science has been allowed to flourish.

One problem with broadening the definition of scientific literacy beyond mere technical competence in, or understanding of, science, is to then withstand pressures for *redemptive* definitions of literacy. Common garden-variety literacy is thought insufficient: teachers are meant to develop "egalitarian" or "liberatory" or "emancipatory" or "powerful" literacy. A teacher is seen as a mere tool of the *bourgeoise* (or whatever is the current preferred equivalent) if this person struggled to help students write clearly, spell correctly, compose an argument, recognize basic logical fallacies, comprehend rudimentary science and scientific method, and understand some key historical episodes. As one critic of garden-variety literacy wrote:

> To get beyond colonized participation in/recruitment to a Discourse and incorporated critique, we need access to standpoints and perspectives from which we can seriously critique that Discourse, and position ourselves to promote and adopt others from an informed and principled base. . . . current educational reform proposals are no more

(or less) than an elaborate strategy for maintaining hierarchies, shoring up particular interests, and reproducing patterns of advantage and disadvantage–albeit in increasingly exacerbated and polarized forms. The ideal of powerful literacy presupposes and agenda of radical democracy, in which possibilities for future ways of doing and being are wilfully kept open . . . (Lankshear, 1998, p. 372)

Such statements of "powerful" literacy (or "emancipatory," "liberatory," "communistic" or "egalitarian") belong to an education-as-redemption genre. The key weakness is that commitment to the social theory, or ideology, is built into the definition of literacy. This, unfortunately, eases the transition from education to indoctrination.

This book defends a liberal account of scientific literacy, where appropriate knowledge *of* and *about* science is required. It is hoped this book will establish that teaching about pendulum motion in a historical, contextual, and integrative manner can contribute both to the learning of science and to understanding of the nature of science. Such understanding is important at the present time as many critics of science raise issues whose resolution depends upon an adequate grasp of the nature of science, an understanding of the procedures of theory choice in science, and an appreciation of the relations of science to its cultural and social milieu. Constructivists down play the "truth seeking" function of science, and emphasize the "interest serving" function: The pendulum story is a nice test case for these alternative interpretations of science. The contextural teaching advocated here can make science classes more engaging and might contribute to the greater participation of students in science programs, thus counteracting the general flight from science occurring in most Western countries.

CHAPTER 2

Navigation and the Longitude Problem

Since the 15th century, when traders and explorers began journeying away from European shores, problems of navigation and position-finding, especially the determination of longitude, became more and more acute. As we will see, accurate time-keeping was recognized as the solution to the longitude problem, and the pendulum played a pivotal role in solving it. Nearly all the great scientists of the 17th century (Galileo, Huygens, Newton, and Hooke) worked intimately with clock-makers and used their analysis of pendulum motion, specifically its isochronism, to create more accurate clocks. As well as being instrumental in the scientific revolution, the pendulum was instrumental in the associated horological revolution.

Position on the earth's surface is given by two coordinates: *latitude* (how many degrees above or below the equator the place is), and *longitude* (how many degrees east or west of a given north–south meridian the place is). When traveling over land, or sailing across oceans, one must know the destination coordinates (latitude and longitude), the present position coordinates, and of course a way of guiding oneself between them. Knowledge of latitude and longitude is essential for accurate mapmaking. Losing one's way on the ocean frequently amounted to "being lost at sea." A lack of knowledge of position made reliable traveling and trading problematic and dangerous.

ERATOSTHENES AND ANCIENT NAVIGATION

The Greek astronomer Eratosthenes (275–194 BC), who not only was a friend of Archimedes but also a librarian of the great collection at Alexandria, provided the first accurate estimation of the circumference of the earth, and hence of the length of a degree of longitude. Eratosthenes realized that the problem could be solved by comparing the angle of the sun's rays at two points, whose distance apart was known, on the same meridian (north–south line), at the same time. The calcula-

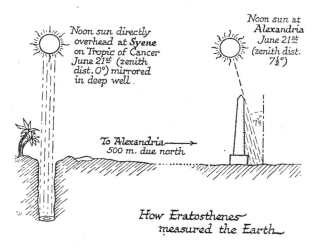

Figure 3. Eratosthenes' Observations at Syene and Alexandria (Hogben, 1940, p. 83)

tion would be simplified, if the time was noon, and at one of the points the sun was directly overhead. He choose the summer solstice (June 21) for the date, and Syene (modern Aswan), because it was on the Tropic of Cancer (lat. 23.5°N), for one location. At noon at Syene the sun was directly overhead and cast no shadow; its light was reflected directly off the water in a deep well (Figure 3).[1] On the same day, at the second location in Alexandria (which he took to be due north on the same meridian), a colleague measured the angle made at noon by the shadow cast by a high vertical obelisk (Figure 4). This shadow angle was 7.5°.

This meant the angle at the center of the earth, between radii from Syene and Alexandria was also 7.5°, or 1/50 of 360°. And because arc lengths are in proportion to the angles that they subtend, then the circumference of the earth must be 50 times the arc distance from Alexandria to Syene. But how to find this distance?

By noting how many days camels took to make the journey between the two centres, and knowing how far camels normally travel in a day, Eratosthenes determined the distance between Syene and Alexandria to be 5,000 stadia. This meant

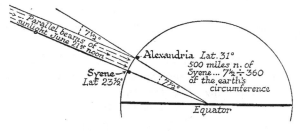

Figure 4. The Calculation of Eratosthenes (Hogben, 1940, p. 83)

the circumference of the earth was then 50 × 5,000, or 250,000 stadia.[2] A stadium was the length of the basic foot-race run in the stadium. Other sources indicate that for Eratosthenes this was equivalent to 157.5 meters (Heath, 1913, p. 339). On this assumption, the earth's circumference can be calculated to be 39,690 km or about 24,662 miles, which is remarkably close to the accepted modern value of about 40,000 km or 25,000 miles! This gives about 60 nautical miles to the degree, a figure that, with some qualifications, was later used by Portuguese, Spanish, and English cartographers.[3] And because circumference is given by the formulae, $C=2\pi r$, Eratosthenes calculation of the circumference allowed an estimate of the radius (3,924 miles), and hence diameter (7,849 miles), of the earth.

The second century BC achievement of Eratosthenes depended upon mathematical and astronomical knowledge. It is a nice early example of mathematics in the service of science, and subsequently of science serving human interests.

DETERMINATION OF LATITUDE AND THE PUZZLE OF LONGITUDE

Latitude determination was relatively straightforward. The inclination of the pole star (when north of the equator), inclination of the sun (when either north or south of the equator), and length of day had all been used with moderate success to ascertain how many degrees above the equator the observer was. Williams relates how Pytheas (ca.300 BC) found, to within the accuracy of a quarter of a degree, the latitude of his sundial's gnomon in his home town of Marseilles by measuring the length of its shadow at noon at the equinox. With the latitude known, the gnomon may be used to find the declination of the sun by measuring the length of its noon-shadow at other times of the year. The theory is that at noon at the equator, the sun is directly overhead and the gnomon's shadow has no length; as one moves progressively north, the gnomon's noon shadow lengthens. The major proviso is that the

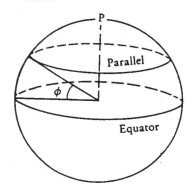

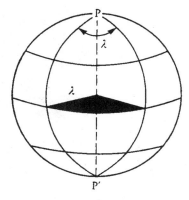

Figure 5. Latitude ϕ (Williams, 1992, p. 11)   Figure 6. Longitude λ (Williams, 1992, p. 11)

gnomon be oriented along the meridian passing through its location—that is, parallel to the earth's axis.

In the second century AD, Ptolemy, one of the foremost early geographers,
produced maps of the then known world with tolerable estimates of latitude for
over 8,000 locations (Ptolemy 1991). Knowing, in theory, that elevation of the
sun at noon, or elevation of the pole star above the horizon, gave a measure of
latitude, still left its measurement to be determined. It is technology and craftskill
that enables this gap between theory and realization to be bridged. Taking sightings
from a heaving deck was one problem, looking into the sun was another, calibrat-
ing instruments was yet another. But these technical problems were progressively
mitigated. The quadrant was the first instrument to provide quantitative readings.
It was introduced to European navigation in the 13th century. Then came the
astrolabe, the cross-staff (designed by Werner in 1514), back-staff (late 16th century), and the sextant in 1757.[4] All of these enhanced the accuracy of latitude
determination.[5] Bartholomeo Diaz at the Cape of Good Hope in 1488 and Vasco
de Gama in the Bay of Sainte-Hélène in 1497 made use of large astrolabes for
determination of their latitude.

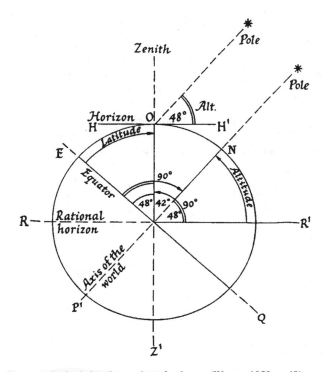

Figure 7. Latitude by observation of pole star (Waters, 1958, p. 49)

In the northern hemisphere the inclination above the horizon of the pole star was the favored method of determining latitude. The North Star is, conveniently, located almost directly above the earth's north pole, and consequently is not seen to rotate in the sky as the other stars do while the earth spins on its axis. For an observer at the north pole, a line between the pole star and the observer is perpendicular to the horizontal. As the observer moves toward the equator, the inclination of the line of sight to the pole star decreases, becoming zero at the equator. Assuming the pole star to be fixed above the pole (which it isn't) and at an infinite distance (which it isn't), each degree of altitude of the star above the horizon represents one degree of latitude above the equator.[6]

In the southern hemisphere, where the pole star was not visible, the altitude of the sun above the horizon at noon was the favored means of determining latitude, with an allowance made for the annual declination of the sun as it moves between the Tropics of Cancer and Capricorn (plus and minus 23°). This variation of the altitude of the sun with seasons meant that Tables of Declination (the "Regiment of the Sun") needed to be developed in order to convert solar altitude readings into location latitude readings (Waters 1958, pp. 48–55; Landes 1998, p. 23). In Figure 8, the sun's altitude is 62°, its declination is 20°, its 'real' or equinoctial altitude is thus 42°, and so the observor's latitude is 90° minus 42° or 48°.

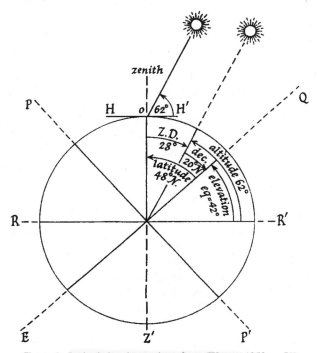

Figure 8. Latitude by observation of sun (Waters, 1958, p. 51)

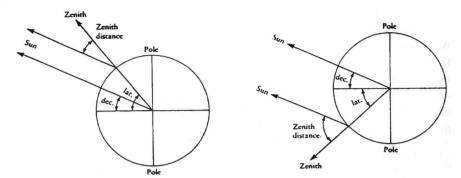

Figure 9. Latitude by noon sight: (a) Summer in northern hemisphere; (b) winter in southern hemisphere

Degrees of latitude are related to distance traveled along the meridian. The length of a degree of latitude, unlike the length of a degree of longitude, does not appreciably change as one moves from the tropics to the poles, the earth being a slightly flattened sphere. Thus provided one traveled north–south, the change in latitude could also be calculated from the distance travelled (approximately 55 miles to the degree).

The Puzzle of Longitude

Longitude was not such a pressing problem for those sailing in the Mediterranean Sea, or close to the coast of Africa or England. Although in the *Acts of the Apostles,* it is written that passengers in Paul's boat had to "abandon all hope" when a storm blew the boat away from the sight of land. However, as soon as the Portuguese, and later the Spanish, in the 15th century sailed into the Bay of Guinea and out into the Atlantic, then it became a matter of life or death, and of profit or bankruptcy. Galleons leaving Europe had to go to islands and harbors in the New World. As they returned home, laden with spoils, their sailors had to know in which direction to sail, how far it was to Lisbon or Seville, and how to ration their food and water. Ships regularly, *Flying Dutchman* like, were blown around with starved and decimated crews until they went aground or went to the ocean's bottom.

Until the problem of longitude was solved, sailors when out of sight of land navigated by a combination of compass bearing, dead-reckoning, and latitude determination. They headed off in the right direction until they reached the latitude on which their destination was located and then sailed along that latitude until they reached the port, fishing ground, island, or wherever else their goal was. This sounds easy, but it was fraught with difficulties.

During the Middle Ages ships' captains and pilots determined their position at sea by depending on approximate latitude determination (as given by the height of the sun, or the pole star), sightings of land, oral and written records of previous journeys, bottom sediments as captured on lead-lines, sea birds, currents, water coloration, water depth, floating weeds, along with other indicators such as prevailing winds. Dead-reckoning was the favored objective method to plot travel routes and location. In this method, direction of travel was ascertained by a compass, and speed was ascertained by noting (with an hour-glass) the time that a log-line, (literally a line with a floating log tied to its end) after being thrown off the ship's bow took to traverse the length of the ship. The direction of water current, and its speed, were just some of the allowances (guesses) that had to be made for dead-reckoning to be approximately correct.

In 1483 a German monk gave a vivid account of his Mediterranean voyage to the Holy Land, an account that nicely describes the state of navigation at the time:

> Besides the pilot there were other learned men, astrologers, and watchers of omens, who considered the signs of the star and sky, judged the winds and gave directions to the pilot himself. And they were all of them expert in the art of judging from the sky whether the weather would be stormy or tranquil, taking into account besides such signs as the colour of the sea, the movements of dolphins and fish, the smoke from the fire, and the scintillations when the oars were dipped into the water. At night they knew the time by an inspection of the stars. And they have as compass a *Stella Maris* near the mast and a second one on the upper poop deck. And beside it, all night long, a lantern burns, and they never take their eyes off it, and there is always a man watching the Star [by which he means the compass-card] and he sings out a sweet tune telling that all goes safely, and with the same chant directs the helmsman. Nor does the helmsman dare alter course in the slightest except at the command of the one watching the *Stella Maris*, from which he sees whether the ship should continue her course. (Taylor and Richey, 1962, p. 8)

The gravity of the problem is well illustrated in the 1599 observation of Simon Stevin, the Flemish polymath: "People going to St. Helena have arrived at the latitude but not known whether to go west or east" (Williams, 1992, p. 88). In a small sailing boat in the middle of the south Atlantic, after a journey of many weeks or months from Europe or the Canary Islands, with limited supplies, and crew and passengers probably dying of scurvy—turning west or east was no light decision (even assuming that the latitude determination was accurate to within half a degree, or thirty miles, which itself made sighting land difficult).

PORTUGUESE NAVIGATION AND EXPLORATION

After Ptolemy's grand achievements in cartography in the second century, the next great European advance in the fields of cartography, navigational science, ship building and maritime technology was made by the Portuguese in the 15th century. In contrast to most of the rest of Europe, the 15th century was relatively calm for Portugal, the small Atlantic-facing kingdom on the Iberian peninsula.

The major powers (Venice, Genoa, Catalan, the Arabs) had locked Portugal out of Mediterranean trade, so it had no choice but to turn toward Africa and the Atlantic.

The Portuguese, under the direction of Prince Henry the Navigator, the deeply religious son of King John I, began exploring the coast of Africa in the early 15th century. In 1419 Henry established the first scientific institute for the study of navigation and chart making at Sagres (initially named by Ptolemy, now Cape Saint Vincent), an ocean promontory in the extreme southwest corner of Portugal. Geographers, mathematicians, astronomers, and engineers were employed irrespective of race or creed.[7] Moors and particularly Jews who were persecuted in Spain, made valued contributions to the work at Sagres. It was an anticipation of Bacon's "commonwealth of learning," and an early form of the modern research and development laboratory.

Exploration of the African Coast

At first, explorers hugged the coast of Africa: The Madeira Islands were reached in 1420, the southern border of Morocco in 1421, Cape Verde in 1445, and Gambia in 1457. This movement south was not all, so to speak, plain sailing. Indeed, initially the Portuguese used galleys, only later was the caravel perfected. About a thousand miles south of the Straits of Gibraltar, on the coast of what is now the Western Sahara, a now insignificant promontory juts out. This was, and is still called, Cape Bojador ("Bulging Cape"). It may be only a fraction of the way down the African coast, but in the early 15th century it was effectively the end of the Portuguese world, a place beyond which even the most intrepid explorers would not venture. One of Henry's captains, Zurara, wrote of why ships would not pass Cape Bojador:

> And to say the truth this was not from cowardice or want of good will, but from the novelty of the thing and the wide-spread and ancient rumor about this Cape, that had been cherished by the mariners of Spain from generation to generation . . . For certainly it cannot be presumed that among so many noble men who did such great and lofty deeds for the glory of their memory, there had been no one to dare this deed. But being satisfied of the peril, and seeing no hope of honor or profit, they left off the attempt. For, said the mariners, this much is clear, that beyond this Cape there is no race of men nor place of inhabitants . . . and the sea so shallow that a whole league from the land it is only a fathom deep, while the currents are so terrible that no ship having once passed the Cape, will ever be able to return . . . these mariners of ours . . . [were] threatened not only by fear but by its shadow, whose great deceit was the cause of very great expense. (Boorstin, 1983, p. 166)

Between 1424 and 1433 Prince Henry sent 15 unsuccessful expeditions to round this cape. It was finally rounded in 1434 by Gil Eannes, who in 1444 pushed 250 miles farther south to Cape Blanco and brought back the first shipload of African slaves for sale in Lisbon. The landing of this human cargo silenced most of Henry's critics who hitherto thought he had been squandering the kingdom's cash on frivolous adventures. Cape Verde was reached in 1445, the Cape Verde Islands and Senegal in 1457.[8]

But Henry also wanted his ships to be able to leave the African coast and efficiently sail into the Atlantic. After improvements in shipbuilding, victualing, and navigation, they did so. The Portuguese colonized the Azores and Madeira islands in 1432. After Henry's death in 1460, John II (until his death in 1495), and later Manuel, continued the program of exploration and the promotion of Portuguese navigational science.

Bartholomeu Dias rounded the Cape of Good Hope in 1488. Vasco da Gama (1460-1524) reached India in 1499.[9] One year later Cabral, sailing west, reached Brazil and claimed it for the King of Portugal. Thus in 80 years one of the smallest and poorest countries of Europe had established colonies, trading posts, and fortifications across the face of the known world. Had the Portuguese not turned down Columbus, who in 1492 sailed west in search of the Spice Islands and dis-

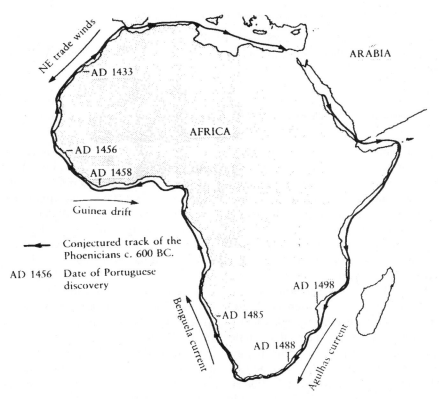

Figure 10. Portuguese exploration of the African coast (Williams, 1992, p.7)

covered the Americas claiming them for the king and queen of Spain, and had they not alienated their own Ferdinand Magellan who also ended up serving Spain, they would also have planted the Portuguese flag in the West Indies and North America. This would have given this insignificant European power a presence in three continents.

Despite the efforts of Henry and his successors at Sagres, navigational science could not solve the problem of longitude. Progress was undoubtedly made in navigation, but the ultimate prize, determination of longitude, was beyond reach.

Medieval Navigation

The ignorance of longitude and dependence upon empirical methods is apparent from this late 15th century book written by Joao de Lisboa for the instruction of Portuguese navigators:

> ... when you are making for the Cape of Good Hope ... you will see a very large round hill ... towards the East, you will see a large mountain with several peaks. Beyond this there is a narrow strip of land with several hills. (Mathew, 1988, p. 23)

Bad luck if it is raining, or a fog has settled in.

In 1597 another pilot, gave advice on how to traverse the route from the Cape to Portugal's Indian settlements (principally Goa and Diu):

> On seeing these signs together with men-of-war birds and winds in the East, you should try to work South-Westwards ... and although gulf weed is found on this coast, it is not found together with sea gulls and men-of-war birds. (Mathew, 1988, p. 23)

Floating weeds and sea water color, indicating depth and river sedimentation, helped tell position, but they could not demonstrate what degree of longitude you were sailing. On each voyage this information was recorded in *roteiros* (itineraries), and was collated back home with that from captains, and improved *regimentos* (rules) issued for subsequent voyages. Thus there was a systematic organization of navigational lore, where previously, as King Alfonso V complained in 1443, navigators' records had been haphazard and slovenly, not "marked on sailing charts or mappe mondes except as it pleased the men who made them" (Boorstin, 1983, p. 162).

The Portuguese also improved astronomical methods of navigation, and they refined astronomical instruments. In 1455 they identified the Southern Cross (called the *Crozier*) constellation and, after observing it, worked out a method for determining the height of the southern pole (Waters, 1958, p. 54). The calculation of the height of the pole at midday required knowledge of the declination of the sun in the earth's annual orbit around it. After years of repeated solar observation, they prepared, a table of declination showing for each day the place of the sun in degrees and minutes. Two surviving handbooks containing this information are

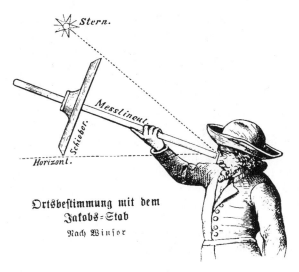

Figure 11. Using Jacob's staff (Mills, 1983, p. 137)

Regiment of Munich (1509) and *Regiment of Evora* (1517), located in the libraries of Munich and Evora. Instruments that assisted astronomical observation (astrolabe, quadrant, Jacob's staff, and compass) were in some cases invented, and in others, improved (Stimson, 1998, Andrewes, 1998c).

The Portuguese paid great attention to that marvellous twelfth century invention, the compass. A contemporary (1601) English writer on Portuguese navigation said:

> Portugalls doe exceed all that I have seen, I mean for their care which is the chiefest in navigation. And I wish in this . . . we should follow their examples. In every ship, . . . upon the half deck or quarter deck, they have a chair or seat, and of which they navigate, the Pilot or his adjutants never depart day or night, from the sight of the compass. (Mathew, 1988, p. 19)

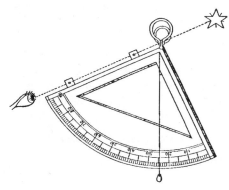

Figure 12. Quadrant, with pinholes for sighting the pole star and a plumb-line which crossed a graduated scale giving the star's altitude (Goodman and Russell, 1991, p. 119)

Francis Drake, who scorned navigational science, took Portuguese navigators with him on his voyagers of plunder (Thrower, 1984).

In 1538 Joao de Castro, during a voyage to India, made an exacting study of the deviation of the compass needle, one of the then hoped-for methods of determining longitude (it was used by Columbus in his 1492 voyage). This "deviation" results from the compass needle pointing to the magnetic north pole, not the geographic north pole. The difference, or deviation, is zero only when the compass is on a line that runs through both the magnetic and the geographic poles. At other points the deviation will vary with the distance from that meridian of zero deviation. In the Mediterranean sea, the deviation is to the east, diminishing to zero at the Azores, and then it is to the west with further voyaging into the Atlantic. Compass deviation thus appeared an attractive way of estimating longitude, given that geographic north was determined by sighting of the pole star.[10] Castro's diligent observations contributed greatly to the study of geomagnetism, but undermined the deviation method for determination of longitude: It was found that deviation varied from year to year, and that the relationship between the magnetic and geographic poles varied.

Cartography was refined in this period. During the early years of the 16th century, Duarte Pacheco sailed in the Portuguese East Indies and produced remarkable maps that were printed in Lisbon. These gave the latitudes of the principal centers: Chaul 22° N, Anjediva 15° N, Cannanore 12° N, Calicut 11° N, and Quilon 8° N.

This accumulated knowledge, navigational lore, and technique was of great worth in Portuguese exploration and exploitation of the New World. And it was the envy of the other European powers. But it fell short of what was needed for reliable navigation and safe transoceanic voyaging (Landes, 1998, p. 23). Knowing a destination's latitude was one thing; knowing whether to go east or west toward it was another.

The magnitude of navigational decisions can be appreciated from this 16th century account of the capacity of the large sailing vessels:

> Imagine for yourself and think what a ship of India is like when put to sea [on a six month voyage] with 600 to 800 and sometimes more than 1,000 persons in it, comprising of men, women, children, slaves, freemen, fidaloges, plebians, merchants, soldiers and sailors. (Mathew, 1988, p. 260)

Being lost in the middle of the ocean, with limited rations being devoured and spoiled, was an unhappy fate. In 1590–1592, seven such ships left Lisbon for Goa, but only two returned. For the period 1500–1610, almost 10 percent of Portuguese ships sailing for India were wrecked.

The Treaty of Tordesillas, 1494

The longitude problem loomed large in the enactment of one of Europe's most famous treaties. Pope Alexander VI, upon the return of Columbus in 1493, tried to

Figure 13. Portuguese caravel (Mendelssohn 1977, p. 27)

balance the competing interests of Spain and Portugal, the major Catholic trading nations, by drawing a meridian line down the globe at a point 100 leagues (about 300 miles) west of the Azores and declaring that all lands to the east of it were to be Portuguese, and all to the west Spanish. That the world could be divided in this way by the Pope at Rome was testimony to the spiritual and temporal power of the Catholic church, a power which 20 years later would be irrevocably challenged by Martin Luther's nailing of his 95 theses to the door of the Wittenberg Cathedral. The Portuguese king, John II, was not happy with the terms of the adjudica-

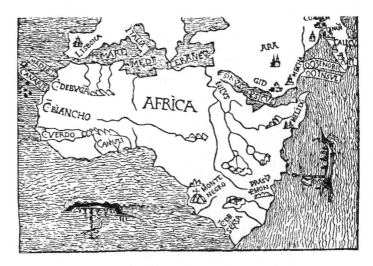

Figure 14. 1508 Portuguese map of Africa (Mendelssohn, 1977, p. 36)

tion; a revised Treaty of Tordesillas, in 1494, divided the Atlantic Ocean at a point 370 leagues (1,200 miles) west of the Cape Verde Islands, or longitude 46 degrees west on current reckoning. This ensured that Brazil, at least as it was then known, fell within the Portuguese sphere of influence—its subsequent expansion westward to the Andes was in violation of the treaty.[11]

Marking the meridian on paper was one thing, *finding* the meridian at sea was altogether different. Locating the anti-meridian in the Pacific as the Portuguese and Spanish sailed to India, the East Indies, and the Philippines was subject to great error. When Magellan crossed the Pacific in 1521 and found the riches of the Spice Islands, the kings of Spain and Portugal could not determine to whom they should belong under the terms of the treaty of 1494. The 46 degrees west meridian, was continued in Asia as the 134 degrees east meridian. But no instruments could ascertain where this line was to be placed, or what territories were to fall either side of it.

The Conference of Badajoz was convened in 1524 to try to sort out these vexatious matters. Magellan made a Solomonic determination, putting three of the Moluccan islands on the Portuguese side of the antimeridian, and two of them on the Spanish side. More committed defendors of Spanish interests put the antimeridian at the mouth of the Ganges in India, thus making all the Moluccas, as well as Java, China, and Japan, fall within the Spanish sphere. Predictably Portuguese interests saw things differently: Their delegates to the Badajoz conference put all the Moluccas 12 degrees within the Portuguese sphere. After two months the conference broke up without coming to a resolution.[12]

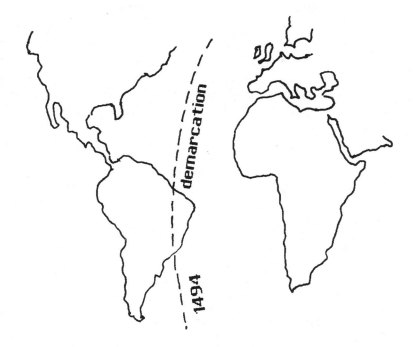

Figure 15. Treaty of Tordesillas demarcation

In 1564 Philip II of Spain sent Andrés de Urdaneta, a Dominican friar and navigator, from Mexico to Asia in search of the Spice Islands. This voyage resulted in Spain's colonization of the Philippines, but not before inciting a dispute about whether they fell on the Portuguese or Spanish side of Tordesillas meridian.

THE LONGITUDE PROBLEM

At the end of the 15th century the hopelessness of the longitude problem can be seen by the fact that Christopher Columbus, although successful in plotting latitude in order to sail a straight path on latitude 28° across the Atlantic in 1492, was so far out in his longitude reckoning that he thought Cuba was a part of China![13] The problem is also evident in a 1499 account by the Florentine explorer Amerigo Vespucci who was working for the Portuguese and following Columbus's route across the Atlantic:

> As to longitude, I declare that I found so much difficulty in determining it that I was put to great pains to ascertain the east–west distance I had covered. The final result of

my labors was that I found nothing better to do than to watch for and take observations at night of the conjunction of one planet with another, and especially of the conjunction of the moon with the other planets, because the moon is swifter than any other planet. (Boorstin, 1983, p. 247)

The 16th century did not add much to the art of longitude determination. The practice of sailing the latitude, and converting distance sailed into longitude degrees (rather than the reverse), was still firmly entrenched. Martin Fernandez de Enciso wrote in 1519 that:

Sailors calculate distance in an East-West direction in nights and days and with an hourglass and the calculation is reasonably correct for those who know their ship well and how much it sails in an hour . . . Because their estimation is approximate . . . they over-estimate rather than under-estimate the number of leagues of their voyage so as to be warned of their approach to land, rather than running upon it suddenly. (Randles, 1985, p. 235)

The practice is well illustrated in a 1577 account by Captain Martin Frobisher of his return to England after his efforts to find the northwest passage:

Having spend foure or five dayes in traverse of the seas with contrarye winde, making oure souther way good as neare as we could, to raise oure degrees to bring ourselves with the latitude of Sylley, we tooke the height the tenth of September, and founde ourselves in the latitude of (cypher) degrees and in ten minutes. . . . And upon Thursday the twelfth of September taking a height, we were on the latitude of (cypher) and a halfe, and reckoned ourselves not paste one hunderd and fiftie leagues [450 miles] short of Sylley . . . then being in the height of Sylley . . . we kept our course east, to run in with the sleeve or channel so called . . . the fifteenth we began to sound with oure lead, and hadde grounde at sixty-one fathome depth . . . the seventeenth . . . we . . . were shotte betweene Sylley and landesendes. (Waters, 1983, p. 144)

The longitude situation, near the end of 16th century, is captured in a journal entry of Abraham Kendall, an English authority on navigation. Crossing the Atlantic in 1594, he wrote:

Lat. 9° 56' N, Long. 337° E. [of the Azores]. We are 640 leagues of a great circle from Cape Blanco. Here we begin to see some birds of the Indies called *forcedos*. . . . Lat. 9°30', Long. 335°. We saw as a sign we were nearing America some great birds like crows, but white with long tails. . . . Birds began to settle at night on the rigging and the sea water was considerably whiter. These are signs of the neighbourhood of the coast. (Taylor, 1956, p. 246)

On his homeward journey, he wrote:

Bermuda was distant 110 leagues of a great circle, and we saw frigate-birds and sea-mews. The current ran to north-east and carried with it weeds from the rocks of the Indies. And when one no longer sees such weeds it is a sign that one passes Cape Race, Long. 344° 10' from Pico of the Azores. (Taylor, 1956, p. 246)

The longitude problem occupied the best navigators and astronomers of the 17th century. Henry Phillippes in his widely used English navigation manual,

Advancement of the Art of Navigation (Phillippes, 1657a), proposed the establishment in London of a navigation college whose chief research occupation would be "the great Master-piece and Mystery of Navigation .. the discovery of the true Longitude of places" (Bennett, 1985, p. 220). In a second publication, *The Geometrical Seaman* (Phillippes, 1657b), Philippes mentions Galileo's discovery of the isochrony of the pendulum as a possible contributor to solving the longitude problem. This was by using the pendulum as a timing device while running out the ship's log, thus obtaining a more accurate estimation of the speed of the ship. That the problem of longitude remained unsolved is indicated in Philippes' advice to seaman to "use dead reckoning, and to avoid sailing due east or west, when no correction can be made by latitude sights" (Bennett, 1985, p. 220). It is noteworthy that Phillippes was hardly asking for too much precision in solving the longitude problem: He was happy to have longitude to within a "degree or two . . . especially in distant places where two or three hundred leagues difference in their account is sometimes no great matter" Bennett, 1985, p. 220.

In 1683 Samuel Pepys, the English diarist, and for a time an official of the Royal Navy, commented on a voyage to Tangiers:

> It is most plain, from the confusion all these people are in, how to make their reckonings, even each man's with itself, and the nonsensical arguments they would make use of to do it, and disorder they are in about it, that it is by God's Almighty Providence, and great chance, and the wideness of the sea—that there are not a great many more misfortunes and ill-chances in navigation than there are. (Sobel, 1995, p. 16)

In the middle of the 18th century (1741) the English commodore George Anson (1697–1762) overpowered the Spanish Acapulco galleon as it was bringing 500 thousand pounds in gold from Mexico. Anson then spent a month rounding Cape Horn, trying to get from the Atlantic ocean to the Pacific. By his reckoning he was 10° west of Tierra del Fuego and so steered his prize due north only to find that he was still *east* of the cape. Food and water supplies declined as scurvy cases and death rates increased. Having finally rounded the cape he sought out the island of Juan Fernandaz, at latitude 35°S and 80°W longitude (about 600 miles off the coast of Chile), so he could land his sick and reprovision his boats. Normally he would have hugged the coast of Chile until he got to the right latitude and then he would "sail the latitude" west until he came to the islands (Chile, being in the control of Spain, was not an option for victualing). With provisions running out, and sickness and death on the rise, Anson decided to sail direct to the islands.

When he got to the correct latitude he did not know whether he was east or west of his destination. Anson spent weeks sailing back and forth across the latitude. On one tack he got within a few hours sail of Juan Fernandaz, only to turn around and sail all the way back to the Chilean coast. Eighty or more of his crew perished.

In 1763, the British Astronomer Royal and navigator, Neville Maskelyne, gave the following less than flattering account of the state of the navigator's art:

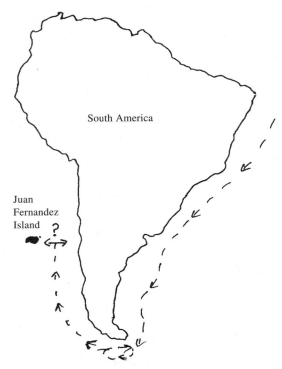

Figure 16. Commodore Anson's 1741 voyage

Daily experience shews the wide uncertainity of a ship's place, as inferred from the common methods of keeping a reckoning, even in the hands of the ablest and most careful navigators. Five, ten or fifteen degrees [of longitude], are errors [of hundreds of miles] which no one can be sure he may not fall into in the course of long voyages. (Waters, 1983, p. 146)

LONGITUDE PRIZES

Governments and kings offered huge rewards for anyone who could solve the problem of longitude.[14] In 1598, Phillip III of Spain offered a perpetual pension of six thousand ducats, a life annuity of two thousand ducats, and a cash prize of one thousand more. In the next 50 years his offer was followed by ones from Portugal, Venice, France, and Holland.

In 1666, King Louis XIV of France established the *Académie des Sciences* and charged it with finding out how to ascertain longitude and improve map-making—perhaps the first case of science in the service of the state. Louis also offered enormous cash prizes for a satisfactory solution to the longitude problem. Any claimant had to satisfy M. Colbert, a lieutenant general in the king's navy

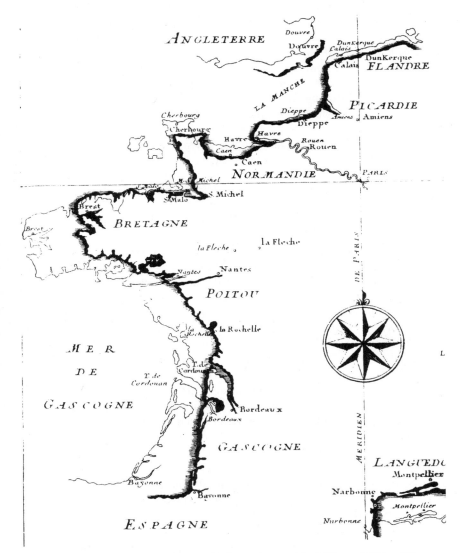

Figure 17. 1681 Survey of France (Paris as prime meridian) (Howse, 1980, p. 17)

and designated representatives of the *Académie des Sciences*. Many came forward with proposals, all went away empty-handed.

On land the problem was not so great. In 1681 the newly published survey of France, commissioned by Louis and conducted by Jean Picard, Philippe de la Hire, and others, using telescopic observations of the eclipses of Jupiter's moons,

moved Ushant and Brittany moved eastwards by 1.5 degrees of longitude, caus-
ing Louis to complain that his "surveyors had lost him more land than his armies
had gained" (Howse, 1980, p. 16).

The English government had long since abandoned Drake's low opinion of
navigational science. When, in 1662, Charles II granted the Gresham College
group its royal charter, making it the Royal Society of London for Improving
Natural Knowledge, a poem celebrating the event included these lines:

> The College will the whole world measure,
> Which most impossible conclude,
> And navigators make a pleasure
> By finding out the longitude.
> Every Tarpalling shall then with ease
> Sayle any ships to th'Antipodes (Howse, 1980, p. 15)

In 1675, when Charles II established the Royal Observatory at Greenwich,[15]
the first astronomer royal, John Flamsteed (1646-1719), was appointed with the
charge: "forthwith to apply himself with the most exact care and diligence to
rectifying the tables of the motions of the heavens, and the places of the fixed
stars, so as to find out the so much desired longitude of places for the perfecting
the art of navigation" (Quill, 1966, p. 2). This charge was the direct consequence
of disastrous failures of the English fleet in the third Anglo-Dutch War, failures
that Sir Jonas Moore, the surveyor-general and director of the Ordinance Office,
attributed to an inability to fix longitude with exactitude (Jardine, 1999, p. 159). It
was Moore who personally installed Flamsteed in the observatory and gave him a
prototype of Hooke's balance-spring watch to better time astronomical events
such as the eclipses of Jupiter's moons upon which longitude determination was
dependent.

But solutions did not come quickly. In 1707, the British admiral Sir Clowdisley
Shovell returning home victorious from Gilbraltar, sailed his fleet, in fog, onto the
Scilly Isles. He and all his navigators had grossly miscalculated their longitude.[16]
Four of his five ships, and 2,000 men were lost. Relentless pressure from trading
merchants, topped by this naval disaster,[17] moved the English parliament in 1714,
during the reign of Queen Anne. With the advice of Isaac Newton they passed the
Longitude Act[18] which provided for £20,000 (approximately a million dollars
today) for a method to determine longitude to an accuracy of a half degree of a
great circle (thirty miles), £15,000 for a method accurate to within two-thirds of a
degree, and £10,000 for a method accurate to within one degree (60 nautical miles
or 68 land miles) on a voyage from England to the West Indies. The act began:

Whereas it is well known by all that are acquainted with the Art of Navigation, That
nothing is so much wanted and desired at Sea, as the Discovery of the Longitude, for
the Safety and Quickness of Voyages, the Preservation of Ships and the Lives of

Men . . . such a Discovery would be of particular Advantage to the Trade of Great
Britain and very much for the Honour of this Kingdom. (Quill, 1966, p. 225)

The act's framers understated their point when they said "such a Discovery
would be of particular Advantage to the Trade of Great Britain." A modern historian has noted, with perhaps some hyperbole, that

the act of 1714 caused the same kind of surge of scientific effort that space research
does today, and was in many ways responsible for the Industrial Revolution that
followed. The invention of the marine chronometer, for which it was directly
responsible, resulted eventually in the domination of the world by the British Fleet,
the expansion of trading, and the acquisition of the British Empire. (Burton 1968, p.
84, quoted in Macey 1980, p. 28)

As one historian put it,

Longitude, then, was the great mystery of the age, a riddle to seamen, a challenge to
scientists, a stumbling block to kings and statemen. Only such will-o'-the-wisps as
the fountain of youth and the philosophers' stone could match its aura of tantalizing
promise—and longitude was real. (Landes, 1983, p. 111)

TIMEKEEPING AS THE SOLUTION TO THE PROBLEM OF LONGITUDE

By the sixteenth century a number of methods had been proposed for ascertaining
longitude. They were all what might be called *astronomical* methods, the fundamental principle of which was that a given astronomical event (a solar eclipse, a
lunar eclipse, a planetry transit) is seen on earth at different times as the earth
rotates. Observers at different *longitudes* see the event at different *times*. If the
time difference for the same observation is known for different locations, then the
longitude separation between the locations can be ascertained. This is because the
earth makes one revolution of 360 degree in 24 hours, thus in one hour it rotates
through 15 degree, or through 1 degree each four minutes. Thus, although the line
of zero longitude is arbitrary, there is an objective relationship between time and
longitude. One such astronomical event was a lunar eclipse, and Hipparchus in
the second century BC and Ptolemy in the second century AD, had both suggested
this as means of determining longitude. If the *local time* of such events in different places were measured and compared, then the difference in longitude straight-
forwardly follows. Measuring local time was the problem! And waiting around
for the event to occur, and communicating with the reference location, made the
method impractical for travellers.

Johann Werner (1468–1522), a priest of Nuremberg, revived the lunar eclipse
method in the sixteenth century. In 1514 he wrote:

For example, I myself saw a lunar eclipse in Rome on the evening of 18 January in the
year 1497. First contact for this lunar eclipse appeared to me, in Rome, to be at 5:24
pm on 18 January. But in Nuremburg first contact for this same lunar eclipse was seen

at approximately 4:52 pm, as computed from the ephemerides of Johannes of Regiomontanus. . . . Hence, it is obvious that the difference in longitude between the city of Rome and of Nuremburg is nearly 32 minutes or 8°. (Murschel and Andrewes, 1998, p. 383)

But lunar eclipses are infrequent, and so the position of the moon relative to some fixed star was also used in attempts to ascertain longitude. This was the *lunar distance* method, and involved measuring the angular displacement of the moon from a star at a given time in one location, and determining the local time at which the same displacement occurred at a second location. The time difference could again be converted into longitude difference by the above calculation. This method was also promoted by Werner (Murschel and Andrewes, 1998, p. 385).

Both of these astronomical methods were beset with severe technical problems—the moon's motion is highly irregular and difficult to predict, the stars' positions were not known with sufficient precision, and angle measuring instruments were inaccurate (Howse, 1998, p. 151). And both depended upon the accurate establishment of local time and some means to keep local time. The third astronomical method later to be introduced by Galileo, which depended upon disparate timings of the eclipses of Jupiter's moons, was also bedeviled by these practical difficulties. Some new method of solving longitude was needed.

Understandably the various Royal prizes, and particularly the British Longitude Act, spawned a host of eccentric would-be solutions to longitude (Gingerich 1998). But the ultimately correct and workable method, a *chronological* not *astronomical* method, was proposed by Gemma Frisius (1508–1555), the Flemish astronomer, professor at Louvain University, and teacher of Mercator the mapmaker. In 1530 he became the first to propose timekeeping as a solution to the longitude problem. In his *De Principiis Astronomiae & Cosmographicae,* published at Louvain, he wrote:

In our times we have seen the appearance of various small clocks, capably constructed, which, for their modest dimensions, provide no problem to those who travel. These clocks operate for 24 hours, in fact when convenient, they continue to operate with a perpetual movement. And it is with their help that the longitude can be found. First of all it is necessary to pay attention, before using them on a voyage, that the clock records with the greatest exactness the time of the place of departure, and that it is not allowed to run down during the voyage. When one is on course for 15 or 20 miles, and wishing to know how distant one is from the point of departure, it would be preferable to wait until the clock is at an exact division of time, and at the same time, with the assistance of the astrolabe, as well as of our globe, to seek the hour of the place in which we find ourselves. If this then coincides to the minute with the time indicated by the clock, it is certain that we find ourselves within the same meridian (as that from which we departed), that is, that we have continued our voyage towards noon or towards the north. In this manner it is possible to find the longitude, even in a distance of a thousand miles, even without knowing where we have passed, and without knowing the distance travelled. (Pogo, 1935, p. 470; see also Murschel & Andrewes 1998, p. 391)

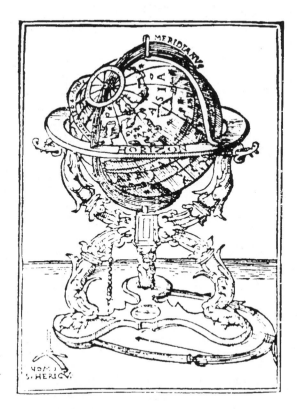

Figure 18. Title page of Frisius'
Cosmographicae (Pogo 1935, p.
474)

Frisius was aware of the deficiencies of mechanical clocks. In the 1553 edition of his *Cosmographicae*, as well as referring to the problem of longitude determination at sea, he said:

> Therefore it would be useful on long journeys, especially sea journeys, to use large clepsydras (that is water clocks) or sand glasses, which will measure a whole day exactly, through which errors of the other clock may be corrected. (Howse, 1980, p. 10)

For each day, provided it was not cloudy or raining, noon could be precisely ascertained by noting the instant at which a gnomon's shadow is at its minimum. An hour-glass or clepsydra could then be set in motion and used as a check on the mechanical clock. Frisius' work was translated into English by Richard Eden in 1555 during the reign of Queen Mary.

The science behind Frisius' suggestion has been noted: the earth makes one revolution of 360 degree in 24 hours. Thus in one hour it rotates through 15 degree, or 1 degree each four minutes. Although the line of 0 longitude is arbitrary,

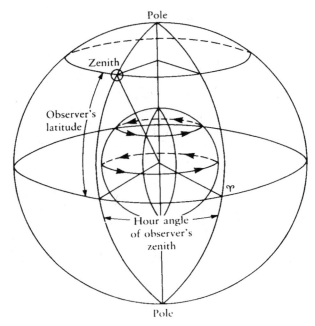

Figure 19. Without time longitude is indeterminate (Williams, 1992, p. 85)

there is an objective relationship between time and longitude.[19] If a reliable clock was set to accurately read 12:00 noon (that is, when the sun is directly overhead) at the outset of the journey, then one only had to look at the clock when the sun was at its highest (noon) throughout the journey to find out how many degrees east or west of the beginning one was (Andrewes, 1998c). If the clock read 2:00 when the sun was highest, then you were 30 degrees west of the starting point; if it read 9:00, you were 45 degrees east of where you started. Once a 0 meridian was chosen, then local time provided an agreed longitude. This decision was rich with political agendas, as just about every major power had the prime meridian on its maps going through its own capital: Munich, Paris, Copenhagen, Rome, Lisbon, Madrid, Washington, Brussels.[20]

On the same principle, the stars can also be used to determine one's longitude. If a traveler knows what time a star appears directly overhead at the prime meridian, then if he sees what time the star appears directly overhead at any other location in a journey he can determine his longitude—provided again there is a way to keep track of time on the prime meridian. Or other regular and common astronomical events can be used—such as the eclipses of Jupiter's moons, or the angular separation of the moon and some fixed star (the 'lunar distance' method). Galileo was the first to suggest using the Jovian moons to determine longitude,

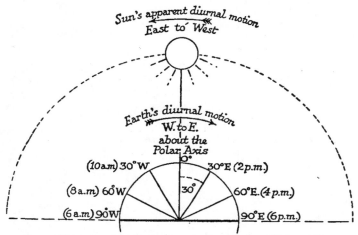

Figure 20. Time and longitude determination (Hogben 1936, p. 169)

and this method was successfully used for terrestrial mapping in the seventeenth, eighteenth and nineteenth centuries (Helden, 1998).

Time and longitude are so co-relative, that the latter is frequently expressed in terms of hours and minutes, instead of degrees.

15 degrees of longitude ≡ one hour (60 minutes) of time
1 degree of longitude ≡ 4 minutes (240 seconds) of time

But wanting to measure time accurately and reliably, and being able to do so, were different things. Frisius's conditions that the clock be set to the hour of departure "with the greatest exactness," and that it not be allowed to run down during the voyage—were, throughout the 16th century, utopian. Early in his reign, Phillip II of Spain (1556–1598) commissioned the royal cosmographer Alonso de Santa Cruz to survey, and report on, the extant methods of longitude determination. Philip was told:

> the longitude is now being sought for in Spain by means of clocks adjusted to register exactly twenty-four hours, and constructed in divers ways: some with wheels, chains, and weights of steel: some with chains of catgut and steel: others using sand, as in sandglasses: others with water in place of sand, and designed after many different fashions: others again with vases or large glasses filled with quicksilver: and, lastly, some, a weight and thereby the chain of the clock, or which are moved by the flame of a wick saturated with oil: and all of them adjusted to measure twenty-four hours exactly. (Gould, 1923, p. 20)

Santa Cruz commented that the best clocks could only get to an accuracy of plus or 30 minutes a day. Little wonder that Phillip II had only a vague idea of the location of his overseas possessions. David Goodman has pointed out that

Figure 21. Hogarth's longitude lunatic (Howse; 1980, p. 55)

from recorded observations of lunar eclipses taken in Panama in 1581 it was calculated that Panama City was 49°15' west of the Canary Islands; the actual difference in longitude is 61°, an error of 11°45', or about 700 miles. And in the Philippines where Urdaneta had asked a fellow Augustinian, Martín de Rada, to take astronomical observations for calculations of longitude, the results indicated that the Philippine town of Cebu was 215°15' west of Toledo; the correct figure is 232°, a large error of just under 1,000 miles. (Goodman, 1991, p. 133)

At the end of the sixteenth century, the best mechanical clocks gained or lost 15 minutes per day. Hardly satisfactory. And that was under ideal land-based conditions, not on storm-lashed heaving ships in the tropics or the Arctic. An hour corresponds to fifteen minutes of longitude, so if at the end of a journey of six weeks we wanted to determine longitude to within one degree (60 miles at the equator) then the error in the timekeeper could not be more than four minutes for the period, or approximately six seconds a day (99.993% accurate). It was thought fanciful to strive for such unheard-of precision.

Swift, Hogarth, and other satirists made great fun of the impossible search for longitude. They thought that the fountain of youth, the philosopher's stone, or the square circle, was more likely to be found. But Frisius' idea was right. And the right idea, along with the king's ransoms in rewards, fueled the search for better timekeeping devices. The search was joined by the most of the great figures of 17th-century science, including Galileo, Huygens, Leibnitz, Hooke and Newton.

CHAPTER 3

Ancient and Medieval Timekeeping

Knowing in theory how to solve the problem of longitude by timekeeping, and implementing this solution were two different things. The fundamental problem was the lack of an accurate and reliable timekeeper. This problem had cultural, scientific, and technical dimensions. Time measurement, as distinct from time-keeping, could not occur until people thought time was something measurable. That is, time measurement depended upon an understanding of time as an independent variable, and as something with metric (measurable) properties. This understanding emerged late in human history. Indeed it did not crystalize until the 16th century, and then, it seems only in Europe. Galileo's law of fall, where he related distance traversed in free fall to the square of the time elapsed ($s \propto t^2$) is arguably the first clear statement of time as an independent quality, being related to another as a dependent quality.

In the ancient world, time was not distinguished from events that took place in time; time was not separate from motion. For Plato and Aristotle, time was identified with the regular motion of the heavenly bodies. They held a *reductionist* account of time: time was constituted by the movement of bodies. Thus for Plato, a world in which there was no movement would literally be a timeless world. Piaget pointed to this as an example of his thesis of ontogenesis recapitulating phylogenesis (Piaget and Garcia, 1989): Young children also do not distinguish time from events that happen in time; the concept of time as something separate from events occurs late in a child's growth (Piaget, 1970).

Time *measurement* needs to be distinguished from time*keeping*. The former requires symbols and units; it is a public activity that also requires some method of *displaying* time. The latter requires neither of these. Animals and people can *keep* time and *estimate* durations in terms of some arbitrary, and even idiosyncratic, standard. A tiger can judge how long it will take prey to reach some point and adjust his own movements accordingly, and a farmer can judge how many days before a crop is ready for harvest. In both cases, the judgments do not require time measurement, only time estimation or timekeeping, and this can be in any "units" one chooses: days, months, or whatever.

Even if the concept of time as something measurable is present, if a society's activities do not require precise timekeeping, then the incentive to produce an accurate timekeeper is lacking. But the concept coupled with social need is insufficient for the production of devices that accurately measure time (clocks). If the science of time measurement is inadequate, then timekeepers will be inadequate. Finally, if technology and craft skills are undeveloped, then timekeepers will be crude and unreliable. All four dimensions (conceptual, social, scientific and technological) have to move forward for there to be accurate and reliable timekeepers, and thus instruments capable of solving the longitude problem.

Since prehistoric time, people have been conscious of the duration of events. But the *measurement* of duration (how long or brief duration is) depended upon the identification of a standard duration so that other durations could be counted as a proportion of the standard. From the earliest days, people were aware of a certain cyclic recurrence in events around them: migration of birds, repetition of seasons, fruiting of trees, rising and setting of the sun, stars and planets, the moon's cycle, and so on. These were all potential standards. And of course standards were relative to a culture's needs. Group standards for time (or length or weight) were satisfactory until the group had dealings with other societies. Ideally a standard should be universally available and reliable. The most commonly used time standard was the daily orbit of the sun, (or rotation of the earth as we would now say).

SUNDIALS

The oldest means of recording the passage of time was the sundial ("dial" comes from the Latin *dies* meaning "day"). The use of sundials for timekeeping probably goes back to inhabitants of the Tigris and Euphrates valleys in 2000 BC[1] Herodotus in the fifth century BC observed that they were widely used in Greece, and mentioned them as coming from Babylonia (Milham, 1945, p. 31 ff). That they measured time only during daylight hours was not a great drawback as this was the important part of the day when people worked and carried on their affairs. Nights were socially less important.

Ancient Egyptians recognized the variation of shadow length over the passage of a year. They saw that the noon shadow was shortest at one time (the summer solstice, June 21, when the sun is highest in the northern hemisphere sky), longest at another (the winter solstice, December 21, when the sun is lowest in the northern hemisphere sky), and in between at another (the vernal, March 21, and autumn, September 23, equinoxes, when the sun rises directly in the east, sets directly in the west, and days and nights are of equal duration). Figure 22 from Hogben (1936, p. 41), indicates the movement, over one year, of an oblesik's

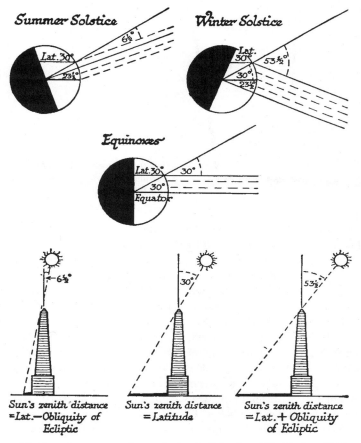

Figure 22. Annual movement of noon shadow in Cairo (Lat. 30°) (Hogben, 1936, p. 41)

noon shadow at Cairo (latitude is 30°) and its relationship to the earth's orientation to the sun.

Early Egyptian shadow clocks and sundials (ca. 1,500 BC) made use of the
relationship between the length of an object's shadow and the position of the sun.
A vertical stick was placed at the end of a long graduated board, and time was
read by noting the position of the shadow as the sun rose and fell (the device had
to have its orientation changed at midday). Egyptians used a decimal system for
daylight hours (ten), a duodecimal system for night hours (12), and had two "hours"
for twilight. Thus their day was broken into 24 "hours" of uneven duration and
uneven distribution between day and night (Neugebauer, 1969, p. 86). Egyptians
were a long way short of the full and complex science that enabled sundials to tell
time in equal hours independently of season and of the location's latitude (Clagett,

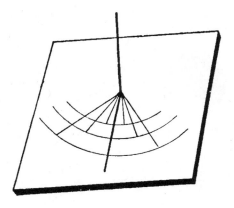

Figure 23. Board with vertical rod for finding meridian and true solar noon (Milham, 1945, p. 32)

1995, p. 97). This knowledge did not develop until the 16th century and was first codified in Sturmy's *The Art of Dialling* (Sturmy, 1683).

Early morning and late afternoon presented problems for flat sundials because the shadow of the gnomon was enormously lengthened. Perhaps the first breakthrough in the technology of sundials was the creation in antiquity of sundials in the form of a hemicycle or the interior of a half-sphere.[2] This overcame the problem of inordinate shadow lengths, and it also allowed, by a system of engraved "longitude" lines, hours to be equalized across the seasons.

Determining the direction of the meridian (see Chapter 1) was a first step in refinement of sundial technology. The direction of the meridian at any point can be easily ascertained by drawing concentric circles around a vertical stick in the ground and bisecting the angles made by the morning and afternoon shadows on each of the circles. When the shadow falls along this line it is noon. At noon, the stick's shadow always falls along the north–south meridian through the location.[3]

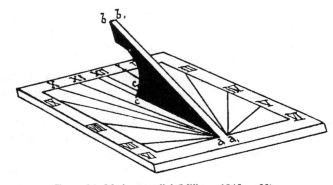

Figure 24. Modern sundial (Milham, 1945, p. 33)

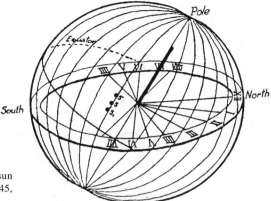

Figure 25. Relation of sundial to sun
(Northern Hemisphere) (Milham, 1945,
p. 34)

For determining time, a mere vertical stick is not sufficient because the sun
changes its height above the horizon throughout the year. The stick has to be
inclined at an angle equal to the location's latitude, and it must lie in the plane of
the meridian, that is point north (in the northern hemisphere). Figure 25 illustrates
how the shadow of a gnomon, which is parallel to the earth's axis, will always
point to the same mark at the same time (in this case VIII), independently of
whether the sun is high S^1, medium S, or low S_1 in the sky.

Thus to make a sundial, a gnomon should be cut such that it is inclined to the
base at an angle, L, equal to the latitude of its location. Then the horizontal base
has to be positioned so that the gnomon is aligned with the meridian (the line
running from pole to pole through the place). All that remains is to graduate the
base, using a fairly simple trigonometrical formula, so the shadow will mark off

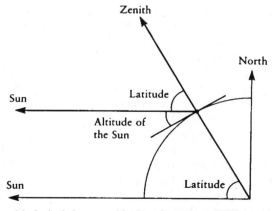

Figure 26. Latitude by noon altitude at the equinox (Williams, 1992, p. 10)

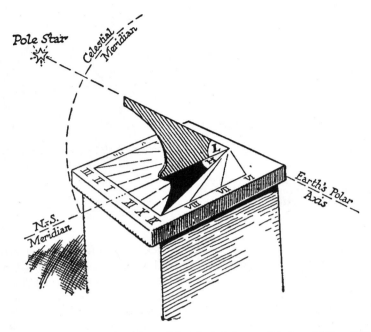

Figure 27. 10th century Moorish sundial, satitude *L* degrees (Hogben, 1936, p. 379)

hours.[4] Spanish Moors, in the 10th century were the first to fully understand the theory of sundials, and they made ones that were aligned to the meridian and whose gnomons were inclined at the appropriate latitude angle.

The Romans were not concerned with precise timekeeping. In the fourth century BC they simply divided their day into before midday (*ante meridiem,* AM) and after midday (*post meridiem*, PM). Only gradually did finer delineations of the day into morning, forenoon, afternoon and evening take hold. Hours, when they were introduced, were "temporary hour's," 1/12 of the daylight or night period.

One noteworthy ancient timekeeping apparatus was the Tower of Winds built in the Agora at Athens by the Greek astronomer Andronicus of Kyrrhos in 50 BC This was an octagonal tower, with sundials on its walls, and it housed an elaborate clepsydra inside. The latter was fed from a stream and was adjusted to fill every 24 hours, with the rising water driving a mechanism that showed the progress of the sun across the heavens (Richards, 1998, p. 55; Price 1967).

Days and nights are of equal duration only at the equinoxes when the sun rises due east and sets due west. Before the 14th century this inequality of "experiential" hours was acceptable and easily lived with. From at least the time of the Romans and onwards, days and nights were split into an equal number, usually

12, of hours (the so-called *horae temporales*). Consequently, hours were of different duration. Summer day hours were longer than summer night hours—there being 12 day hours in a long duration, and 12 night hours in a short duration. At the winter solstice, there are just under nine "modern" hours of daylight in Rome, while at the summer solstice there are just over 15 "modern" hours of daylight.

Chaucer in his *Canterbury Tales* (The Parson's Prologue, 5–9), written about the year 1400, made reference to shadow length as a rule-of-thumb for telling time:

> It was four o'clock according to my guess,
> Since eleven feet, a little more or less,
> My shadow at the time did fall,
> Considering that I myself am six feet tall.

These lines also well express the inexactitude of timekeeping in Chaucer's period. Of course, the length of Chaucer's shadow at 4 PM changed throughout the year: in mid-winter it was longest, in midsummer shortest. In June, at 41 degrees north latitude, someone who is 6 foot tall has a 72 foot shadow—a practical problem for Chaucer's method of telling time by length of his shadow!

The use of sundials represents an important move toward objectivity in measurement, which is a prerequisite for all science. It is the position of the shadow line that determines the time of day, not one's feelings, wishes or subjective experience. This move from subjective to objective measurement was not universally approved. The Roman poet Platus, in the second century BC, complained that

> The gods confound the man who first found out
> How to distinguish hours—confound him, too
> Who in this place set up a sun-dial,
> To cut and hack my days so wretchedly
> Into small pieces! When I was a boy,
> My belly was my sun-dial – one more sure,
> Truer, and more exact than any of them.
> This dial told me when 'twas proper time
> To go to dinner, when I ought to eat;
> But, now-a-days, why even when I have,
> I can't fall to, unless the sun gives leave.
> The town's so full of these confounded dials,
> The greater part of its inhabitants,
> Shrunk up with hunger, creep along the streets.
> (Milham, 1945, p. 37)

Nor did external indicators ensure agreement. Initially the Romans simply brought sundials to Rome from conquered lands. They did not appreciate that the time read at one latitude will not be the same as that read at another latitude (with the same dial), that the angle of inclination of the gnomon needs to be the same as the latitude of Rome for it to give the same time each day as the sun moves up and down in its annual path, and that the gnomon needs to be aligned with the earth's axis. The frustrations of the situation are well captured by Seneca who, spoke of Roman clocks at the very beginning of the Christian era when he said: "philosophers agree more readily among themselves than clocks" (Milham, 1945, p. 85). It was as late as 164 BC when, at the direction of Marcius Philippus, a properly oriented sundial was set up in Rome for the telling of time.

WATER CLOCKS

Even the best sundials could not tell the time at night, or on cloudy days. For these reasons the hour-glass, and the water clock (or clepsydra, "water thief"), with all their faults, were widely used to measure the passage of time.

One of the earliest clepsydra was an earthen jar with a hole in the bottom from which water dripped. The jar would empty in equal periods of time. Sometimes there would be a graduated receiving vessel with "hours" marked off. In more complex clepsydra, the flow of water was stabilized by providing a constant head of water, and the outflow activated a dial of some form. With technological ingenuity, the rate at which a vessel emptied or filled could be made steady, and with suitable calibration, it could measure equal periods of time. But dividing the hours of day or night into equal portions proved difficult because the length of the day, and hence the night, varied with the season. Initially the people's interest was not with abstract time (Newton's absolute time that "flows equably without relation to anything external") but with the daylight period and being able to divide that into equal periods. Such periods, or hours, were called "temporary" or "temporal" hours because their duration varied from one day to the next as the seasons took their course. Water clocks in antiquity and the Middle Ages were set up, and calibrated, to display these unequal hours.

Aristophanes in his 400 BC *Comedies* often referred to clepsydra, particularly in the context of timing speeches by orators. At one point Demosthenes accused a man of "talking in my water"; while at another time, when interrupted, he called for the court official to "stop the water."

Ctesibius (Ktesibios), the Alexandrian mathematician (ca. 300 BC), had a wide reputation for making refined clepsydra. He introduced the syphon principle into his clocks, whereby when the water container emptied, it automatically refilled. He employed geared wheels in the mechanisms, and he used jewelled bearings. He successfully tackled the vexing problem of summer daylight hours being longer than winter daylight hours by having the falling water level mark out the hours on a rotating cylinder which was divided into narrower and wider bands. In midwinter, the narrowest bands were presented to the marker; in midsummer, the widest bands were presented. The cylinder was so geared as to make one complete revolution in 365 days.[5]

Vitruvius, the first century BC commentator, described Ctesibius' water clock;

> The water, flowing regularly through this orifice [a tank located above the reservoir containing the float] causes the rise of an inverted vessel. On this [the float] a rod is placed beside a drum that can turn. They [the rod and drum] are provided with equally spaced teeth, which teeth impinging on one another cause suitable turnings and movements. Also other rods and other drums toothed in the same way driven by a single movement caused by their turning effects and varieties of movements; that causes statues to be moved, obelisks to be turned, pebbles or eggs to be thrown, trumpets to sound, and the other by-works. (Sleeswyk and Hulden, 1990, p. 34)

Figure 28. Ctesibius' water clock, 250 B.C. (Bond, 1948, p. 59)

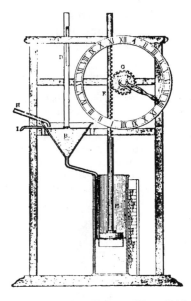

Figure 29. An ancient clepsydra (Milham 1945, p. 49)

In the second century, Archimedes wrote a treatise *On the Construction of Water Clocks* (Hill, 1976). In 807, the King of Persia presented the emperor Charlemagne with a most complex water clock that struck hours. The Islamic world perhaps contributed most to the development of water clocks from the ancient period up to the late medieval time (Hill, Hill, and al-Hassan, 1986, Hill, 1993), and its technology slowly diffused into Europe. But it appears that the fundamental jump to mechanical clocks, originally with foliot and verge escapements, was not made in the Islamic world.

A form of clepsydra figured in the physics of the mature Galileo. In the *Dialogue Concerning the Two New Sciences*, he detailed his inclined-plane experiments for relating time of fall to distance travelled:

> For the measurement of time, we employed a large vessel of water placed in an elevated position; to the bottom of this vessel was soldered a pipe of small diameter giving a thin jet of water, which we collected in a small glass . . . the water thus collected was weighed, after each descent, on a very accurate balance. (Galileo, 1638/1954, p. 179)

He obtained remarkably accurate measurements with this weighing-of-water method of timekeeping.

MECHANICAL CLOCKS

For 2000 years after the sundial's gnomon was aligned with the meridian, there was little advance in the science or technology of timekeeping (horology). While time was not standing still, timekeeping had certainly slowed. Mechanical clocks, the creation of which marked the next major advance in timekeeping, did not appear until the 13th century. Horologists say their first appearance was some time between 1277 and 1310. Richards believes the first mechanical clock was "made in 1283 for Dunstable Priory in England" (Richards, 1998, p. 58). These early mechanical clocks were not thought of as time measurers; they were seen as timekeepers. Even the most grand of the early mechanical clocks were set up primarily to mimic, or reproduce, heavenly motion: They were not built to measure time.

One historian said about medieval technology that:

> The ambition of inventors was unlimited, their imagination boundless, but of all the extraordinary machines they conceived and sometimes built, one above all symbolizes the inventiveness of the age: the mechanical clock. (Gimpel, 1992, p. 149)

while another historian, D.S. Cardwell, opined:

> There can be little doubt, however, that of all the great medieval inventions none surpassed the weight-driven clock and the printing press, measured by the scales of inventive insight on the one hand and social, philosophical, even spiritual, importance on the other. (Cardwell, 1972, p. 12)

Cardwell went on to say: "the clock and the printing press are, in fact, the twin pillars of our civilization and modern organized society is unthinkable without them" (Cardwell, 1972, p. 12).

There is a debate among historians on the origin of clocks as timekeepers. On the one hand, Derek de Solla Price claims that timekeeping clocks resulted incidently from centuries of building models and devices that imitated heavenly movements. As he expressed it, 'The mechanical clock is naught but a fallen angel from the world of astronomy" (Price, 1975). On the other hand, scholars such as Landes, maintain that, "It was the clock, in other words, that facilitated and thereby fostered the automated planetarium or astrarium, not the reverse" (Landes, 1983, p. 58).

One key matter that this debate hinges upon is whether the concept of time had been separated from the indicators of time; that is, whether the idea of time as something measurable in itself had emerged in society. Landes points out, against the Solla-Price thesis, that both Europe and China had an abundance of planetaria and other amazingly complex machines to imitate the movement of planets, but it was only in Europe that the mechanical clock emerged. For Landes, this was because in Europe the concept of time emerged as an independent entity or variable; and thus, "The clock did not create an interest in time measurement; the interest in time measurement led to the invention of the clock" (Landes, 1983, p. 58).

Monastic Timekeeping

The church, and more specifically the monastery, where monks were required to say communal prayers at seven set times through the day and night, provided the impetus, or social need, for the creation of the mechanical clock. [6] One historian observed of the earliest clocks that:

> They served not to indicate the passage of time to the public eye, but to give within the monastery a signal at proper intervals for sounding the hour or for ringing the bell to summon the monks to prayer by day and night. (Robertson, 1931, p. 24)

Monastic clocks did not display time; they rang time. The time they rang was *temporal,* or varying, time. The monastery's bell rang "canonical hours," that is, it rang seven times during the day, but not at night. The canonical hours, as with the hours of the ancient world and of Rome, were temporary or unequal hours. The monks, following the rules of St. Benedict and of other founders, had to pray seven times a day; and as the day varied in length from summer to winter, and from low to high latitudes, the time between bells varied accordingly.

For instance, the following are among the long list of rules (custumals) governing Cistercian monasteries in the 12th century:

> CXIV The sacristan shall set the clock and cause it to sound before lauds. And also in order to awaken himself before vigils every day. And after having arisen, he shall light up the dormitory and church.

> LXXIV The clock being heard they shall attend to their necessities so that when the bell shall be rung, they shall be ready to go into the choir.

> LXXXIII But if it shall be a fast day, [they] shall pause from their reading until the eighth hour. And then the sacrist, aroused by the sound of the clock shall ring the bell. (Drover, 1954, p. 55)

Our word "clock" reflects this monastic heritage: the Middle English *clok,* the old German *Glocke,* the French *cloche*, and the Middle Latin *clocca* all meant "bell." "Clock" initially referred only to timekeepers that rang the hours on bells. It was a species word, not a generic word as it is today. Thus in the early medieval period, sundials were not referred to as clocks; they, and water clocks and sand clocks, were referred to as *horologia*.

The first monastic clocks were water clocks (Drover, 1954) where the striking bell alerted an attendant, who then proceeded to manually ring the big bell that alerted the monks in their cells. These water clocks had their own inherent problems: maintaining a water supply, keeping even pressure, preventing impurities from blocking the apertures, having enough force to ring loud bells, and so on. It was in everyone's interest to automate and regularize this process of bell ringing, the more so when clocks began to move out of the monasteries into the towns. For example, in 1403, the keeper of the belfry in Montpellier was accused of having too high wages, sleeping on the job, and being irregular in striking the time. The town authorities then ordered a large clock from Dijon which had an automatic striking mechanism (Robertson, 1931, p. 25). At some time between 1270 and 1320 the monastic water clocks began to be replaced with their newly created mechanical relatives, the first mechanical clocks.

The Coming of Equal and Secular Hours

Monastic clocks *marked* time; they did not *measure* it. That was sufficient for their purpose. The process of sounding canonical hours sufficed for the monk's needs, but eventually monastic clocks would be replaced by secular hours, usually 24 in the day.[7] When clocks became public, that is placed in church spires, towers, town halls, then the latter civil time rapidly replaced ecclesiastical time.

In the ancient and the medieval world, people usually divided the daylight period into 12 hours (six and ten were other conventions). Little, if any, effort was spent on dividing or measuring night time. Temporal hours (the period of daylight divided by twelve) vary with season and latitude. For instance, in northern Egypt at latitude 30° north, the time between sunrise and sunset varies between 10–14 hours from winter to summer; in London, at latitude 51½ degrees north, the duration of daylight varies between 7¾ and 16½ hours from winter to summer. Thus temporal hours in London vary in duration between about 38 minutes in winter to 82 minutes in summer (Gimpel, 1992, p. 168). It was clearly easier for mechanical clocks to mark equal, or equinoctial,[8] hours than to mark temporal hours. This would require adjustments each day, and with each move north or south in latitude. The concept of equal hours was necessary for advance in timekeeping, and for advances in physics that depended upon measuring changes as a function of time.

The change from church time and unequal hours to secular time and equal hours began in the early14th century—1330 is one date mentioned (Boorstin, 1983, p. 39). It was a significant event in the development of European culture and society, and according to Boorstin:

> There are few greater revolutions in human experience than this movement from the seasonal or "temporary" hour to the equal hour. Here was man's declaration of independence from the sun, new proof of his mastery over himself and his surroundings. Only later would it be revealed that he had accomplished this mastery by putting himself under the dominion of a machine with imperious demands all of its own. (Boorstin, 1983, p. 39)

Another historian explained that

> . . .the regular striking of the bells brought a new regularity into the life of the workman and the merchant. The bells of the clock tower almost defined urban existence. Timekeeping passed into time-saving and time-accounting and time-rationing. As this took place, Eternity ceased gradually to serve as the measure and focus of human action. (Mumford, 1934, p. 14)

Along with equal hours came the division of the hour into 60 minutes and, at least theoretically, the minute into 60 seconds. Crombie noted that this fine-scale division of time was "fairly common as early as 1345" (Crombie, 1952, p. 213). However these fine distinctions were not widely utilized in everyday life: A 14th century cookbook, for instance, directed novices to boil eggs "for the length of time that you can say a *Miserere*" (Crosby, 1997, p. 32). According to Crombie,

> the adoption of this system of division completed the first stages in the scientific measurement of time, without which the later refinements of both physics and machinery would scarcely have been possible. (Crombie, 1952, p. 213)

Clock Components

When Robertus, in 1271, commented on the standard astronomical text, *The Sphere of Sacrobosco*, he was among the first to formulate the principles of mechanical clockwork:

> Clockmakers are trying to make a wheel which will make one complete revolution for every one of the equinoctial circle [sun's rotation], but they cannot quite perfect their work.... The method of making such a clock would be this, that a man make a disc of uniform weight in every part so far as could possibly be done. Then a lead weight should be hung from the axis of that wheel, so that it would complete one revolution from sunrise, minus as much time as about one degree rises according to an approximately correct estimate. (Gimpel, 1992, p. 153)

He realized part of the problem was that the weight accelerated as it fell, and so the wheel speeds up, hence the early hours would pass faster than the later hours. Robertus sought to balance the weight drive prefectly, so the wheel would turn evenly. Unfortunately his handiwork—technology—was not equal to his mental work. Recognizing the problem, and solving it in practice, were, as they usually are, two different things. Whatever retarded the falling weight was subject to wear and tear, and so there was no even rotation of the wheel.

Robertus identified the basic components of all mechanical clocks:

- a *driving* mechanism (falling weight, spring, etc.)
- a *transmitting* mechanism (axles or arbors, and toothed wheels)
- a *controlling* or *regulating* mechanism, usually called the *escapement*
- an *indicating* mechanism (dial and hands)

Improvement of mechanical clocks depended upon suitable advances in metallurgy and metal-working skills. Although some timber clocks, and even watches, were made,[9] timber had inherent limitations for use in clockwork. Clocks are constructed so that one axle or arbor will turn once in one hour or once in 24 hours. Apart from ordinary technological problems and challenges (refinement of metal work, precision tooth-cutting, use of jewels in the axle pinions, and better lubricants), the major challenge facing mechanical clocks was the escapement. One horologist comments that:

> The heart of any mechanical clock is its escapement, the device which through a repetitive mechanical motion regulates the running down of the motive power. The history of mechanical clocks is, to a large extent, the history of improvement in their escapements. (Macey, 1980, p. 20)

Eventually it was the adaptation of the pendulum as an escapement that brought the necessary increase in reliability and accuracy to mechanical clocks and thus allowed them to (almost) solve the problem of longitude, and enabled precision time measurement in the physics and astronomy of the scientific revolution.[10]

The early 14th century witnessed the first major breakthrough in controlling mechanisms when an unknown clockmaker used a *foliot* and *verge* escapement as

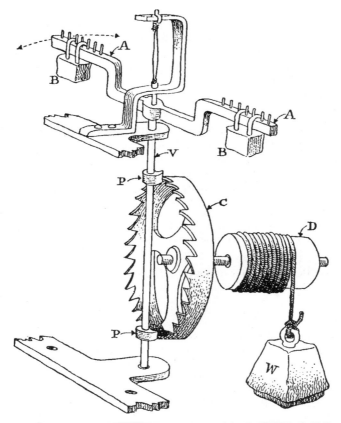

Figure 30. Foliot controlling power from a falling weight (Bell and Bell, 1963, p. 57)

a regulating device to arrest and equalize the turning power of the falling weight. The weight (W) turned a vertical crown wheel (C), the motion of which was checked by pallets (P) on a vertical verge, or rod (V). A horizontal foliot (A) was attached to the verge, and balance weights (B) were hung off this. As the inertia of the foliot and balance weight was overcome by the turning action of the crown wheel, one pallet was turned away, allowing it to escape. Simultaneously the second pallet was moved into the crown wheel, arresting its motion. The ease with which the pallets could turn on the verge was determined by the mass of the suspended weights on the foliot and their radial distance from the verge.

After trial-and-error adjustment, the clockmaker determined the appropriate combination of falling weight, hanging weight, radial distance from verge, diameter of the crown wheel, and number of its teeth. This knowledge was part of clockmaking lore. The escapement had three functions: It retarded and regularized the fall of the weight; its oscillating foliot divided time into equal beats; and the verge counted the beats (Burnett, 1998, p. 66). Thus the verge and foliot converted an accelerating falling motion into a regular oscillating motion. This second motion could then, by wheels, gears, and so on, move an indicating wheel, or a pointer on an indicating face.

Dante described such a mechanism in his *Divine Comedy:* "And the wheels in the clock array turn in such a way that the first appears to the observer in rest while the last to fly" (Dante, 1310/1967, Canto 24). There is some debate about the exact dating of the *Comedy*, but 1310 is a reasonable estimate, thus setting a minimum earliest date for the presence of mechanical clocks in Europe. A contemporary chronicler, Froissart, coined the term *foliot* (from the French for a quivering leaf or, a more rare translation, a woman's temperament) to name the ingenious escapement device. Heinrich Halder, who built a clock for the city of Lucerne in 1385, described the foliot's regulating action:

> And if the woman's temperament goes faster than you think it should, hang the lead blocks away from the wheel, and if it goes too slowly hang them closer to the wheel, in this way you will make it go forward and backward as you wish. (Rossum, 1996, p. 49)

Galileo in the Fourth Day of his 1633 *Dialogue* wrote of the operation of the foliot:

> In order to regulate time in wheel clocks, especially large ones, the builders fit them with a certain stick which is free to swing horizontally: At its ends they hang leaden weights, and when the clock goes too slowly, they can render its vibrations more frequent merely by moving the weights somewhat toward the center of the stick. On the other hand, in order to retard the vibrations, it suffices to draw these same weights out towards the ends, since oscillations are thus made more slowly and in consequence the hour intervals are prolonged. Here the motive force is constant—the counterpoise— and the moving bodies are the same weights; but their vibrations are more frequent when they are closer to the center; that is, when they are moving along smaller circles. (Galileo, 1633/1953, p. 449)

The importance of the verge and foliot mechanism should not be underestimated: It represented an enormous leap forward in the art of timekeeping. As Carlo Cipolla writes:

> If a broad historical point of view invites one to stress the continuity and gradualness inherent in the process of technological change, a strictly technological point of view forces one to emphasize the fundamental difference between water clocks, bell-ringing mechanisms, and the like on the one hand and the mechanical clock on the other. Historians of technology and science have good cause to dramatize the sharp break in the history of horology brought about by the invention of the verge escapement with foliot. . . . In the mechanical clock, the time-keeping is governed by an oscillator or escapement which controls the unidirectional movement of the motive power and transforms it into a slow, steady and regular motion whose meaning appears on the face of the clock. (Cipolla, 1967, p. 39)

However, the great weakness of the verge and foliot mechanism also needs to be stressed; the foliot had no *natural* frequency of oscillation. Horizontal foliots when set swinging, oscillated in a nonregular, nonuniform manner. There was no universal law of oscillation that described their movement. The movement could be made faster or slower, but there was no set pattern that all foliots displayed. In 1675, 300 years after their introduction, J. Smith (author of *Horological Dialogues*) wrote an account of the verge and foliot that indicated how little this crucial piece of horological technology had developed since its creation:

> These movements going with weights must be brought to keep true time by adding to or diminishing from them; if they go too slow you must add thin shifts of lead to the weights to make it go faster, but if it go too fast then you must diminish the weight to make it go slower; for that whensoever you find either to gain or lose, you must thus, by adding or diminishing, rectifie its motion; note that these Balance movements are exceedingly subject to be altered by the change of weather, and therefore are most commonly very troublesome to keep to a true time. (Welch, 1972, p. 31)

It was the pendulum's natural frequency, ascertained by Galileo, that was to make it the core of accurate timekeeping. In the ideal situation, all pendulums of the same length vibrated with the same frequency independently of mass or amplitude of swing. Gravity alone, which was thought to be the same in all countries, controlled the pendulum's oscillation.

The Spread of City Clocks

In the 300 years between the early 14th and mid-17th centuries, the use of mechanical clocks spread throughout Europe.[11] It was a matter of personal prestige for nobility to possess one, and a matter of civic pride for towns to have one in their church or town hall. In 1481 a petition presented to the Town Council of Lyon stressed that in the town:

> is sorely felt the need for a great clock whose strokes could be heard by all citizens in all parts of the town. If such a clock were to be made, more merchants would come to

the fairs, the citizens would be very consoled, cheerful and happy and would live a more orderly life, and the town would gain in decoration. (Cipolla, 1967, p. 42)

At the time, François Rabelais commented that "a city without bells is like a blind man without a stick" (Crosby, 1997, p. 77). Among the better known early cathedral, palace, or town hall clocks were: Dunstable (1283), London (1286), Canterbury (1292), Milan (1335), Cluny (1340), St. Albans (1344), Padua (1344), Dover (1348), Strasboug (1352), Genoa (1356), Bologna (1356), Perpignan (1356), Chartres (1359), Frankenberg (1359), Ferra (1362), Ghent (1370), Pavia (1370), Munich (1371), Cologne (1372), Salisbury (1386), and Wells (1392).[12]

In 1335 a chronicler described the Milan cathedral clock:

> A very large hammer . . . strikes one bell twenty-four times according to the number of the twenty-four hours of the day and night; so that at the first hour of the night it gives one sound, at the second, two strokes, at the third, three, and at the fourth, four; and thus it distinguishes hour from hour, which is the highest degree necessary for all conditions of men. (Boorstin, 1983, p. 39[13])

In 1350 Giovanni de Dondi constructed a clock in Padua that was widely acknowledged as the masterpiece of medieval clockwork. He wrote a 130,000-word manual and included 180 diagrams, detailing its construction and operation. The clock had seven sides, in the upper part of which were dials giving the position of the Primum Mobile, the moon, and the five planets then known—Venus, Mercury, Saturn, Jupiter, and Mars. In the lower parts there were dials for the 24-hour clock, and for the fixed and movable feasts of the church. The exactitude of his reporting can be seen in his description of the clock train:[14]

> The hour circle revolves in 24 hours, 144 teeth, pinion of 12 carrying a wheel of 20 meshing with a wheel of 24 on the great wheel. Thus the barrel rotates 10 times in 24 hours. Great wheel 120 teeth mesh with pinion of 12 carrying a second wheel of 80, which therefore revolves 100 times a day. The second wheel meshes with a pinion of 10 carrying the escape wheel of 27 teeth, which therefore makes 800 revolutions a day, each calling for 54 oscillations of the balance, i.e. 43,000 a day or 2 second beat. This is the standard beat. (Gimpel, 1992, p. 161)

A contemporary, Philippe de Maizières, wrote of Dondi's clock that:

> This master John of the clocks has produced famous works in the three sciences [philosophy, medicine and astronomy], works that are held in great repute by the great scholars in Italy, Germany and Hungary. Among other things he has made an instrument, called by some a sphere or clock for the celestial movements; which instrument shows all the movements of the signs and the planets with their circles and epicycles, and differences, and each planet is shown separately with its own movement in such a way that at any moment of the day or the night one can see in which sign and to which degree the planets and the great stars appear in the sky. The sphere is constructed in such a subtle way that in spite of the fact that there are so many wheels that they cannot be counted without taking the clock to pieces, all goes with one weight. ... In order to have his sphere done as he had it in his subtle mind, the said master John actually forged it with his own hands out of brass and copper, without the help of anyone, and he did nothing else for sixteen years. (Cipolla, 1967, p. 45)

Figure 31. The Strasbourg cathedral clock (woodcut, 1574) (Hoskin, 1997, p. 138)

Dondi was both a scholar and an artisan, a common combination that was important to medieval and renaissance horology and, later, a combination so well exemplified by Galileo and Huygens.

Medieval clocks and horologies[15] were elaborate creations, and utility was merely one consideration in their construction. They frequently amounted to public puppet shows performed on the hour, half-hour, and quarter-hour. The falling weight activated wheels, gears, levers, and trip devices so that throughout the day a panel would open regularly and out of it would appear knights jousting, chariots racing, wild animals fighting, and so on.

Clocks had an entertainment function in addition to their timekeeping function. Improvement in clocks usually meant making their performances more elaborate, not making them more reliable timekeepers. As the foregoing chronicler wrote, accuracy to the nearest hour was the "highest degree necessary for all conditions of men." During the 14th century, dials were seldom found on public clocks. These clocks sounded rather than showed time. In 1500, the clock at Wells Cathedral in England was considered advanced because it rang the quarter-hours.

For some appreciation of the extra-timekeeping function of medieval clocks, consider the following description of the 1352 Strasbourg cathedral clock (Fig. 31) which stood eighteen meters high and was nine meters wide:

> Enormous in size, it included a moving calendar and an astrolabe whose pointers indicated the movements of the sun, moon and planets. The upper compartment was adorned with a statue of the Virgin before whom at noon the Three Magi bowed while a carillon played a tune. On top of the whole thing stood an enormous cock which, at the end of the procession of the Magi, opened its beak, thrust out its tongue, crowed and flapped its wings. (Cipolla, 1967, p. 44)

It is easy to appreciate why the early mechanical philosophers constantly had recourse to metaphors of "clock work" to explain their mechanistic view of the universe, and it is also easy to appreciate why this metaphor was so believable and convincing for people used to watching the operations of elaborate clocks, like the one at Strasbourg.[16] Robert Boyle, for instance, writing 400 years after the clock was constructed, said this of his mechanistic conception of nature:

> . . .it is like a rare clock, such as may be that at Strasbourg, where all things are so skilfully contrived, that the engine once set a-moving, all things proceed according to the artificer's first design, and the motions . . . do not require the particular interposing of the artificer, or any intelligent agent employed by him, but perform their functions upon particular occasions, by virtue of the general and primitive contrivance of the whole engine. (Boyle, 1772, vol. 5, p. 163; in Whitrow, 1988, p. 122)

Following Jewish and canonical practice, one day ended and the next commenced at dusk, that is, about a half hour after sunset. The Italian clockmakers divided the period between dusk and dusk into 24 hours, with their bells striking up to 24 to signify the final hour of the day. This convention became known as Italian time and held its ground in Italy even into the 19th century (Robertson,

1931, p. 36). The convention had its obvious problems as the commencement of the day depended upon the caprice of the clock attendant, the latitude of the town, the season of the year, and whether sunset and dusk could be ascertained—storms, rain, and snow not making the task any easier! Also striking up to 24 times was not ideal for quick signalling of time to citizens. These problems led gradually to the adoption of the 12-hour dial, and the day being defined as the period from midday to midday, or midnight to midnight.

Dante provides, some time between 1316 and 1321, one of the earliest literary references to these new inventions. In his *Paradiso* he wrote:

> As clock, that calleth up the spouse of God
> To win her Bridegroom's love at matin's hour,
> Each part of other fitly drawn and urged,
> Send out a tinkling sound, of note so sweet,
> Affection springs in well-disposed breast;
> Thus saw I move the glorious wheel; thus heard
> Voice answering voice, so musical and soft,
> It can be known but where the day endless shines.
>
> (Gimpel, 1992, p. 155)

The year 1360 is important in the early history of mechanical timekeeping: This was the year Charles V of France employed Henry de Vick of Würtemberg to make a clock for the royal palace. This was Paris' first public clock: It was a true

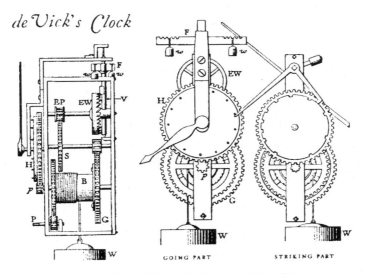

Figure 32. De Vick's clock (Brearley, 1919, p. 81)

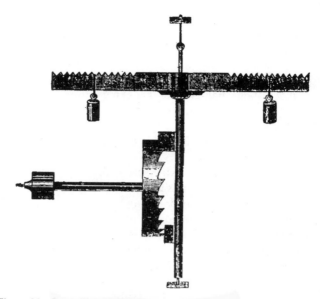

Figure 33. Controlling mechanism of De Vick's clock (Milham, 1945, p. 85)

mechanical clock that broke with the tradition of sundials and water clocks, and in virtue of the verge and foliot escapement, it was a comparatively accurate time-keeper. The clock took eight years to complete. It combined the crafts of the blacksmith and locksmith with every wheel and tooth being individually cut. The driving weight of 500 pounds (227 kg) fell through 32 feet (10 m) in 24 hours. It struck the 24 hours using a hammer to hit a 1,500 pound (680 kg) bell. It was a marvel because it kept time accurately to within *two hours* per day.[17] Social life—meeting times, wage labour, dinner appointments, commencement of religious services, etc.—obviously adjusted to the accuracy available. The growth of astronomical knowledge, which depended upon the timing of eclipses, timing the rising and setting of the moon and planets, and so on, was handicapped by the inaccuracy of these clocks.

Jean Froissart was inspired by de Vick's masterpiece to write a much reproduced poem "Li Orloge Amoureus." The poem, of 1,174 lines, bears witness to the effect of the clock on late 14-century thinkers.

> I well myself may liken to the Clock,
> For when that Love, that dwells within my heart,
> Directs my mind thereto and bids me think,
> I see therein is a similitude,
> Wherefor I must rejoice and pride myself;

For the clock, if well we ponder, is
An instrument most fine and notable,
And pleasant is, besides, and servicable;
For night and day the hours it lets us know,
Through the subtlety that it possesses,
Even in the absence of the sun;
Whence all the more its mechanism should be prized,
As that which other instruments do not do,
However much by art and compass they be made:
Wherefor I hold him for a mighty man and wise
Who first of all the use of it discovered,
When of his intellect he took in hand and made
A thing so noble and of so great profit.

Next it behoves us of the "dial" to speak;
And this dial is the journal wheel
Which only in one natural day
Is moved and makes one turn exactly,
Like as the sun doth make a single turn
Around the earth in just one natural day;
Upon this "dial", whereof the merit's great,
The four and twenty hours are inscribed.

And by reason that the clock cannot
Go of itself, nor in any way could move,
Were there no one to keep it and to tend,
Therefore it needs for its special care
To have a clockman, who both late and early
Shall diligently attend and regulate it,
Draw up the weights and set them to their task;
'Tis thus he makes them orderly to move,
And both controls and regulates the wheels,
And for the striking gives them governance.

Besides the clockman accurately sets
The foliot, so that it may never stop,
The spindle too and likewise all the pins,
The wheel, moreover, so that all the bells
Whose hours, which are set upon the dial,
May right truly be arranged to sound,
When at the appointed time the detent's raised.

King Charles V, having installed de Vick's clock in his palace, decreed that the churches and monasteries of Paris set their time by his clock (Rossum, 1996, p. 217). This edict is the beginning of state control of time and the weakening of the Church's control.

The move from church dominance of time finishes with the altogether impersonal time of Newton, a time that "flows equably without relation to anything external." But this "desacralisation" of time did not occur quickly. The Roman Catholic church in western Europe readily accepted Charles V decree and supported the installation of mechanical clocks in church towers. It was different with the Orthodox church in eastern Europe. There secular timekeeping was rejected, and it was not until the 20th century that mechanical clocks were allowed to be installed in Orthodox churches.[18]

The late 14th century operation of the Nuremberg clock illustrates the limitations of mechanical clocks. The Nuremberg clock kept "day hours" and "night hours," so in December the short winter day had eight hours and the night 16 hours.

From December to June the day was lengthened by one hour approximately every three weeks and the night shortened accordingly. From June to December the procedure was reversed. All public civic activities that were fixed by the hours, and work times in particular, had to take this adjustment of hours into consideration. This was a "natural" but complicated system. In 1489 the clock was corrected by astronomers, and it was decided that there would be 52 eight-hour days, 17 13-hour days, and 22 15-hour days (Rossum, 1996, p. 115).

Figure 34. German city in 1497, showing tower clock, (Rossum, 1996, p. 149)

Accuracy of Mechanical Clocks

Although during the 13th and 14th centuries mechanical clocks became more common, there was little improvement in their accuracy. Metal working skills were refined. Tooth-cutting became more accurate. Better lubricants were used. And from 1500 on, the coiled spring was used in the driving mechanism as an alternative to falling weights thus enabling chamber clocks to be made—but the heart of the clock (the verge and foliot escapement) was retained. This governed the accuracy of the clock. No one had devised a more accurate and reliable way of controlling the driving mechanism. There is some evidence of a semicirculus, and more accurate, escapement being made, but only one such device is known. Also it needs to be remembered that degrees of accuracy were judged against personal and social needs: Monks needed to be awakened during the night, but the exact time was not important. Likewise wage labor had a sense of when clocks were running slow, but an accuracy of 15 minutes in a paid work-day of 8–12 hours was probably sufficient for the purpose. Up till the end of the 16th century clocks did not have minute hands. The "idea" of modern minutes was present—that is, a sixtieth part of the hour—but there was no impetus to measure them. "Seconds" had a purely theoretical existence. The recitation length of prayers and chants was a common measure of brief time intervals. For example, a chronicle of Constance said about an earthquake on August 13, 1295 that "around midday there came the greatest earthquake . . . and it lasted about as long as it takes someone to say a Paternoster and an Ave Maria" (Rossum, 1996, p. 42). It is easy to appreciate how the development of quantitative science was held back for lack of accurate time measurement. The world of precision, characteristic of post pendulum-clock science, was not yet possible.

Astronomers wanted accurate clocks because they realized the connection between degrees of separation of astronomical events, and time elapsed in their observation. With the earth rotating one degree each four minutes, angular separation and time were interchangeable. Early use of this interconnection was recorded in 1484 by the Nuremburg astronomer Bernhard Walther:

> I observed Mercury with a well regulated clock, which gave the time precisely from noon the day before to noon [today]. I saw Mercury in the morning in contact with the horizon, and simultaneously I put a weight on the clock, which had 56 teeth in the hour wheel. It made one complete turn and 35 teeth more, by the time the Sun appeared on the horizon, whence it followed that Mercury on that day rose one hour 37 minutes before the Sun, which almost agrees with calculation. (Andrewes, 1985, p. 69)

Tycho Brahe, the great observational astronomer and contemporary of Galileo, sought to get clocks to keep accurate time so as to refine and test astronomical theories.[19] In a much-reproduced plate of Tycho in his observatory, his assistant is avidly looking at *two* clocks as Tycho notes astromomical events. Tycho used two clocks because he could not have confidence in the reliability of a single clock.

In 1641 the town council in Dijon, desperate because none of the public clocks agreed with each other, gave an order that they should all be put in accordance with the town sundial—an indication of how limited was horological progress since ancient Egyptian days (Cipolla,1967, p. 43). In the middle of the 14th century, an accurate clock was one that gained or lost just 15 minutes per day. By the middle of the 16th century, improvements in craft skills brought this time down to 12 minutes per day. By the middle of the 17th century, *prior* to the use of the pendulum as a regulator, it was as low as eight minutes per day (Bertele 1953). When in 1660 the pendulum was first used in mechanical clocks, accuracy improved to ten *seconds* per day, shortly after it was one second per day.

CHINESE WATER CLOCKS

Whatever the problems and limitations of European mechanical clocks, they were far superior to what had been achieved in China. As China was a sophisticated and literate society, it is useful, for technical and cultural reasons, to briefly comment on the achievements and limitations of Chinese timekeeping.[20]

There is documentary evidence of the use of clepsydra in China from the seventh century BC (Needham, Ling, and Solla Price, 1960, p. 85). Water clocks were progressively refined over the centuries to eliminate the effect of falling water pressure as resevoirs emptied and variation in the seasonal length of the day. The culmination of this clepsydra tradition was the great *Tower for the Water-powered Sphere and Globe* built by Su Sung in 1092 for the Emperor Che Tsung. There has probably never been built a water clock to rival Su Sung's in grandeur and complexity. However his magnificient creation was essentially private technology. It did not spawn the growth of similar public clocks as the first mechanical clocks of Europe did 200 years later. This was largely because the emperor's authority was affected by the accuracy of his timekeeping corps and their ability to forecast astronomical events such as eclipses, the movement of planets, and phases of the moon. The emperor was covetous of this knowledge. It was never part of the public domain.

A powerful domestic reason for precision in an emperor's timekeeping had to do with affairs of the bedroom, which were of course affairs of state. Su Sung, as with emperors before and after, had hundreds of concubines and wives of various status, and there was a rigid timetable for their sharing of the emperor's bed. As auspiciousness of offspring, and hence recognition of the next emperor, was determined by planetary conjunctions at conception rather than birth, astronomers and timekeepers were integral to the imperial office. The complexity of nocturnal arrangements is evident from the following courtly instruction:

> The lower ranking [women] come first, the higher-ranking come last. The assistant concubines, eighty-one in number, share the imperial couch nine nights in groups of

nine. The concubines, twenty-seven in number, are allotted three nights in groups of nine. The nine spouses and the three consorts are allotted one night to each group, and the empress also alone one night. On the fifteenth day of every month the sequence is complete, after which it repeats in reverse order. (Boorstin, 1983, p. 76)

Despite the imperial imperative for accurate timekeeping, and despite China being an advanced literate, numerate and technological civilization, the Chinese never made reliable clocks. Their science and scientific organization was defective.

Su Sung's *Heavenly Clock* was between 30–40 feet (9–12m) high. The driving wheel was 11 feet in diameter and carried 36 scoops on its circumference, into each of which in turn water poured at a uniform rate from a constant-level tank. The cycle of scoops filled every nine hours, consuming in the process about half

Figure 35. Pictorial reconstruction of Su Sung's heavenly clockwork (Needham, Ling & Solla Price, 1962, p. i)

TRIPS
WHEN
CUP IS
FULL

Figure 36. Su Sung's Escapement Mechanism
(Jespersen & Fitz-Randolph 1982, p. 26)

a ton of water. The axle of the water wheel articulated with a vertical 20-foot
timber transmission-shaft which by a system of gears and wheels finally drove a
time-dial and a constantly-focused star-sighting viewer (Needham, Ling, and Solla
Price, 1960, p. 48).

Technically the most noteworthy thing about Su Sung's clock is its escapement mechanism, this being the soul of any timekeeping machine and, as has
been mentioned, the crucial component that the pendulum was eventually used to
regulate.

Clepsydra traditionally had no escapement device. They had continuously
acting mechanisms: graduated bowls filled with, or emptied of, water; floats with
pointers constantly rose or fell in cylinders; and so on. Su Sung's scoop-laden
wheel was held stationary while a scoop filled. A tripping mechanism, designed
so that the weight of the filled scoop was sufficient to release the wheel, allowed
the next scoop to come under the water source and simultaneously prevent it from
further movement (Needham, Ling, and Solla Price, 1962, pp. 55–59). The scoop
and the tripping mechanism constituted the escapement.

Needham and colleagues estimated each scoop took about 15 minutes to fill
before the tripping mechanism was activated, this then caused a few seconds
movement in the mechanism. Thus the clock was stationary, for perhaps 59.5
minutes in each hour. This lack of activity perhaps accounted for its longevity—
it apparently ran without interruption from 1092 to 1126. Thus Su Sung's escapement "constitutes an intermediate stage between the time-measuring properties of
liquid flow and those of mechanical oscillation" (Needham, Ling, and Solla Price
1962, p. 59).

Culturally the Su Sung clock is important because it seems to have led no
where. It did not represent a stage in the development of Chinese horology, rather,
it was a terminus. In the 16th century, when Jesuit missionaries brought mechanical clocks to China they were greeted with universal amazement. There was nothing comparable in China. In 1600, the well respected missionary Father Matteo

Ricci, a friend of Galileo and former student of the Jesuit astronomer Clavius, wrote that timepieces used in China were:

> ... made to operate by the means of water or fire, and in those which are operated by fire, time is measured by odoriferous ashes of a standard size. Besides the Chinese make also other clocks with wheels that are made to turn with sand. But all such instruments are very inaccurate. Of sundials, they have only the equinoctial ones but do not know well how to adjust them for the position [latitude] in which they are placed. (Cipolla, 1967, p. 80)

There were no mechanical clocks in China when Ricci arrived in 1590 (Cipolla, 1967, p. 77). This despite the fact that mechanical clocks had appeared around 1300 in Europe, and that commercial and social intercourse between China and the West, both along the Silk Road and by ocean, had been occurring for centuries—Marco Polo's *Travels* having been published in 1295. The failure of Chinese society to master timekeeping is especially remarkable as the office of the emperor and the administration of empire had long been bound up with astrology and knowledge of astronomical events, for both of which timekeeping was vital.

In 1590, Father Ricci (1552–1610) and a Jesuit companion became the first Jesuit missionaries to enter China. They journeyed via Portuguese Goa and Macao "to garner into the granaries of the Catholic Church a rich harvest from this sowing of the gospel seed" (Boorstin, 1983, p. 56).[21] It is likely they were also the first people to bring western clocks to China. For 20 years they travelled back and forth across the land without the emperor granting them an audience, an audience necessary for conversions to be sanctioned. Finally it was the emperor's interest in Ricci's clocks that brought the long sought audience. Actually it was not an audience for, since time immemorial, Chinese emperors had kept themselves isolated, even from their own subjects and ministers. The emperor had artists paint a portrait of Ricci and bring it to the emperor's chamber. This constituted the audience, from which permission for missionary activity was given.

Jesuit horological and astronomical knowledge provided the key for the "harvest of souls." The emperor's astronomers predicted a two hour eclipse of the sun for 10:30 AM on June 21, 1629. The Jesuits predicted it would occur at 11.30 AM and last for only two minutes. The heavens behaved exactly as the Jesuits forecast. The imperial astronomers with their arcane timekeeping were cast aside, and the Jesuits installed in their place with an order to revise the Chinese calendar. They then translated into Chinese, European works on astronomy, optics, mathematics, music, and other subjects.

A Chinese historian writing in the century after Su Sung's work suggested a reason for the backwardness of Chinese astronomy and timekeeping: "Barbarians had no restrictions on astronomical and calendrical study," whereas the official court was highly protective and exclusionst about such study. Astronomers were operating under an imperial decree of 840 that had all the marks of a modern "state security" regulation. The decree warned:

> If we hear of any intercourse between the astronomical officials or their subordinates and officials of other government departments or miscellaneous common people, it will be regarded as a violation of security regulations which should be strictly adhered to. From now onwards, therefore, the astronomical officials are on no account to mix with civil servants and common people in general. (Boorstin, 1983, p. 75)

A later Chinese historian wrote:

> Thus astronomical instruments have been in use from very ancient days; handed down from one dynasty to another, and closely guarded by official astronomers. Scholars have therefore had little opportunity to examine them, and this is the reason why unorthodox cosmological theories were able to spread and flourish. (Needham, Ling, and Solla Price, 1962, p. 6)

Joseph Needham said of this situation:

> In Europe, unlike China, there was some influence at work ... that pushed forward to make the junction between practical knowledge and mathematical formulations ... Part of the story undoubtedly concerns the social changes in Europe which made the association of the gentleman and the technician respectable. (Cipolla, 1967, p. 35)

The history of Chinese timekeeping provides gist for the epistemological mill of Marx, Dewey, and others who stressed that progress in science, that knowledge, comes from the unity of mind and hand, from *praxis* (Suchting, 1986a; Matthews, 1980 Chap. 6). It also provided support for Merton's thesis that the development of science requires genuinely democratic participation by scientists and open and accountable dissemination of research results.[22]

CHAPTER 4

Galileo and
the Pendulum Clock

Although Galileo's achievements are most frequently thought of as purely intellectual or scholarly, this is not a complete picture.[1] Galileo united technological competence with scholarship. He bridged the worlds of manual and intellectual labor, and he united the traditions of workshop and university. He was the most illustrious member of the newly emerging class of scientist–engineers that formed in Italy in the 16th century. Galileo made great contributions to physics, astronomy, and applied mathematics. It is not for nothing that he is regarded as the founder of modern science. Additionally he made contributions to philosophy[2] and theology.[3] However Galileo also had a lifelong engagement with technology and its utilization in personal and commercial life. He developed the prototypes of many scientific instruments—the *pulsilogium, thermoscope, military* and *geometrical compass, lodestone, telescope*—and he drew a plan for the first *pendulum clock.* When these instruments were refined by others, they brought about the precision and objectivity of measurement that characterized the European scientific revolution.[4]

Against the ancient Platonic, and Chinese mandarin's disdain of manual labor and dismissal of craft skills, Galileo praised William Gilbert when he said:

> To apply oneself to great inventions, starting from the smallest beginnings, and to judge that wonderful arts lie hidden behind trivial and childish things is not for ordinary minds; these are concepts and ideas for superhuman souls. Galileo, 1633/1953, p. 406

In his early 20s Galileo studied in Florence with Ostilio Ricci, a noted engineer and mathematician who introduced him to the work of Archimedes, "whose name ought never be mentioned except in awe" as Galileo later wrote. At this time Galileo developed an improved form of hydrostatic balance and spent time in the Venice arsenal, where he met its director Guidobaldo del Monte, who was to become his first patron and collegial investigator of topics in mechanics and mathematics. In this early period he also mastered machine design, canal construction, dyking, and the building of fortifications. In 1597 he patented a water-hoisting engine.

Galileo was an energetic participant in the quest for the solution to the longitude problem. As Silvio Bedini remarked, Galileo sought "over a period of 30 years for a solution to the problem of longitude, and it was the final project with which he became involved before his death" (Bedini, 1991, p. 6). Bedini added:

> It was in fact for a means of achieving a precise measurer of time segments for scientific use that Galileo Galilei experimented with the mechanical clock, and not as an improvement of the domestic clock; he was seeking a means of accurately determining the longitude at sea. (Bedini, 1991, p. 1)

The pendulum features in Galileo's physics from his earliest writings in 1590 through to his mature *Dialogues* of 1633 and 1638 and up to the time of his death in 1641 when he was engaged in the planning of a pendulum clock. He was the first person to recognize the isochronic properties of the pendulum, the dependence of the period of the pendulum upon its length, and the independence of period from amplitude of swing. He was also the first to recognize the timekeeping properties of the pendulum and to detail how the swinging pendulum could be utilized in clockwork. He used the pendulum to illustrate the conservation of "force" (roughly our momentum), to dissolve the Aristotelian distinction between natural and violent motions, to formulate his theory of the tides, and to propose his theory of freefall. From beginning to end, the pendulum played a major role in Galileo's physics, and in his overturn of Aristotelian philosophy. He got certain details wrong (for instance, the supposed isochrony of circular pendulums, the supposed complete independence of period from amplitude, and the supposed equality of oscillations of different weight pendulums). He dramatically overstated other properties (for instance, claiming that pendulums remain swinging for thousands of oscillations), but he put the pendulum onto the centre stage of the new science of the 17th century. Within 50 years of his death the work of Huygens, Hooke, and Newton filled out and corrected Galileo's basic findings about pendulum motion. These refined properties were then used in the creation of the pendulum clock (the world's first accurate and reliable time keeper), in nearly solving the longitude problem, in unifying celestial and terrestrial mechanics, in developing gravitational theory, and in a multitude of other achievements in mechanics.

As with most other things of intellectual and social importance, some historical perspective is necessary for appreciating Galileo's achievements in the science and practice of timekeeping: Specifically, some understanding of the achievements and limitations of the medieval science of motion and horology is required. Understanding the Aristotelian and medieval background, and its transformation by Galileo, is also useful for teachers whose students so often think about projectiles and pendula in Aristotelian ways (Eckstein and Kozhevnikov, 1997).

THE MEDIEVAL PENDULUM

Thirty years ago, a historian remarked, "There is no adequate history of the [medieval] pendulum" (White, 1966, p. 108). Unfortunately this is still the case.[5] Some elements, however, of the theoretical, technical, and cultural aspects of the medieval pendulum story can be discussed.

Pendulums, of one form or another (natural, as with fruit swaying at the end of vines, or created, as with children swinging on rope or lamps oscillating on chains) have existed since the dawn of time. These real objects do not, however, have a history. What has a history is their conceptualization. The history of the pendulum begins when real objects become theoretical objects. That is, when people begin to describe, conceptualize, explain, and ultimately theorize about pendulums. The pendulum as a theoretical object first emerged in late medieval Europe. As Thomas Kuhn explained: "Until that scholastic [impetus] paradigm was invented, there were no pendulums, but only swinging stones, for the scientist to see" (Kuhn, 1970, p. 120).

The Aristotelian Problematic

Medieval impetus theory[6] was the theoretical milieu which prepared the way for Galileo and Newton's 17th-century-analysis of pendulum motion. Impetus theory was developed to save a key feature of Aristotle's mechanics, namely the conviction that *nothing moves violently without a mover;* or as the medievals said: *Omne quod movetur ab alio movetur.* This doctrine, which states a commonplace of everyday experience, illustrates why Aristotelianism has been so often called "the philosophy of commonsense."[7]

Aristotle divided all motion into either *natural* (motion derived from the nature of the body, for example, heavy bodies falling) or *violent* (motion derived from outside the body, for example, thrown stones or spears). In *Physics* Book VIII, Chapter 4 he wrote:

> If then the motion of all things that are in motion is either natural or unnatural and violent, and all things whose motion is violent and unnatural are moved by something, and something other than themselves, and again all things whose motion is natural are moved by something. . . . then all things that are in motion must be moved by something. (Barnes 1984, p. 427)

It is important to recognize that for Aristotle, the *Omne quod movetur ab alio movetur* doctrine applied only to bodies that are moved violently, not to bodies moving naturally.[8] As he said in the opening sentence of *Physics* Book VII, "Everything that is in motion must be moved by something. For if it has not the source of its motion in itself it is evident that it is moved by something other than itself, for there must be something else that moves it" (Barnes, 1984, p. 407). For

natural motion (freefall towards the centre for heavy terrestrial bodies and upward movement toward the heavens for light bodies; circular motion about the
earth for celestial bodies) there is no mover, external or internal, required; the
bodies just naturally move downwards,[9] upwards, or in heavenly circles. Their
source of motion is within themselves. It is their *nature*.

Since first formulated by Aristotle, this doctrine had been contradicted by
the everyday phenomenon of projectile motion: What is the mover which is responsible for motion after a stone leaves the thrower's hand? Projectile motion
was a stumbling block for Aristotle's theory of motion. As Alexandre Koyré ob-
serves: "Aristotelian physics thus forms an admirable and perfectly coherent theory
which, to tell the truth, has only one flaw . . . that of being contradicted by everyday practice, by the practice of throwing" (Koyré, 1968, p. 26).

Aristotle recognized this, and on at least three occasions addressed the "problem" of projectiles.[10] For Koyré, his "genius" is displayed in the explanation he
provided for this seemingly fatal flaw in this physics of motion. What is the saving explanation? Aristotle in *Physics* Book VIII, Chapter 10, interrupted a discussion about how the fact of motion ultimately required that there be a first mover
that is not moved—an argument picked up by medieval catholic philosophers to
establish God's existence[11]—and said:

> But first it will be well to discuss a difficulty that arises in connection with locomotion.
> If everything that is in motion with the exception of things that move themselves is
> moved by something else, how is it that some things, eg., things thrown, continue to
> be in motion when their mover is no longer in contact with them? (Barnes, 1984, p.
> 445)

He answered by saying that "the original mover gives the power of being a
mover either to air or to water or to something else of the kind, naturally adopted
for imparting and undergoing motion" (Barnes, 1984, p. 445). Aristotle is stating
the thoroughly improbable view that the mover moves the medium, which then
moves the projectile.[12] The thrower's hand moves the air, which then continues to
move the thrown stone. This was the *antiperistasis* theory. It is the Archilles heel
of his physics. The history of medieval mechanics can, in large part, be seen as an
effort to provide a better, yet still Aristotelian, account of what moves a moving
projectile after it leaves a thrower's hand.

Medieval Impetus Theory

It was fully one thousand years before the first published dissent from Aristotle's
account of projectile motion appeared. The sixth century Christian Platonist John
Philoponus of Alexandria,[13] restated the more obvious problems with Aristotle's
antiperistasis theory: Why can a stone be thrown further than a feather? Why
does a thin arrow fly further than a blunt stick? Why can't the thrower just wave
air at a projectile and hence make it move? When a wheel or top is set spinning, is

it believable that the surrounding air keeps it in motion? Philoponus, in his commentary on Aristotle's *Physics*, said:

> From these considerations and from many others we may see how impossible it is for
> forced motion to be caused in the way indicated. Rather it is necessary to assume that
> some incorporeal motive force is imparted by the projector to the projectile, and that
> the air set in motion contributes either nothing at all or else very little to this motion
> of the projectile. (Clagett, 1959, p. 509)

Thomas Kuhn in his landmark *Structure of Scientific Revolutions* regarded
medieval impetus theory as a "paradigm change" from Aristotelian science (Kuhn,
1970, p. 119). In some sense it was, but it needs to be remembered that Philoponus
accepted that Aristotle's physics was basically correct. He thought the physics
just needed improvement on this one point concerning the motor for projectile
motion. Philoponus agreed that the projectile did need a mover to keep in motion,
but he thought that the thrower did not communicate this power of movement to
the medium (as in Aristotle), but communicated it directly to the projectile.
Aristotle's was a systematic, interconnected philosophy or world view. The parts
(physics, biology, cosmology, ethics, and politics) all cohered. Philoponus can be
seen as pulling just one thread of the Aristotelian tapestry, and that not very strongly,
yet the pull on that one thread eventually resulted in the whole fabric unravelling.
Once Philoponus put the mover into the body for violent motion, then the standard Aristotelian natural–violent motion distinction dissolved. The idea of objects having a natural place was then threatened, and the idea of efficacious natures became problematic. Thus ideas fundamental to Aristotle's physics were
gradually rejected.[14] It is arbitrary to ask where in this unravelling, the paradigm
change may be said to begin and end.

Philoponus' idea of an incorporeal motive force (*virtus impressa*), although
subject to diverse interpretations and reworkings, was influential among Arab
philosophers, particularly Avempace (a Spanish Arab who died in 1138) and his
successors Averroes and Avicenna. Through these Arab philosophers Philoponus
influenced the 14th century impetus theory developed in Paris by Franciscus de
Marchia, Jean Buridan, Nicole Oresme, and others, and in Merton College, Oxford by William of Ockham, Thomas Bradwardine, John Dumbleton, William of
Heytesbury, and colleagues.[15]

Philoponus did have his critics, and it is illustrative to briefly consider one of
foremost in the 13th century, Thomas Aquinas. The "Angelic Doctor" wrote against
Philoponus' idea of absorbed or impressed impetus:

> However, it ought not to be thought that the force of the violent motor impresses in
> the stone which is moved by violence some force (virtus) by means of which it is
> moved...For [if] so, violent motion would arise from an intrinsic source, which is
> contrary to the nature (ratio) of violent motion. It would also follow that a stone
> would be altered by being violently moved in local motion, which is contrary to sense.
> (*Quasetiones Supra Libros Octo Physicorum Aristotelis*, in Hanson, 1965a, p. 137)

Aquinas was a great medieval Aristotelian, indeed one of the profoundest thinkers of all time, and the influence of Aristotle's core natural–violent motion distinction can be seen in his comment. If impetus is truly *internal*, and if it is the source of the projectile's motion, then violent motion has become converted to a natural motion. *Per impossible*. The kind of investigation in physics that engaged the medievals is also illustrated. Medieval physics was largely philosophical and conceptual, not empirical and much less experimental. The medievals were concerned with qualitative, not quantitative relations, with the "why" of motion, not the "how."

We now know that an adequate science requires *both* conceptual and experimental investigation. One conceptual question about Philoponus' impressed force was: Does it diminish of its own accord, or is it diminished only when it "works" against a resistance? As early as 1323, Franciscus de Marchia, who rehabilitated impressed force theory after its dismissal by Aquinas, proposed that the impressed force (*vis derelicta*) communicated to a projectile dissipated of its own accord, independently of any "work" (overcoming resistance of the medium) that it might have to do. Thus a thrown body, even in a void, would come to a halt.[16] The alternative view, elaborated most clearly by Jean Buridan (ca. 1295–1356),[17] was that *impetus* (Buridan seems to be the first to have introduced this term) was only used up when it did work against resistance. If a thrown body met no resistance, then it would continue in motion indefinitely. Buridan introduced this anti-Aristotelian idea in a very tangible context:

> [The impetus also explains why] one who wishes to jump a long distance drops back a way in order to run faster, so that by running he might acquire an impetus which would carry him a longer distance in the jump. Whence the person so running and jumping does not feel the air moving him, but [rather] feels the air in front strongly resisting him. (Clagett, 1959, p. 536)

Buridan proceeded, again in a most anti-Aristotelian fashion, to bridge the terrestrial–celestial divide by attributing impetus to the heavenly bodies, saying that "God when He created the world, moved each of the celestial orbs as He pleased, and in moving them impressed in them impetuses which moved them" (Clagett, 1959, p. 536). He then speculated on what the subsequent history of such divinely moved bodies would be, and took a position that subsequently would be argued over by Leibniz and Newton. Buridan said:

> And these impetuses which He impressed in the celestial bodies were not decreased nor corrupted afterwards, because there was no inclination of the celestial bodies for other movements. Nor was there resistance which would be corruptive or repressive of that impetus. (Clagett, 1959, p. 536)

Elsewhere, in his commentary on Aristotle's *Metaphysics*, Buridan wrote:

> You know that many hold that a projectile, after it has left the hand of the thrower, is moved by an impetus given to it by the thrower, and that it is moved for as long a time as the impetus remains stronger than the resistance. And this impetus would endure

for an infinite time, if it were not diminished and corrupted by an opposed resistance
or by something tending to an opposed motion. (Moody, 1975, p. 267)

Buridan's decisive shift from a naturally degenerating impetus to one that is
enduring and only diminished when it has to overcome obstacles has frequently
been hailed as the "precursor" to Newtonian inertia.[18] Stanley Jaki used this shift
to support his thesis about the Christian origins of modern science:

> [Buridan's] Belief in a Creator whose powers were not limited to the features of the
> actually observed world contributed, therefore, most effectively to the liberation of
> critical thinking from the shackles of Aristotelian science. (Jaki, 1974, p. 232)

Nicole Oresme's Introduction of the Pendulum

Nicole Oresme (ca. 1320–1382), Buridan's most illustrious pupil, introduced discussion of the pendulum into Western science when, in his 1377 *On the Book of
the Heavens and the World of Aristotle* he developed the thought experiment of a
body dropped into a well that had been drilled from one side of the earth, through
the centre, and out the other side. Oresme said the dropped body would eventually
come to a halt at the centre of the earth.[19] Oresme likened this imaginary situation
to that of a weight which hangs on a long cord and swings back and forth, each
time losing a little of its initial height (Clagett, 1959, p. 570). He said this of
impetus:

> And it is not weight properly [speaking] because if a passage were pierced from here
> to the center of the earth or still further, and something heavy were to descend in this
> passage or hole, when it arrived at the center it would pass on further and ascend by
> means of this accidental and acquired quality, and then it would descend again, going
> and coming several times in the way that a weight which hangs from a beam by a long
> cord [swings back and forth]. . . . And such a quality exists in every movement—
> natural or violent—as long as the velocity increases . . . And such a quality is the
> cause of the [continued] movement of projected things when they are no longer in
> contact with the hand or the instrument [of projection]. (Clagett, 1959, p. 570)

It is impetus that carries the falling body *past* the centre of the earth, and the
mass on a suspended string *upwards* after it reaches its nadir. Galileo used almost
the exact words[20] when, in his discussion of projectile motion in the *Dialogue*
(1633), he said:

> If two strings of equal length were suspended from that rafter, with a lead ball attached
> to the end of one and a cotton ball to the other, and then both were drawn an equal
> distance from the perpendicular and set free, there is no doubt that each would move
> toward the perpendicular and, propelled by its own impetus, would go beyond that by
> a certain interval and afterwards return. (Galileo, 1633/1953, p. 151)

Oresme, in his commentary on Aristotle and projectile motion, said:

> We can understand this more easily by taking note of something perceptible to the
> senses: if a heavy object *b* is hung on a long string and pushed forward, it begins to

move backward and then forward, making several swings, until it finally rests absolutely perpendicular and as near the center as possible. (Menut and Denomy, 1968, p. 573)

This first recorded mention of a pendulum in a theoretical context is note-worthy because it is an accurate description. The *several* swings that Oresme sees when he looks at a suspended heavy weight pushed at the end of a cord, are the same as most people, including physics students, see.[21] This is in contrast to the *hundreds* and sometimes *thousands* that, later, Galileo will claim to have seen.[22]

Oresme's observation is subject to two different *theoretical* interpretations. One can look at the gradually slowing pendulum and "see" its initial impetus draining away and thus the pendulum coming to a stop because without a mover (impetus) it is no longer capable of movement. This is the self-decaying impetus that Marchia, Oresme, and Galileo at Pisa, saw. Alternatively, one can look at the same pendulum and "see" a constant impetus that is only used up because there is work being done, that is, friction to be overcome. This is what Buridan saw, and what Galileo saw at Padua after abandoning the impetus theory of his early *De Motu*. For both observers, *impetus* has ontological reality; it is something pos-sessed by the body. The disagreement is whether or not this entity dissipates of its own accord or only when used up in doing work.

This is another case of different theoretical orientations resulting in different observation statements.[23] The perspective (theory) of Buridan, the mature Galileo, and further down the line, Newton, is, manifestly, not easy to acquire. Marchia, Oresme, and most people, see what is happening. Friction, air resistance, impedi-ments, and imperfections, are part of the world around us. It is difficult to see them *at work*. It is easier to "see" the initial impetus that was clearly given to the pendulum, draining away as the pendulum slowly comes to a stop. Buridan, then Galileo, and especially Newton, all "rise above the everyday" and try to see "the heart of the matter." For them, natural motion is not—contrary to the whole em-piricist tradition of Aristotelianism—what actually happens to bodies, but what would happen if the impediments and accidents were removed. This process of abstraction and generalization is the beginning of modern science. It is an ability to see the forest, and not just the trees. Galileo's "thousands of swings" is clearly not what he saw, but what he would see if the impediments were removed. Simi-larly Newton did not see inertial bodies continuing to move in a straight line indefinitely. This is what he would have seen if all resistance were removed. Fermi and Bernardini, in their biography of Galileo, emphasized this innovation:

> In formulating the "Law of Inertia" the abstraction consisted of imagining the motion of a body on which no force was acting and which, in particular, would be free of any sort of friction. This abstraction was not easy, because it was friction itself that for thousands of years had kept hidden the simplicity and validity of the laws of motion. In other words, friction is an essential element in all human experience; our intuition is dominated by friction; men can move around because of friction; because of friction they can grasp objects with their hands, they can weave fabrics, build cars, houses,

etc. To see the essence of motion beyond the complications of friction indeed required a great insight. (Fermi and Bernardini, 1961, p. 116)

Leonardo da Vinci's Pendulum Investigations

The limitations of observation in understanding the pendulum are illustrated by Leonardo da Vinci, acknowledged as one of the most acute observers of all time. In the late 15th century Leonardo examined pendulums and clocks but he appears

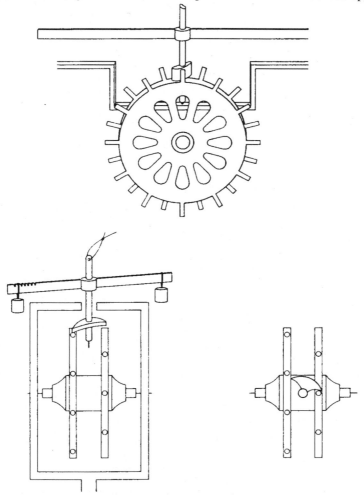

Figure 37. Leonardo da Vinci's sketches for a mechanical escapement with foliot (Edwardes, 1977, p. 227)

not to have put the two together, nor to have seen the isochronic properties of the pendulum.[24] He suggested improved spring mechanisms for driving clocks. He was the first to suggest using jewels in the pivot of the clock pins or axles to reduce friction and wear, and he was aware of the problems of "regulating" the effect of the driving mechanism. In the late 1490s he sketched two pendulums, one on a reciprocating pump, the other on what appears to be a clock. He recognized that the descent along the arc of a circle is quicker than that along the shorter corresponding chord, an anticipation of Galileo's later Law of Chords.

Leonardo was thorough in his treatment of clocks and timekeeping. He wrote of escapements:

> It is customary to oppose the violent motion of the wheels of the clock driven by their counterweights with certain devices called escapements, as they keep the timing of the wheels which move it. They regulate the motion according to the required slowness and the length of the hours. The purpose of the device is to lengthen the time, a most useful thing. They are made according to several different methods, but all have the same nature and value. Their variety is born out of the necessity of taking into account the space for the collocation. (Bedini and Reti, 1974, p. 254)

In discussing the problem of some hours being "longer" than others, he commented:

> This [lengthening] happens as a result of the humidity inherent in bad weather; the air becomes thick and the escapements of the clocks find more resistance in such air. As a consequence, the clocks slow down and the hours become longer. . . . Therefore, the lengthening of the hours is caused solely by the clock escapements and has no other possible cause. (Bedini and Reti, 1974, p. 256)

Leonardo suggested using vanes as a regulator. His regulator was made by fixing plates to an axle so that as the driving mechanism speeds up, the axle rotates and the vanes impede the motion in a way that is directly proportional to the speed of rotation. In all his writing on, and drawing of, clocks Leonardo did not stumble on to a more effective way of regulating their driving mechanism, and never realized, as Galileo later would, that a pendulum could perform this task.

Leonardo utilized the pendulum in discussing the fundamental Aristotelian distinction of natural and violent motion. His discussion is interesting because it demonstrates not only his powers of observation, but how these powers, when wedded to an Aristotelian empiricism about how one gains knowledge of the world, impeded science. Leonardo discussed the proposition that "Accidental motion will always be shorter than the natural," and proceeded as follows:

> In order to demonstrate the truth of this statement, we shall propose as example a round weight suspended from a cord, and let the weight be *a*. And it shall be lifted as high as the point of suspension of the cord. This point will be *f* and the cord will be *fa*, which should always be pulled along a straight line. I affirm thus, that if you let this weight fall, all the motion made from *a* to *n* will be called natural, because it moves in order to stop at *n*, that is, beneath its point of suspension *f*, seeking to approximate itself as near to the center of the world as possible. After reaching the desired site, that

is, *n*, another motion takes place, which we will call accidental, because it goes against the desire. Such accidental motion will be from *n* to *m* and will always be less than the natural *an*. Consequently natural motion, that more it approaches its end, increases its velocity. Accidental motion does the contrary. However, the end of accidental motion well be that much weaker than the beginning of the natural motion as the former is less than the latter. ... The smaller is the natural motion of a suspended weight, the more the following accidental motion will equal it in length. (Bedini and Reti, 1974, p. 262)

Leonardo made many drawings to illustrate the movement of the pendulum, or oscillating motion as he called it. These included conical and inverted pendulums.

Despite Leonardo's unrivaled observational skills and his enormous intellectual ability, "he failed, however, to recognize the fundamental properties of the pendulum, the isochronism of its oscillation, and the rules governing its period" (Bedini, 1991, p. 5). Given that we expect beginning high school students to see these things, Leonardo's failure to see them is worthy of some examination.

Leonardo saw that the weight *does not* rise to the exact height from which it was released. This is precisely how things are in the world. Leonardo did not make the abstractions from *impediments* and *accidents* that Galileo made, and Newton perfected. These abstractions make possible the birth of modern science. In terminology to be introduced in Chapter 10, he does not distinguish the real and theoretical objects of science.

Understandably, lesser lights than Leonardo also failed to see the pendulum's isochronism. In 1569 Jacques Besson published a book in Lyon detailing the use of the pendulum in regulating mechanical saws, bellows, pumps, and polishing machines. The book contained numerous drawings of pendulums, but made no mention of isochronism.

Figure 38. Leonardo's pendulum drawing (Bedini and Reti, 1974, p. 262)

At least two historians of horology have remarked on this peculiar circumstance. Ernest Edwardes, in his *The Story of the Pendulum Clock*, related how

> the late Lord Grimthorpe, in his *Clocks, Watches and Bells* (8th ed., London, 1903), has an amusing comment on this point, to the effect that it is strange how the isochronism should only be discovered in the later sixteenth century, with so many pendulums swinging about all over the world! (Edwardes, 1977, p. 18)

For his part, Edwardes added: "With this remark the present writer is inclined to concur, for we have ample proof that the capacity for intelligent observation is not limited to our modern era, nor even to post-medieval times."

GALILEO'S PULSILOGIUM

This now brings us back to Galileo's revolutionary analysis of the pendulum. This analysis had a humble enough beginning. While the youthful Galileo was briefly a medical student at Pisa he utilized the pendulum to make a simple diagnostic instrument for measuring pulse beats. This was the *pulsilogium*. It had been long known that the pulse rate was indicative of various illnessess and of the recovery process. Yet until Galileo there had been no effective and objective measure of this important medical phenomenon. Physicians spoke of pulses being very fast, quick, slow, or normal, but agreement on even these broad categories was difficult: How fast was fast? How slow was slow? Pulse rate required that time be measured. But tower clocks, water clocks, sundials, and primitive house clocks were hardly usable for this purpose. We are familiar with doctors walking along wards, stopping to feel the pulse of patients, looking at their watch to determine whether the beat is 50, 70, 100 or whatever, and pronouncing a verdict on the state of the patient. Doctors in Galileo's day did not have pocket watches, yet they realized that pulse rate was of great medical significance. It was a matter of frustration that objective, let alone accurate, measurement of pulse beat was beyond them.

Galileo's answer to the problem was ingenious and simple: He suspended a lead weight on a short length of string, mounted the string on a scaled board, set the pendulum in motion, and then moved his finger down the board from the point of suspension (thus effectively shortening the pendulum) until the pendulum oscillated in time with the patient's pulse. As the period of oscillation depended only on the length of the string and not on the amplitude of swing or the weight of the bob, the length of string provided an objective and repeatable measure of pulse speed.[25] As the string was finely divisible, accurate and precise measurements could be made—although the instrument involved no units of time, time was measured by length of the pendulum. Doctors could communicate, patient charts could be kept, and the possibility of research on conditions governing pulse rate was opened.

The *pulsilogium* was developed further in the 1620s by Santorio Santorio for measuring breathing rates and other medical applications. Santorio worked at the University of Padua, where 50 years earlier Andreas Vesalius lectured and where Galileo would soon take an academic position. In 1690 John Foyer was the first to use a seconds pendulum for measuring pulse-beat.

The *pulsilogium* provides an interesting epistemological lesson. Initially something subjective, the pulse, was used to measure the passage of time: Occurrences, especially in music, were spoken of as taking so many pulse-beats. With Galileo's *pulsilogium* this subjective measure itself becomes subject to an external, objective, public measure—the length of the *pulsilogium's* string. This was a small step in the direction of objective and precise measurement upon which scientific advance in the 17th century would depend.

GALILEO AND THE LONGITUDE PROBLEM

Galileo was aware of King Phillip III's longitude competition, and between 1609 and the late 1620s he had correspondence back and forth with the Spanish court on the matter (Bedini, 1991, pp. 9–18). In 1627 he was advised of the prize offered by the states-general of Holland for solving the longitude problem. Up until 1640 he forwarded, through his friend and agent Elio Diodati, proposals for its solution.

On August 15, 1636, Galileo wrote to the states-general of Holland, who had offered a reward for a means of determining longitude, saying that he had a

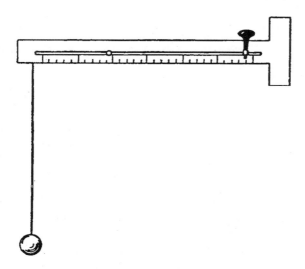

Figure 39. Bar pulsilogium (Wolf, 1950, p. 433)

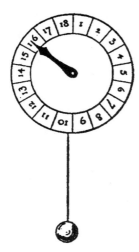

Figure 40. Circular pulsilogium (Wolf, 1950, p. 433)

marvellously accurate timepiece capable of determining longitude on long voyages, and that "their construction is very simple and far less subject to outside influences than are other instruments which have been invented for a similar purpose." He said of these instruments, in a piece of typical Galilean hyperbole, that if one let four or six of these chronometers run simultaneously, they would not differ by one second, even in a period of several months (*Opere*, vol. XVI, pp. 463–469). He repeated the claim a year later (June 6, 1637) in a letter to the Dutch Admiral Lorenzo Reael (*Opere*, vol. XVII, pp. 96–105):

> It would be a waste of time to occupy Your Lordship's attention any longer with the details. You can command artists of the utmost skill in the manufacture of clocks and other excellent mechanisms. They have only to know that the pendulum gives vibrations of exactly equal duration, whether the arc be great or small, in order to devise methods of construction of greater precision than any that I could devise. (Robertson, 1931, p. 86)

Galileo's basic idea was to use timing of the eclipses of his recently discovered (1610) moons of Jupiter to ascertain longitude. The moons as they circle the planet are frequently and regularly eclipsed, and the regularity is independent of the location of the observer. Thus if the time for the eclipse of a moon is known for one location, say Venice, and if local time is known for its eclipse at another location then, by the difference in times, the longitude of the second location can be calculated. For example if the Venetian time of an eclipse is 9 PM and it is seen to occur at 11 PM at another location, then the second place is:

$$360° \times 2h/24h \text{ west of Venice.}$$
$$\text{That is, } 30° \text{ west of Venice}$$

The method depends upon (1) having tables of the eclipse times at Venice (or where ever else one wants), (2) making accurate observations of the eclipses (almost impossible on a rocking boat, but practical on land) (3), and crucially, having an accurate means of telling local time. The benchmark for local time was local noon, easily enough found as it was the moment that the sun was over the local meridian and hence when a stick's shadow was at its minimum, (or alternatively at night in the northern hemisphere, by noting the position of the circumpolar stars). Once noon was established, then an accurate means had to be found of telling how many hours, minutes, and seconds had elapsed since noon. Galileo saw the pendulum-regulated clock as providing the last.

Galileo's correspondence on this method was passed through Constantijn Huygens, secretary to Prince Frederick Henry Stadtholder of Orange and father of the Dutch astronomer Christiaan Huygens, who was to become, for historians, the rival of Galileo for title of the "discoverer of the pendulum clock."[26] The Dutch court thought so highly of Galileo and his suggestion that they sent two commissioners to Galileo to deliver a gold chain as a mark of esteem. Unfortunately by the time they arrived he was blind and ill, and nothing more had been achieved with his idea.

In September 1639 he wrote to Giovanni Battista Baliani in Genoa, who also experimented on falling bodies and inclined planes:

> That the use of the pendulum for the measurement of time is a most exquisite thing I
> have stated many times; as a matter of fact I have brought together diverse astronomical
> operations, in which by means of this measurer I have achieved a precision infinitely
> more exact than that which was possible with any other astronomical instruments, as
> well as even with quadrants, sextants, armillaries, and whatever else, having diameters
> of not only the two or three *braccia* such as those of Tycho, but even 20, 30 or 50,
> divided evenly not only into degrees and minutes, but into parts of minutes as well.
> (*Opere* vol. XVIII, pp. 93–95; in Bedini, 1991, p. 24)

GALILEO'S PROPOSAL FOR A PENDULUM CLOCK

In 1639 Galileo published in Paris a work entitled *L'usage du cadran ou de l'horloge physique universel*. In this book, after explaining the movements and properties of a pendulum, its construction and the means of making it swing faster or slower, he dealt with its use for ascertaining longitudes and for observing eclipses. For these purposes, Galileo recognized, it is necessary to count the vibrations, and said that "this method was valueless so long as the pendulum was not used for regulating the movement of clockwork" (Robertson, 1931, p. 93).

By 1636 Galileo's basic idea for a clock was to attach a stylus to a solid pendulum. Each time the pendulum passed the vertical, this stylus would strike an elastic bristle fixed at one end and resting in the tooth of a horizontal crown wheel, so that it would move the crown wheel forward one tooth and come to rest in the next tooth. This was his counting device, and he immediately went on to suggest that it be connected to the workings of a regular weight-driven clock (Robertson, 1931, p. 85, Edwardes, 1980).

The best account we have of Galileo's conception and execution of a pendulum clock comes from a letter of Vincenzo Viviani to Prince Leopold de Medici dated August 20, 1659:

> One day in the year 1641, whilst I was living with him [Galileo] in his country house at Arcetri, I recollect that it came into his mind that the pendulum could be adapted to clocks driven by weight or spring, in the hope that the perfect natural equality of its motion would correct the imperfections of mechanical construction. But, being deprived of sight and unable himself to execute the plans and models which would be required to ascertain which would be best adapted for carrying out this project, he communicated his idea to his son Vincenzio, who had come out one day from Florence to Arcetri. They had several discussions on the subject, with the result that they fixed upon the method shown by the accompanying drawing, and they decided to proceed

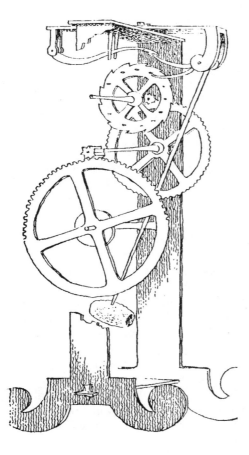

Figure 41. Galileo's original drawing for pendulum clock (Bell and Bell, 1963, p. 58)

at once with its execution, in order to determine what were the difficulties, which, as a rule, in the construction of machines, a theoretical design does not reveal. But Vincenzio, being desirous to construct the instrument with his own hands, for fear lest the artificers who might be employed should divulge it before it had been presented to the Grand Duke and to the States-General of Holland for the measurement of longitudes, kept putting off its execution, and a few months later Galileo, the author of this admirable invention fell ill, and died on January 8, 1642. As a consequence, Vincenzio's enthusiasm cooled down, so that it was not until the month of April 1649, that he took in hand the manufacture of the present clock made in accordance with the conception which his father had already imparted to him in my presence. (*Opere,* vol. XIX, pp. 655–657; Edwardes, 1977, p. 21)

Righly, or wrongly, Viviani distanced Galileo from the more vulgar motives of many who pursued the longitude solution:

For a long time Galileo [was aware] that the greatest obstacle and the most important objection that his proposal probably would have encountered might be that others would believe he had demonstrated it for that rich reward and the honors promised by all the kings of Spain and the other princes to the man who made the invention. He wished, on the contrary, to make known that he was never moved by such a vulgar stimulus but instead was inspired by confidence in his discovery. His only desire was to enrich the world with knowledge that was so necessary and profitable to human commerce, and to cover himself with the glory which was therefore due to him. He decided to offer it freely and generously to the most powerful States General of the United Provinces. (*Opere,* vol. XIX, pp. 599–632; Bedini, 1991, p. 21)

Galileo recognized—or worked out—the isochronic properties of the pendulum, and furthermore, he conceived the idea of using the pendulum in a new escapement mechanism that would replace the 300 year-old verge and foliot. This was a technical achievement of the highest order and of immense consequence. Figure 42 shows details of Galileo's escapement mechanism. The pivoted pawl, or bar A, falls against tooth B to stop the crown wheel. C and D are finger bars rigidly attached to the pendulum rod F. When the pendulum swings left, the upper finger lifts the pawl to release the wheel, while the lower finger rises to intercept pin E, allowing the crown wheel to turn the distance of one tooth only. Thus the pendulum provides intermittent motion of the wheel. Galileo's plans for a pendulum clock were first materialized by the clockmaker Johann Philipp Tefler (1625–1697) in the late 1650s (Bedini, 1991, Chap. 3).

Galileo's invention of the dead-beat form of the pinwheel escapement was a major technical step in the evolution of accurate timekeepers. In 1954, the president of the British Horological Society reported using a clock built to Galileo's design, and said "the variation from day to day was only a few seconds" (Edwardes, 1977, p. 22). Although a major technical breakthrough, Galileo's pin-wheel escapement design was not immediately utilized. His son Vincenzio apparently did little to promote or publicize his father's pendulum clock and design. Huygens was seemingly oblivious to the new escapement, or at least to its advantages. He kept using the verge escapement that had remained virtually unchanged from the

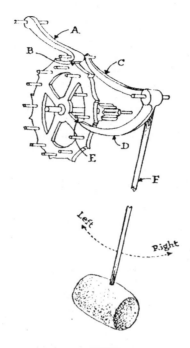

Figure 42. Galileo's escapement mechanism (Bell and Bell, 1963, p. 58)

14th century. Galileo's pin-wheel was reinvented in 1753 by P.A. Caron, and then developed by Lepaute (Robertson, 1931, p. 120).

Galileo achieved more than enough to firmly establish his place in the pantheon of illustrious contributors to knowledge and human self-understanding; perhaps it was best that he fell just short in his efforts to make what would have been the world's first accurate mechanical clock. This distinction would go, within a few years of Galileo's death, to the only slightly less illustrious Christiaan Huygens.

CHAPTER 5

Galileo's Analysis of Pendulum Motion

Galileo's contribution to horology hinged on his analysis of pendulum motion. It was his proposed use of the isochronic pendulum as an escapement that made his clockwork truly novel. As developed in the next half-century by Huygens, Hooke, and others, the pendulum clock revolutionized timekeeping and enabled major strides to be made in astronomy, mechanics, and other sciences. But the *sine qua non* of this development was a correct scientific account of the oscillating pendulum. This in turn hinged on Galileo's overturning of the long dominant Aristotelian epistemological position concerning what constituted a "correct scientific account." That is, the new technology of timekeeping resulted from a new scientific analysis of pendulum motion, which in turn was possible because of a new philosophical theory of knowledge. The scientific revolution was not just a change in science, it was also a change in philosophy of science.[1]

Galileo at different stages made four claims about pendulum motion:

1. Period varies with the square root of length; the Law of Length
2. Period is independent of amplitude; the Law of Amplitude Independence
3. Period is independent of weight; the Law of Weight Independence
4. For a given length all periods are the same; the Law of Isochrony

Contrary to the textbooks, it was not observation but mathematics, and experiment guided by mathematics, that played the major role in Galileo's discovery and proof of the properties of pendulum motion.

GALILEO AT PISA

In 1588, Galileo was appointed to a lectureship in mathematics at the University of Pisa. He quickly became immersed in the mathematics and mechanics of the

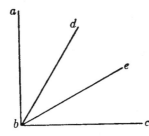

Figure 43. Galileo's 1590 inclined plane construction (Galileo, 1590/1960, p. 63)

"superhuman Archimedes," whom he never mentions "without a feeling of awe" (Galileo, 1590/1960, p. 67). Archimedes, above all else, utilized Euclidean geometry in the service of mechanics. He gave geometrical solutions to practical problems in using common machines: lever, pulley, wheel and axle, wedge, and screw.

Galileo's major Pisan work is his *On Motion* (1590/1960).[2] In it he dealt with the full range of problems then being discussed among natural philosophers: freefall, motion on balances, motion on inclined planes, and circular motions. Physical circumstances are depicted geometrically, and mathematical reasoning is used to establish various conclusions in physics: Galileo here began the mathematizing of physics so characteristic of modern science. Galileo's genius lay in his ability to see that all of the above motions could be dealt with in one geometrical construction. That is, motions which appeared so different in the world could all be depicted and dealt with in a common manner.

Galileo began with inclined plane motion and inquiries about the different amount of "force" [effort] required to pull bodies up planes of different inclinations. By working with the diagram in Figure 43 he decided:

> . . . a heavy body tends downward with as much force as is necessary to lift it up . . . If, then, we can find with how much less force the heavy body can be drawn up on the line *ba*, we will have found with how much greater force the same heavy body descends on line *ab* than on line *db*. . . . We shall know how much less force is required to draw the body upward on *bd* than on *be* as soon as we find out how much greater will be the weight of that body on the plane along *bd* than on the plane along *be*.

Galileo related these different forces to the different vertical heights that the bodies are drawn up, not the incline distances. This is a major conceptual step forward in understanding the motion of bodies. He proceeded to investigate the differences in weight of the body on the two planes *bd* and *be* and, in so doing, laid the conceptual and geometrical foundation for his pendulum investigations. He constructed the drawing (shown in Figure 44), and reasoned thus:

> Let us proceed then to investigate that weight. Consider a balance *cd*, with center *a*, having at point *c* a weight equal to another weight at point *d*. Now, if we suppose that the line *ad* moves towards *b*, pivoting about the fixed point *a*, then the descent of the body on the line *ef* will be a consequence of the weight of the body at point *d*. Again,

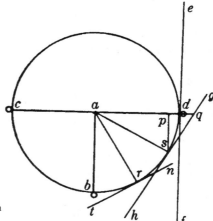

Figure 44. Galileo's 1590 composite construction
(Galileo, 1590/1960, p. 64)

when the body is at *s*, its descent at the initial point *s* will be as if on line *gh*, and hence its motion on *gh* will be a consequence of its weight at point *s*.

Now it is clear that the body exerts less force at point *s* than at *d*. For the weight at point *d* just balances the weight at point *c*, since the distances *ca* and *ad* are equal. But the weight at point *s* does not balance that at *c*. For if a line is drawn from point *s* perpendicular to *cd*, the weight at *s*, as compared to the weight at *c*, is as if it were suspended from *p*. But a weight at *p* exerts less force than at *c*, since the distance *pa* is less than the distance *ac* ... It is obvious, then, that the body will descend on line *ef* with greater force than on line *gh*. . . .

But with how much greater force it moves on *ef* than on *gh* will be made clear as follows. Extend line *ad* beyond the circle to meet line *gh* at point *q*. Now since the body descends on line *ef* more readily than on line *gh* in the same ratio as the body is heavier at point *d* than at point *s*, and since it is heavier at *d* than at *s* in proportion as line *da* is longer than *ap*, it follows that the body will descend on line *ef* more readily than on *gh* in proportion as line *da* is longer than *pa* ... And as *da* is to *pa*, so *qs* is to *sp*; that is, as the length of the oblique descent is to the vertical drop. And it is clear that the same weight can be drawn up an inclined plane with less force than vertically in proportion as the vertical ascent is shorter than the oblique. Consequently the same heavy body will descend vertically with greater force than on an inclined plane in proportion as the length of the descent is greater than the vertical fall. (Galileo, 1590/1960, p. 64–65)

This 1590, early Galilean, construction is remarkable in the history of science. He is saying that if one considered horizontal motion on a perfectly smooth plane, then as the plane was progressively tilted, one had inclined plane motion and, finally, when the plane was vertical, this was the situation of freefall motion. The progressive situations could be represented geometrically. It was only a small intellectual leap to see pendulum motion in the diagram (*a* the point of suspension, and the radius as length), and hence to analyze horizontal, inclined plane, freefall and pendulum motion in a unified manner. This is an important aspect of Galileo's break with Aristotelian–empiricist epistemology. As Winifred Wisan, an important analyst of Galileo's early work, wrote: "It is this exploitation of the creative power of mathematics that lies at the heart of the phenomenal acceleration of the mathematical sciences beginning with Galileo's discovery of a way to solve problems involving terrestial motions taking place in given ratios of time" (Wisan, 1980, p. 332).

For the Aristotelian tradition, things that *looked* the same were treated the same. They were taken as having the same essence or nature; they were members of the same *species*. This reflected the primacy that the senses played in Aristotle's epistemology. Above all, the eye was the key to knowledge of the world. As Aristotelians were fond of saying: "If the eye cannot be believed, what can?" This epistemological commitment is maintained by the 17th-century British empiricists, even though they were trenchant critics of Aristotle. Twentieth-century positivists and logical empiricists, with their commitment to "observation statements" being the foundation of science, have maintained this Aristotelian position. There are good grounds, some of which will be mentioned in Chapter 10, for thinking that this long empiricist tradition is simply mistaken about the role of observation in science.

Galileo introduced a classificatory system where very different looking things (balance motion, inclined plane motion, the lever, pendulum motion ,and freefall) were regarded as all belonging to the same category, and hence were analyzable in a coherent and comparable manner. They were seen as instances of the same thing. In much the same way as a moving compass needle, patterned iron filings, and induced current in a moving conductor—observationally all very different— are seen as indicators of one thing, a magnetic field.

But there is more than just the methodological innovation. Initially Galileo claimed that the downward tendency is measured by the upward resistance, a statement in keeping with what Newton maintained, one hundred years later, in his third law of motion. Galileo in his argument reduced the different motions to the problem of the lever, an Archimedean topic, that previously had been dealt with as an isolated topic in statics. Here Galileo used the principles of the lever to analyze a dynamic situation. He also used the notion of virtual, or infinitesimal, displacements in reasoning about the situation. This is something Huygens and Newton made more formal and mathematical. Finally, Galileo made vertical motion the common measure of all motion on inclined planes. This common measure enabled him to generalize his results to other simple machines.[3]

Galileo developed this line of analysis in an unpublished work, *On Mechanics* (Galileo, 1600/1960), a work that followed in the decade after his *On Motion*. It is in the *Mechanics* that the pendulum situation which is implicit in the 1590 construction, is made explicit. He dealt with all the standard machines (the screw,

plane, lever, pulley) and made the very un-Aristotelian observation, or rather theoretical claim, that:

> ... heavy bodies, all external and adventitious impediments being removed, can be
> moved in the plane of the horizon by any minimum force. (Galileo, 1600/1960, p. 171)

This is un-Aristotelian because the core of Aristotle's philosophy is that physics deals with the world as it is, not with an idealized or mathematical world
where, *contra-reality*, all external and adventitious impediments are removed. For
Aristotelian scientists, this is fantasy land. This claim of Galileo's put him on the
track toward a doctrine of circular inertia. He said Pappus of Alexandria "missed
the mark" in his discussion of forces on bodies, because he made the assumption
that "a weight would have to be moved in a horizontal plane by a given force"
(Galileo, 1600/1960, p. 172). Galileo believed that this assumption was false "because no sensible force is required (neglecting accidental impediments which are
not considered by the theoretician)" (Galileo, 1600/1960, p. 172). He used the
following construction (Fig. 45) and argument to make his point, and in so doing
set up a situation that enabled him to analyze pendulum motion in terms of circular motion and motion along chords of a circle.

> Consider the circle *AJC* and in this the diameter *ABC* with center *B*, and two weights
> at the extremities *A* and *C*, so that the line *AC* being a lever or balance, movable about
> the center *B*, the weight *C* will be sustained by the weight *A*. Now if we imagine the
> arm of the balance as bent downward along the line *BF*...then the moment of the
> weight *C* will no longer be equal to the moment of the weight *A*, since the distance of
> the point *F* from the line *BJ*, which goes from the support *B* to the center of the earth,
> being diminished.
>   Now if we draw from the point a perpendicular to *BC*, which is *FK*, the moment of
> the weight at *F* will be as if it were hung from the line *KB*; and as the distance *KB* is
> made smaller with respect to the distance *BA*, the moment of the weight *F* is accordingly
> diminished from the moment of the weight *A*. Likewise, as the weight inclines more,
> as along the line *BL*, its moment will go on diminishing, and it will be as if it were
> hung from the distance *BM* along the line *ML*, in which point *L* a weight placed at *A*
> will sustain one as much less than itself as the distance *BA* is greater than the distance
> *BM*.
>   You see, then, how the weight placed at the end of the line *BC*, inclining downward
> along the circumference *CFLJ*, comes gradually to diminish its moment and its impetus
> to go downward, being sustained more and more by the lines *BF* and *BL*. But consider
> this heavy body as descending and sustained now less and now more by the
> radii *BF* and *BL*, and as constrained to travel along the circumference *CFL*, is not
> different from imagining the same circumference *CFLJ* to be a surface of the same
> curvature placed under the same moveable body, so that this body, being supported
> upon it, would be constrained to descend along it. For in either case the moveable
> body traces out the same path, and it does not matter whether it is suspended from the
> center *B* and sustained by the radius of the circle, or whether this support is removed
> and it is supported by and travels upon the circumference *CFLJ* . . . Now when the
> moveable body is at *F*, at the first point of its motion it is as if it were on an inclined
> plane according to the tangent line *GFH*, since the tilt of the circumference at the

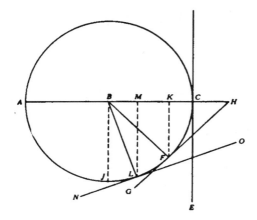

Figure 45. Galileo's 1600 construction
(Galileo, 1600/1961, p. 173)

point F does not differ from the tilt of the tangent FG, apart from the insensible angle
of contact. (Galileo, 1600/1960, pp. 173–174)

This is a most fruitful construction. It allowed Galileo to analyze pendulum
motion as motion in a circular rim and as motion on a suspended string. By considering initial infinitesimal, motions, he was able to consider pendulum motion
as a series of tangential motions down inclined planes. Two years later he wrote,
in the letter reproduced below, to Guidobaldo del Monte about these propostions.

GALILEO'S 1602 LETTER TO GUIDOBALDO DEL MONTE

The most significant opponent of Galileo's nascent views about the pendulum
was his own academic patron, the distinguished Aristotelian Guidobaldo del Monte
(1545–1607). The 1602 correspondence between the two nicely incarnates the
grand-historic tension between the old and new science.   ·

Del Monte was one of the great mathematicians and mechanics of the late
16th century. He was a translator of the works of Archimedes, the author of a
major book on mechanics (Monte 1581/1969), a book on geometry (*Plani-
spheriorum universalium theorica*, 1579), a book on perspective techniques
Perspectiva (1600), and an unpublished book on timekeeping *De horologiis* that
discussed the theory and construction of sun dials. He was a highly competent
mechanical engineer as well as a director of the Venice arsenal. Additionally he
was an accomplished artist, a minor noble, and the brother of a prominent cardinal. And of course he was a patron of Galileo who secured for Galileo his first
university position as a lecturer in mathematics at Pisa University (1588–1592),
and his second academic position as a lecturer in mathematics at Padua University (1592–1610).[4]

Figure 46. Guidobaldo del Monte (Ronan, 1974, p. 77)

Del Monte was not only a patron of Galileo, but from at least 1588 to his death in 1607, he was also actively engaged in Galileo's mechanical and technical investigations. They exchanged many letters and manuscripts on broadly Archimedean themes. Del Monte believed theory should not be separated from application, that mind and hand should be connected. As he said in the preface of his *Mechanics*: "For mechanics, if it is abstracted and separated from the machines, cannot even be called mechanics" (Drake and Drabkin, 1969, p. 245).

Del Monte was concerned with the longstanding Aristotelian problem of how mathematics related to physics. In his *Mechaniche* he said:

> Thus, there are found some keen mathematicians of our time who assert that mechanics may be considered either mathematically, removed [from physical considerations], or else physically. As if, at any time, mechanics could be considered apart from either geometrical demonstrations or actual motion! Surely when that distinction is made, it seems to me (to deal gently with them) that all they accomplish by putting themselves forth alternately as physicists and as mathematicians is simply that they fall between stools, as the saying goes. For mechanics can no longer be called mechanics when it is abstracted and separated from machines. (Monte, 1581/1969, p. 245)

The methodological divide between del Monte and Galileo, between Aristotelian science and the embryonic new science of the scientific revolution, was signalled in del Monte's criticism of contemporary work on the balance, including perhaps drafts of Galileo's first published work, *La Bilancetta* (Galileo, 1586/1961). Del Monte cautioned that physicists are:

> . . . deceived when they undertake to investigate the balance in a purely mathematical way, its theory being actually mechanical; nor can they reason successfully without

the true movement of the balance and without its weights, these being completely physical things, neglecting which they simply cannot arrive at the true cause of events that take place with regard to the balance. (Drake and Drabkin, 1969, p. 278)

In a 1580 letter to Giacomo Contarini, del Monte said:

> Briefly speaking about these things you have to know that before I have written anything about mechanics I have never (in order to avoid errors) wanted to determine anything, be it as little as it may, if I have not first seen by an effect that the experience confronts itself precisely with the demonstration, and of any little thing I have made its experiment. (Renn et al., 1988, p. 39)

This then is the methodological basis for del Monte's criticism of Galileo's mathematical treatment of pendulum motion. The crucial surviving document in the exchange between Galileo and his patron is a November 29, 1602 letter where Galileo wrote of his discovery of the isochrony of the pendulum and conveyed his mathematical proofs of the proposition.[5] As the letter is a milestone in the history of timekeeping and in the science of mechanics, and as it illustrates a number of other things about Galileo and his scientific style, it warrants reproduction in full.

> You must excuse my importunity if I persist in trying to persuade you of the truth of the proposition that motions within the same quarter-circle are made in equal times. For this having always appeared to me remarkable, it now seems even more remarkable that you have come to regard it as false. Hence I should deem it a great error and fault in myself if I should permit this to be repudiated by your theory as something false; it does not deserve this censure, nor yet to be banished from your mind—which better than any other will be able to keep it more readily from exile by the minds of others. And since the experience by which the truth has been made clear to me is so certain, however confusedly it may have been explained in my other [letter], I shall repeat this more clearly so that you, too, by making this [experiment], may be assured of this truth.
>
>   Therefore take two slender threads of equal length, each being two or three braccia long [four to six feet]; let these be *AB* and *EF*. Hang *A* and *E* from two nails, and at the other ends tie two equal balls (though it makes no difference if they are unequal). Then moving both threads from the vertical, one of them very much as through the arc *CB*, and the other very little as through the arc *IF*, set them free at the same moment of time. One will begin to describe large arcs like *BCD* while the other describes small ones like *FIG*. Yet in this way the moveable [that is, movable body] *B* will not consume more time passing the whole arc *BCD* than that used up by the other moveable *F* in passing the arc FIG. I am made quite certain of this as follows.

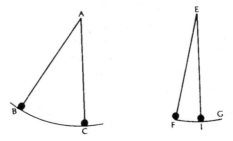

The moveable B passes through the large arc BCD and returns by the same DCB and then goes back toward D, and it goes 500 or 1,000 times repeating its oscillations. The other goes likewise from F to G and then returns to F, and will similarly make many oscillations; and in the time that I count, say, the first 100 large oscillations BCD, DCB and so on, another observer counts 100 of the other oscillations through FIG, very small, and he does not count even one more—a most evident sign that one of these large arcs BDC consumes as much time as each of the small ones FIG. Now, if all BCD is passed in as much time [as that] in which FIG [is passed], though [FIG is] but one-half thereof, these being descents through unequal arcs of the same quadrant, they will be made in equal times. But even without troubling to count many, you will see that moveable F will not make its small oscillations more frequently than B makes its larger ones; they will always be together.

The experiment you tell me you made in the [rim of a vertical] sieve may be very inconclusive, perhaps by reason of the surface not being perfectly circular, and again because in a single passage one cannot well observe the precise beginning of motion. But if you will take the same concave surface (Fig. 47) and let ball B go freely from a great distance, as at point B, it will go through a large distance at the beginning of its oscillations and a small one at the end of these, yet it will not on that account make the latter more frequently than the former. Then as to its appearing unreasonable that given a quadrant 100 miles long, one of two equal movables might traverse the whole and [in the same time] another but a single span, I say that it is true that this contains something of the wonderful, but our wonder will cease if we consider that there could be a plane as little tilted as that of the surface of a slowly running river, so that on this [plane] a moveable will not have moved naturally more than a span in the time that on another plane, steeply tilted (or given great impetus even on a gentle incline), it will have moved 100 miles. Perhaps the proposition has inherently no greater improbability than that triangles between the same parallels and on equal bases are always equal [in area], though one may be quite short and the other 1,000 miles long. But keeping to our subject, I believe I have demonstrated that the one conclusion is no less thinkable than the other.

Let BA be the diameter of circle BDA (Fig. 48) erect to the horizontal, and from point A out to the circumference draw any lines AF, AE, AD, and AC. I show that equal movables fall in equal times, whether along the vertical BA or through the inclined planes along lines CA, DA, EA, and FA. Thus leaving at the same moment from points B, C, D, E, and F, they arrive at the same moment at terminus A; and line FA may be as short as you wish.

And perhaps even more surprising will this, also demonstrated by me, appear: That line SA being not greater than the chord of a quadrant, and lines SI and IA being any whatever, the same moveable leaving from S will make its journey SIA more swiftly than just the trip IA, starting from I. This much has been demonstrated by me without transgressing the bounds of mechanics. But I cannot manage to demonstrate that arcs SIA and IA are passed in equal times, which is what I am seeking.

Do me the favor of conveying my greetings to Sig. Francesco and tell him that when I have a little leisure I shall write to him of an experiment that has come to my mind for measuring the force of percussion. And as to his question, I think that what you say about it is well put, and that when we commence to deal with matter, then by reason of its accidental properties the propositions abstractly considered in geometry commence to be altered, from which, thus perturbed, no certain science can be assigned – though the mathematician is so absolute about them in theory. I have been too long and tedious with you; pardon me, and love me as your most devoted servitor. (Drake, 1978, p. 69–71)

Figure 47. Ball in rim.

Thus in 1602 Galileo claimed two things about motion on chords within a circle:

1. That in a circle, the time of descent of a body freefalling along all chords terminating at the nadir is the same regardless of the length of the chord.
2. In the same circle, the time of descent along a chord is longer than along its composite chords, even though the latter route is longer than the former.

This gets him tantalizingly close to a claim about motion along the *arcs* of the circle, the pendulum case, but not quite there. He was not prepared to make the leap, saying "But I cannot manage to demonstrate that arcs *SIA* and *IA* are passed in equal times, which is what I am seeking." Galileo "sees" that they are passed in equal times; he also had empirical proof—if one can take hypothetical behavior in ideal situations as empirical proof, but he lacked a "demonstration." This is something that he believed only mathematics could provide.

Figure 48 presents a circle with its chords that had great potential in Galileo's physics. Galileo asserted that the time of freefall along *BA* is the same as the motion along the inclines *DA*, *EA*, *FA*. Two years later, he formulated his "times-square" law, saying that distance fallen varies as the square of the time elapsed ($s \propto t^2$); or, as we can say, time elapsed varies as the square root of distance fallen ($t \propto \sqrt{s}$). In Figure 48, the circle is traced out by a pendulum whose length is the semidiameter (half *BA*), and he intuits (but could not prove) that times of descent along the arcs *DEFA*, *EFA*, *FA* are equal. Galileo then has a proof for isochronic motion of a pendulum whose length is the semidiameter of *BA*. Further, the period *t*, will be as the square root of the pendulum's length, $\sqrt{l}$.

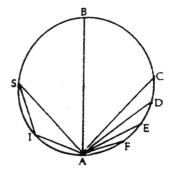

Figure 48. Chords and planes construction.

This 1602 letter is interesting for a number of reasons, one being that it illustrates the working of the patronage system which was important in early modern European science.[6] Mario Biagioli, a cultural historian of science, wrote of this system:

> In short, the legitimation of the new science involved much more than an epistemological debate. The acceptance of the new worldview depended also on the sociocognitive legitimation of the disciplines and practitioners upholding it. . . . high social status was the password to cognitive legitimation. (Biagioli, 1993, p. 18)

Biagioli's comment is, as commonly occurs in cultural histories of science and in the sociology of scientific knowledge, ambiguous about the duties performed by the "password." One can see that a password or an influential supporter might give "sociocognitive legitimation" to a new field of endeavour; it might get the new field a publisher or the new author a position. However, the interesting epistemological, and also surely historical point, is whether possession of the password truly gives "cognitive" legitimation, or only "socio" legitimation. Running the two terms together confuses the issue. In most senses, "cognitive legitimation" implies truth, or being productive of truth. History is littered with systems of ideas that had the right password and support from influential circles (Aristotelianism in Galileo's day, creationism in the 19th century, Nazism, Stalinism, Lysenkoist genetics, Islamic biology, Maoism and so on) but none of this "socio legitimation" rendered the ideas true, or the research progams cognitively productive.

Nevertheless Biagioli is correct in saying that the sociological and political realities of scholarly life was not unnoticed by the insightful, and ambitious, Galileo. In 1610 while at Padua University under the patronage of one prince, Galileo wrote about his desire for a position at the court of a still higher prince:

> Regarding the everyday duties, I shun only that type of prostitution consisting of having to expose my labor to the arbitrary prices set by every customer. Instead, I will never look down on serving a prince or a great lord or those who may depend on him, but, to the contrary, I will always desire such a position. (Biagioli, 1993, p. 29)

Galileo was solicitous of del Monte's opinions and he displayed a certain flattery. He apologized for not having explained things clearly enough in earlier letters and signed off as a "devoted servant." All of this was good manners. It was expected, and it was not unwarranted. Del Monte was the first to materially assist Galileo's academic career. He was able to do this through the agency of his brother Francesco Maria del Monte, who in 1588 was made a cardinal by Ferdinand de Medici, the powerful grand duke. While Francesco was a mere *monsignore* all of Guidobaldo's efforts to secure advancement for Galileo came to naught. But when Guidobaldo's brother donned the cardinal's cap, Galileo's job fortunes changed: He was given a lectureship at Pisa in 1588. When he wanted advancement, he was given a lectureship at Padua in 1592–both universities were under the authority of the doge of Venice. Guidobaldo died in 1607, but by then he had put Galileo into the higher echelons of the patronage circuit. Indeed so much so, that Galileo was, after del Monte's death, able to turn his back on the de Medici's patronage and successfully seek an appointment as "mathematician and philosopher" in the court of the grand duke of Tuscany.

However Galileo was far from being a sycophant. He had a point to make about pendulum motion and a new epistemology and methodology for its analysis. He was determined to make the point, even though it was at odds with the view of his patron.

A second reason for the letter's interest is that it illustrates Galileo's style of argument, specifically his clever use of rhetoric.[7] He clearly thought that the point of argument was not only to establish the truth, but also to have the reader believe it. He wrote, for instance, of "500 or 1,000 oscillations." This sounds impressive, but no ordinary pendulum will make anywhere near this number of oscillations before coming to a halt. When finely set up, heavy aerodynamic pendulums will, if released from a large amplitude, struggle to perhaps 300–400 oscillations before stopping. Simple weights on the end of strings will not get near to 100 before halting. Likewise, when Galileo said that large and small amplitude swings remained in synchrony after 100 oscillations, it should be remembered that when two pendulums with initial amplitudes of 60° and 10° are released, they are usually out of synchrony within 30 oscillations. Galileo's claims are rhetorical. There is an element of truth and a considerable element of exaggeration. Perhaps this exaggeration is understandable given the novelty of the claims—things that are novel frequently need more "beating up" just to get a hearing.

Likewise, Galileo's appeal to the theorem about the equality of areas of triangles having the same base when drawn between parallel lines is irrelevant to his claim about tautochronism. The theorem might well be amazing and counterintuitive, but it does not lend any epistemic support to the proposition being advanced by Galileo—rhetorical support, but not epistemic. It is like someone advancing the claim that they are as bright as Einstein, and when listeners express amazement, the person says that it is equally amazing, but true, that triangles drawn between parallels on the same base, even if the sides of one are hundreds of miles long, have the same area. The claim about triangles is amazing and true, but it simply has no bearing on the truth or falsity of the first claim about being as bright as Einstein. This type of argument is a clever, and a much used, ploy of debaters, politicians, and demagogues. Galileo's use may perhaps be legitimate. He was well aware that the new science, specifically Copernicus' claims for a rotating earth, required people to abandon deeply entrenched common sense beliefs, that were supported by masses of everyday evidence. In this context, drawing attention to other truths, that initially seem at odds with common sense, is sensible.

A third reason why the 1602 letter is interesting is that it illustrates, in embryonic form, the role of abstraction, idealization, and mathematics in Galileo's

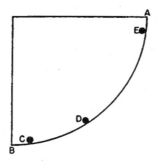

Figure 49. Ball in Hoop (Galileo, 1633/1953, p. 451)

new science. This makes it of particular interest to philosophers of science because this Galilean strategy came to typify the new science of the scientific revolution. Galileo and del Monte had had an earlier exchange of letters (they were
lost by the time Galileo's *Opere* was collected) about motion in a semicircle, and
it was Galileo's belief that such motion was tautochronous.[8] Del Monte could not
believe these claims and found them wanting when he rolled balls inside an iron
hoop. He was a scientist-engineer and enough of an Aristotelian to believe that
tests against experience were the ultimate adjudicator of claims in physics. Galileo's
claims failed the test. But Galileo replied that *accidents* interfered with del Monte's
test: his wheel rim was not perfectly circular and, as he stated elsewhere, the rim
was not smooth enough. These are perfectly understandable qualifications, yet it
needs to be appreciated that they are *modern* qualifications. Galileo introduced
this, now well established, process of abstracting from real circumstances to ideal
ones.

Thirty years later, in his *Dialogue Concerning the Two Chief World Systems*
(1633), Galileo returned to this example of del Monte's, when he said, in defense
of his claims for the tautochronism of circular motion that:

> Take an arc made of a very smooth and polished concave hoop bending along the
> curvature of the circumference *ADB*, so that a well-rounded and smooth ball can run
> freely in it (the rim of a sieve is well suited for this experiment). Now I say that
> wherever you place the ball, whether near to or far from the ultimate limit *B* . . . and
> let it go, it will arrive at the point *B* in equal times. . .a truly remarkable phenomenon.
> (Galileo, 1633/1953, p. 451)

Jürgen Renn and colleagues, after detailed analysis of Galileo's early works
and of the process of his discovery of the law of fall, maintain that

> . . . it was primarily the contact with Guidobaldo del Monte which, in a decisive moment
> of Galileo's intellectual development, encouraged him to take up the life-perspective
> of the risky but rewarding career of an engineer-scientist. (Renn et al., 1998, p. 41)

A not insignificant influence for a Tuscan nobleman.

GALILEO AND MEASUREMENT

It should not be inferred from the foregoing that Galileo was indifferent to experimental evidence. He was a most careful experimenter. His insight was to have a mathematical model of motion and then to compare real motions with this model; his "world on paper" preceded his "real world." This of course was the method employed by astronomers since before Plato's time. The philosophically interesting point is what adjustments, if any, and for what reasons, one makes to the mathematical model in the light of experimental results. How much does reason and how much does measurement contribute to theory development?

Galileo, as with other natural philosophers since Aristotle, was interested in understanding free fall, the most obvious terrestrial "natural" motion. Thus in Day Three of his *New Sciences,* Salvati (Galileo's alter ego) said: "thus we may picture to our mind a motion as uniformly and continuously accelerated when, during any equal intervals of time whatever, equal increments of speed are given to it." (Galileo, 1638/1954, p. 161). Sagredo, the supposedly impartial bystander to the discussion, then said: "I may nevertheless without offense be allowed to doubt whether such a definition as the above, established in an abstract manner, corresponds to and describes that kind of accelerated motion which we meet in nature in the case of freely falling bodies." (Galileo, 1638/1954, p. 162). As has often been pointed out, Galileo was interested in the "how" of such natural motions, as well as the "why" questions which preoccupied the Aristotelian tradition. But freefall was too fast for direct investigation and measurement, so Galileo turned to two motions that embody freefall, but in which the motion can be more easily manipulated and investigated: motion on an inclined plane and pendulum motion.

Scientific measurement was, to put it mildly, problematic in Galileo's day. Galileo's unit of length, the *punto* (about 0.94 mm), was arbitrary and idiosyncratic. He chose it simply because he had a brass ruler that was finely divided into 60 parts. These parts were his *punto*. His initial time measure was equally arbitrary and idiosyncratic: He weighed collected, water that flowed at the rate of 3 fluid ounces per second from a master container. He recorded those weights in grains, 1 ounce being 480 grains. He used the same container with the same hole and used the weight of water collected in a second vessel as a measure of the time elapsed. He thus measured *time* elapsed by *weight* of water collected, with his time unit, the *tempo*, being 16 grains weight. Because he worked with ratios—the time of travel for such a distance compared with the time for some other distance—the actual units chosen for both time and distance were of no consequence, the units canceled out.[9] By opening and closing a cock on the outlet of a water container, his "water clock," and weighing the amount of water that flowed out, he was able to measure time intervals down to one tempo, or 1/92 of a second. An almost modern Olympic-games degree of accuracy! But of course no one else in the world used the *tempo*, nor did Galileo ever relate it to the second.

In Galileo's famous inclined plane experiment, he noticed, initially by chanting, that the distances travelled along the plane in equal time (chant) intervals were (taking the first distance to be one unit) 1, 3, 5, 7, 9, 11, etc. He then saw that the sums of these distances (total distance travelled—1, 4, 9, 16, 25, 36) were in the proportion to the squares of the times elapsed. These chant-based, hence fixed-interval measurements were later replaced by his ingenious water-weighing measure of time.[10]

In 1604 he tried to time accurately pendulum swings of different amplitude using his weight of flowing water method. He took a 1740 *punto* (1.635 m) pendulum and let it swing to a vertical board from 90° and from 10°, while taking his finger off the outflow (when releasing the pendulum) and then putting it on the outflow tube of the water bowl (when the pendulum struck the board). For the full quadrant of fall he got 1043 grains (2.17 oz) of water, for the small amplitude he got 945 grains (1.96 oz). The difference in time (9.4%) for large and small amplitude swings is very close to the 10% that can now be calculated for this particular length (Drake, 1990, p. 22). Galileo was meticulous about his experimental results: In this case it was his interpretation that was flawed. He attributed the difference in time to *impediments*, not realizing that it was the *circular* arc that was the fundamental disturbing cause. Huygens would discover this geometrically and instead use the *cycloidal* curve for his isochronic pendulum.

Stillman Drake, who thoroughly researched Galileo's working papers on the pendulum and fall (Drake, 1978, 1990), wrote of finding the following Latin inscription in Galileo's folio 189vl: "The vertical whose length is 48,143 [*punti*] is completed in 280 *tempi*." Drake comments that: "In metric units, that says that fall through 45¼ meters takes 3.07 seconds; I calculate the time at Padua [for such a fall] to be 3.04 seconds" (Drake, 1990, p. 27).

One most important measurement that Galileo performed in these early years (1600–1604) was the ratio of the distance fallen from rest to the length of a pendulum timing that fall by swinging to the vertical through a small arc. A pendulum was held out from a vertical board and released similtaneously with the dropping of a weight. He adjusted the pendulum length until the "thud" of the pendulum hitting the wall coincided with the "thud" of the weight hitting the floor. If t is the time from release to "thud," then $4t$ will be the period of the pendulum (from release to vertical is one quarter of its period). This ratio is a constant for all heights of fall, at all places on earth and even on the moon, and it is equal to $\pi^2/8$.

Galileo measured the time of freefall through a given length to the time of the swing of a pendulum of the same length from release to a vertical board (one quarter of its period, T). He let a ball drop 2000 *punto* (1.88 m) and timed its fall using his weight of water method (850 grains). He then took a 2000 *punto* pendulum and timed its quarter-oscillation (942 grains). The ratio of the two times was 1.108. And it is a constant for all lengths of fall. That is, for all heights, the time of fall compared to the time for a quarter oscillation of a pendulum whose length is equal to the particular height, is constant. Using modern methods, we can calculate this ratio to be equal to $\pi/2\sqrt{2}$, or 1.1107. Galileo's result of 1.108, by water weighing on a beam balance, is not too bad.[11]

Galileo's reasoning and his calculations can be rendered in modern form. (However it must be remembered that Galileo never in his life wrote an equation nor used an algebraic expression.) For a pendulum of length l, having a quarter period of T (time from release to vertical) we have

$$T = \pi/2 \times \sqrt{l/g}$$

For a freefall of distance l in time t, we have

$$l = g/2 \times t^2$$

Rearranging this formula to give time in term of distance, we have

$$t = \sqrt{2l/g}$$

Thus if the body falls a distance equal to the length of the pendulum, then

$$t/T = 2/\pi \times \sqrt{2}$$

Galileo, in his 1602 notes, gave this ratio, by water-weighing, as 850/942, or 0.902. The above formula has the ratio as 0.900. An impressive degree of accuracy for someone without so much as a stopwatch, much less an electronic timer such that can be found in most modern schools.

THE PENDULUM IN GALILEO'S LATER WORK

We can divide Galileo's scholarly life into three periods: his early period (1581–1592) when he was first a student then a professor at the University of Pisa,[12] a period culminating in his first systematic work on motion, *De Motu* (Galileo, 1590/1960); his middle period (1592–1610), when he was a professor at the University of Padua, and his mechanical work was encapsulated in *On Mechanics* (Galileo, 1600/1962), and in letters to friends, particularly del Monte (1602) and Sarpi (1604); and his mature period (1611–1642), during which time he defended Copernician astronomy and published his major treatises *Dialogue Concerning the Two Chief World Systems* (Galileo, 1633/1953) and *Dialogues Concerning Two New Sciences* (Galileo, 1638/1954).[13] The year 1609, when he began his telescopic observations and turned his attention from physics to astronomy, is a pivotal year.[14] The essentials of his pendulum analyses were complete before then, though not published until in his late *Dialogues*.[15]

Galileo's law of length—that period is dependent upon length—first appeared in his 1590 *De Motu*, and returned in his 1633 *Dialogue,* but not in its refined "square root" form. In 1633, in discussing periodic tidal motion,[16] he said that "one true, natural, and even necessary thing is that a single moveable body made to rotate by a single motive force will take a longer time to complete its circuit along a greater circle than along a lesser circle. This is a truth accepted by all, and in agreement with experiments, of which we may adduce a few" (Galileo, 1633/ 1953, p. 449). Among the experiments adduced is this:

> Let equal weights be suspended from unequal cords, removed from the perpendicular, and set free. We shall see the weights on the shorter cords make their vibrations in shorter times, being things that move in lesser circles. Again attach such a weight to a cord passed through a staple fastened to the ceiling, and hold the other end of the cord in your hand. Having started the hanging weight moving, pull the end of the cord which you have in your hand so that the weight rises while it is making its oscillations. You will see the frequency of its vibrations increase as it rises, since it is going continually along smaller circles. (Galileo, 1633/1953, p. 449)

Five years later, in the First Day of his 1638 *Discourse*, Galileo refined this law of length when, in discussing the tuning of musical instruments:

> As to the times of vibration of bodies suspended by threads of different lengths, they bear to each other the same proportion as the square roots of the lengths of the thread; or one might say the lengths are to each other as the squares of the times; so that if one wishes to make the vibration-time of one pendulum twice that of another, he must make its suspension four times as long. In like manner, if one pendulum has a suspension nine times as long as another, this second pendulum will execute three vibrations during each one of the first; from which it follows that the lengths of the suspending cords bear to each other the [inverse] ratio of the squares of the number of vibrations performed in the same time. (Galileo, 1638/1954, p. 96)

The law of amplitude also runs through Galileo's work from beginning to end. It is equivalent to saying that the circle is tautochronous—that is, the time taken by a freely moving body to reach the nadir is the same irrespective of where on the circle's circumference it is released. In his early work, *On Motion* (1590), as we have seen, Galileo strove for a mathematical proof of this property. This was his Law of Chords—the time taken to descend any chord in the lower quarter of a circle terminating at the nadir was a constant. He needed a Law of Arcs—the time taken to descend any arc in the lower quarter of a circle terminating at the nadir was a constant. Galileo recognized that he did not have a mathematical proof of this property. Del Monte, as we have seen, was quick to point that he lacked an empirical proof.

In the fourth day of the 1633 *Dialogue* where Galileo discusses pendula, he commented that it is:

... truly remarkable ... that the same pendulum makes its oscillations with the same frequency, or very little different—almost imperceptibly—whether these are made through large arcs or very small ones along a given circumference. I mean that if we remove the pendulum from the perpendicular just one, two, or three degrees, or on the other hand seventy degrees or eighty degrees, or even up to a whole quadrant, it will make its vibrations when it is set free with the same frequency in either case. (Galileo, 1633/1953, p. 450)

At this stage Galileo still lacked, or at least did not offer, a proof. For Galileo, merely showing that it happened did not constitute a proof. A proof is mathematical. It is something that reason, using the tools of mathematics, establishes. Experiment may suggest the proof, or it may confirm or illustrate the proof, but it is not itself the proof. If experiment does not confirm the proof, then there is always the "accidents" and "imperfections" of matter to consider. Galileo is confident that the real world and his world on paper, when stripped of accidents and imperfections, will coincide.

It was in his 1638 *Discourse* that Galileo provided what he thought was the mathematical proof of his amplitude law. In Theorem VI of the *Discourse* he provided a geometric proof of the law of chords. The theorem stated:

> If from the highest or lowest point in a vertical circle there be drawn any inclined planes meeting the circumference the times of descent along these chords are each equal to the other. (Galileo, 1638/1954, p. 188)

This Law of Chords is close to a proof for isochrony of pendulum motion. The law has shown that the time of descent down inclined planes (chords) is the same, provided the planes are inscribed in a circle and originate at the apex or terminate at the nadir. This means that amplitude does not affect the time taken. This is highly suggestive of a Law of Arcs where amplitude should not affect the time of descent or time of swing. The Law of Arcs is proved later in Theorem XXII of the *Discourse* that also demonstrates the counterintuitive proposition that the quickest time of descent in freefall is not along the shortest path:

> If from the lowest point of a vertical circle (*C*), a chord (*CD*) is drawn subtending an arc not greater than a quadrant, and if from the two ends of this chord two other chords be drawn to any point on the arc (*B*), the time of descent along the two latter chords (*DB, BC*) will be shorter than along the first, and shorter also, by the same

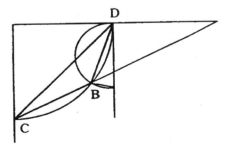

Figure 50. Galileo's brachistochrone proof (Galileo, 1638/1954, p. 237)

amount, than along the lower of these two latter chords (*BC*). (Galileo, 1638/1954, p. 237)

This is the beginning of Galileo's proof that the arc of a circle is the brachistochrone, or the line of quickest descent in freefall. In the Scholium that follows the above demonstration, he said (Fig. 51):

> From the preceding it is possible to infer that the path of quickest descent from one point to another is not the shortest path, namely, a straight line, but the arc of a circle. In the quadrant *BAEC*, having the side *BC* vertical, divide the arc *AC* into any number of equal parts, *AD, DE, EF, FG, GC*, and from *C* draw straight lines to the points *A, D, E, F, G*; draw also the straight lines *AD, DE, EF, FG, GC*. Evidently descent along the path *ADC* is quicker than along *AC* alone or along *DC* from rest at *D*...descent along the three chords, *ADEC*, will take less time than along than along the two chords *ADC*. ...along the five chords, ADEFGC, descent will be more rapid than along the four, *ADEFC*. Consequently the nearer the inscribed polygon approaches a circle the shorter is the time required for descent from A to C. (Galileo, 1638/1954, p. 239)

Galileo made a false move at this point. Having established that for a circle, the arc *not* the chord, is the path of quickest descent in freefall, he concluded the circle is the brachistochrone.[17]

He is confident of his proof, and in this *Discourse* stated that:

> It must be remarked that one pendulum passes through its arcs of 180°, 160° etc in the same time as the other swings through its 10°, 8°, degrees. . . . If two people start to count the vibrations, the one the large, the other the small, they will discover that after counting tens and even hundreds they will not differ by a single vibration, not even by a fraction of one. (Galileo, 1638/1954, p. 254)

The law of length, and the law of amplitude together almost amount to the law of isochrony. To claim that a pendulum's motion is isochronous is to claim that every swing takes the same time. That is, that all subsequent oscillations take the same time as the first oscillation. This does not follow from the law of amplitude. It almost follows from the law of length: if period depends only upon length, then, ideally, each swing will take the same time as any other swing. But friction at the fulcrum, the air's resistance to the bob, and the dampening effects of the

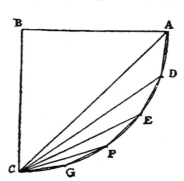

Figure 51. Galileo's brachistochrone proof (continued) (Galileo, 1638/1954, p. 237).

weight of the string or wire—will all contribute to a slowing of the pendulum. Galileo addressed the isochrony question (and the law of weight independence) in his 1638 *Discourse*.[18] When discussing his proposition that bodies of different weight fall with the same speed, and in attempting to overcome friction effects when demonstrating this proposition on inclined planes, he turned to his trusty pendulum for vindication:

> Accordingly I took two balls, one of lead and one of cork, the former more than a hundred times heavier than the latter, and suspended them by means of two equal fine threads, each four or five cubits long. Pulling each ball aside from the perpendicular, I let them go at the same instant, and they, falling along the circumferences of circles having these equal strings for semi-diameters, passed beyond the perpendicular and returned along the same path. This free vibration repeated a hundred times showed clearly that the heavy body maintains so nearly the period of the light body that neither in a hundred swings nor even in a thousand will the former anticipate the latter by as much as a single moment, so perfectly do they keep step. We can also observe the effect of the medium which, by the resistance which it offers to motion, diminishes the vibration of the cork more than that of the lead, but without altering the frequency of either; even when the arc traversed by the cork did not exceed five or six degrees while that of the lead was fifty or sixty, the swings were performed in equal times. . . .
>
> But observe this: having pulled aside the pendulum of lead, say through an arc of fifty degrees, and set it free, it swings beyond the perpendicular almost fifty degrees, thus describing an arc of nearly one hundred degrees; on the return swing it describes a little smaller arc; and after a large number of such vibrations it finally comes to rest. Each vibration, whether of ninety, fifty, twenty, ten or four degrees occupies the same time: accordingly the speed of the moving body keeps diminishing since in equal intervals of time, it traverses arcs which grow smaller and smaller. . . . Precisely the same thing happens with the pendulum of cork. (Galileo, 1638/1954, p. 84)

PROBLEMS WITH GALILEO'S PENDULUM CLAIMS

Although when Galileo's *Dialogue* was published in 1633, the novel geometrical analysis of the pendulum was widely and quickly embraced, there was genuine scientific ill-ease about the details of this analysis. Because history books, and more particularly science textbooks, are usually written from the modern, victor's point of view, contemporary opposition or dissent from Galileo's findings have been glossed over, or dismissed as the complaints of obscurantists. This is not only poor history, but it is grossly unfair to Galileo's pendulum critics. The situation is similar to that, 200 years later, of Darwin and his critics. There were empirical and conceptual holes all over the place in Darwin's grand evolutionary picture. In the case of both Galileo and Darwin, effort should be made to sympathetically evaluate the opposition. It is only by so doing that serious lessons can be learned from the history of science.[19]

It is easy to appreciate the empirical reasons for opposition to Galileo's law. The overriding argument was that if the law were true, pendulums would be perpetual motion machines, which clearly they are not. An isochronic pendulum is one in which the period of the first swing is equal to that of all subsequent swings. This implies perpetual motion. We know, and so did Galileo as seen in the above extract from the *Discourse*, that any pendulum when let swing will very soon come to a halt; the period of the last swing will be by no means the same as the first. Given isochrony, if the first swing takes *n* seconds, then all subsequent swings should also take *n* seconds. This means that the pendulum never stops. But it does stop. Furthermore it was plain to see that cork and lead pendulums have a slightly different frequency, and that large amplitude swings do take somewhat longer than small-amplitude swings for the same pendulum length. All of this was pointed out to Galileo, and he was reminded of Aristotle's basic methodological claim that the evidence of the senses are to be preferred over other evidence in developing an understanding of the world.

Marin Mersenne

Marin Mersenne (1588–1648)[20] and René Descartes (1596–1650) corresponded at length about Galileo's pendulum theory.[21] They established that, despite Galileo's claims, a pendulum's period was not independent of its amplitude, and hence not isochronic, which is the crucial property for its accurate timekeeping.

Mersenne in his *Harmonie Universelle* (1636), published five years before Galileo's death, related an experiment in which two identical pendulums were set swinging, one with an amplitude of two feet and one with an amplitude of one inch. He claimed (Proposition XIX) that the former is retarded by one oscillation after the passage of 30 oscillations. In the following proposition he said that there is a difference in period for a pendulum of eight pounds and one of half a pound. Mersenne not only agreed with del Monte, but doubted whether Galileo had ever conducted the experiments (Koyré, 1968, pp. 113–117).

In 1639 Mersenne, in his *Les nouvelles pensées de Galilée*, wrote:

> If [Galileo] had been more exact in his trials, he would have noticed that the string [of the pendulum] takes a sensibly longer time to descend from the height of a quarter-circle to the vertical position than when it is pulled aside only ten or fifteen degrees. This is shown by the *two sounds* made by two equal strings when they strike against a plank placed in the vertical position.
>
> And if he had merely counted to thirty or forty oscillations of the one pulled aside twenty degrees or less and of the other eighty or ninety degrees, he would have known that the one having the shorter swings makes one oscillation more in thirty or forty oscillations. And if you made one always swing to eighty degree, while that of ten or twenty degrees continued to slacken, this latter would gain one oscillation in ten or twelve. (MacLachlan, 1976, p. 182)

Mersenne, like del Monte, was no obscurantist. He endeavoured to be as accurate an observer and as good a scientist (natural philosopher) as he could be. He certainly valued the pendulum for its contribution to mechanics and to music;

and he was the first person to explicitly identify and state the law that the period of a pendulum is proportional to the square root of the length, $T \propto \sqrt{l}$ (Dear, 1988, p. 166). For a long time Galileo was vague on the specific relationship between a pendulum's period and its length: He stated that period varied with length, but as late as 1632 in correspondence with Baliani, who was questioning him on the relationship, he did not provide a more accurate statement of this relation (Crombie, 1994, p. 868). Galileo did not state the square root relationship until his final work of 1638. It was Mersenne who quantified this key relationship. He communicated this finding in a 1634 letter, and reproduced the proof in his 1637 *Harmonie universelle* (Fig. 52):

> As for the length that the string *AB* must have to make its returns faster or slower according to any proportion one wishes, it must be in proportion to the square of the slowness one wants, for example if the string *AB*, one foot long, makes each return in half a beat of the pulse, it must be four feet long to make each return in one beat of the same pulse; and if it makes each return in one second when it is 4 feet long, it must be 16 feet long to make each return in 2 seconds; and so on to infinity. (Crombie, 1994, p. 869)

Mersenne used Galileo's circular pendulum, and its theoretical underpinnings, to ascertain the value of the gravitational constant **g**. Neither Galileo, nor other 17th century scientists, used the idea of **g** in its modern sense of acceleration. *Their* gravitational constant was the distance that all bodies fall in the first second after their release—this is, numerically, half of the modern constant of acceleration.[22]

Mersenne's earlier investigations pointed to three Parisian feet being the length of the seconds pendulum,[23] and so in a famed experiment (1647) he held a 3 foot pendulum out from a wall and released it along with a freefalling mass (Fig. 53). He adjusted a platform under the mass until he heard both the pendulum strike the wall and the mass strike the platform at the same time. He reasoned that this should give him the length of freefall in half a second (a complete one-way swing

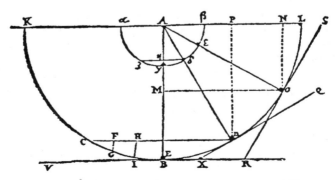

Figure 52. Mersenne's $T \propto \sqrt{l}$ Proof (Crombie, 1994, p. 869).

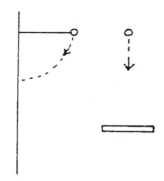

Figure 53. Mersenne's gravitational constant experiment (based on Westfall 1990, p. 71)

of the pendulum taking one second), and so, by the times-squared rule, he calculated the length of freefall in one second, the gravitational constant.

Mersenne did not get consistent results. He was frustrated by his experiments and became convinced that the circular pendulum was not isochronic. He was not however able to proceed beyond this point of frustration.[24] Huygens thought the ear could not separate the pendulum and freefall sounds to better than six inches of free fall. This was a significant enough "experimental error" to make the empiricist Mersenne despair of the possibility of there ever being an exact and certain "science" of mechanics. Exactitude and certainty were the qualities required at the time by both followers of Galileo and of Aristotle.[25] Modern notions of probabilism and fallibilism, as characteristics of science, were at least two centuries away, ushered in with Darwin's transformation of 19th-century philosophy of science.

Descartes also suspected that pendulum swings were not isochronous. In one of many letters to Mersenne he said:

> I believe that the vibrations will be a little slower toward the end than at the beginning,
> because the movement, having less force, does not overcome the hindrance of the air
> so easily. However about this I am not entirely certain. (*Works,* vol. I, p. 77)

Giambattista Riccioli

Mersenne's frustration was shared by another scientist, the anti-Copernican Jesuit Giambattista Riccioli, who in Italy in 1651 also tried to obtain the value of the gravitational constant—or more to be more precise, the distance that a freefalling body drops in the first second of fall. Riccioli's unflagging efforts to obtain the result are a monument to empiricism, and another stark exception to the popular image of anti-Copernicans being readers of the "book," rather than of nature.[26]

Riccioli realized that to ascertain the distance fallen in one second he first had to determine the duration of a second, for which, following Galileo's lead, he tried to construct a seconds pendulum. It was first necessary to establish a standard against which to measure his second. Sandglasses, water clocks, and mechanical clocks could not provide a standard. For a standard he looked to the heavens, this being the only standard for time then recognized. He took a "seconds" pendulum of length 98.5 cms (3 Roman feet, 4 inches), set it swinging, and kept giving it impulses. Riccioli enlisted the assistance of nine of his Jesuit colleagues, and they counted the oscillations over a period of 24 hours (from noon April 2, 1642 to noon April 3). They got 87,998 oscillations, whereas the number of seconds elapsed was 86,640. He slightly lengthened the pendulum chain, and used the sidereal day, rather than the solar day, for his standard. Again his colleagues were called in to do the counting, and again he did not get a correct result: 86,999 oscillations, against the 86,400 seconds in the sidereal day that he should have gotten had his pendulum been truly a seconds pendulum. Riccioli made a third attempt, this time slightly shortening the pendulum. Only two fathers carried on with their tiresome counting task, and on three different nights they got results of 3,212, 3,214, and 3,192 oscillations for the transit of the same star. His pendulum thus seemed to be both unreliable and inaccurate, but it was the best on offer.

Using this pendulum, he refined Mersenne's experiment and obtained a value of 15 Roman feet (4.43 m or 14.6 feet) for the distance traversed by a freefalling body in its first second of fall—the "gravitational constant." This compares favorably enough with the modern value of 4.9 m or 16 feet.

But did all bodies fall with the same speed? Riccioli used his pendulum, actually a drastically shortened version of Mersenne's, to settle the then vexing question of the speed of fall of bodies. Aristotelians maintained that bodies of different weight fell with different speeds; neo-Aristotelians maintained that bodies of different density fell with different speeds, but those of the same density, irrespective of their weights, fell with the same speed. Galileans maintained that all bodies, irrespective of weight and density, fell with the same speed. Riccioli dropped bodies from a high hill, the Torre degli Asinelli, and used his pendulum to record the time of fall. After 15 experiments his results were conclusive: Heavy balls fell quicker than light ones, the difference being 12–40 feet!

CONTEMPORARY REPRODUCTIONS

Beginning with his first biographer Vicenzio Viviani (1622–1703), through the work of Mach in the late 19th century, and up to Stillman Drake's compendious studies this century, Galileo has been depicted as a patient experimentalist who examined nature rather than books about nature, as Aristotelians were supposedly doing. This empiricist interpretation has been the dominant tradition in Galilean historiography. Alexandre Koyré's rationalist, intellectualist, neo-Platonic interpretation burst in the 1930s and 1940s like a thunderclap over this empiricist tradition (Koyré, 1939/1978, 1943a, 1943b). This was not surprising, because in 1960 Koyré wrote of Galileo's work:

> I have tried to describe and justify Galileo's use of the method of imaginary experiment concurrently with, and even in preference to, real experiment. In fact, it is an extremely fruitful method which incarnates, as it were, the demands of theory in imaginary objects, thereby allowing the former to be put in concrete form; and enables us to understand tangible reality as a deviation form the perfect model which it provides. (Koyré, 1960, p. 82)

Against this contested historiographical background, it is also not surprising that scholars have scrutinised Galileo's pendulum experiments and have endeavoured to replicate them (Ariotti, 1968; MacLachlan, 1976; Naylor, 1974a, 1980, 1989; Settle, 1961). Concerning Galileo's claims for weight independence, Ronald Naylor, for instance, found that;

> using two 76 inch pendulums, one having a brass bob, the other cork, both swinging initially through a total arc of 30°, the brass bob was seen to lead the cork by one quarter of an oscillation after only twenty-five completed swings. (Naylor, 1974a, p. 33)

Concerning amplitude independence, Naylor realized:

> the repetition of Galileo's pendulum experiment shows that after some thirty swings the pendulum completing large arcs has lagged one complete swing behind that completing small arcs. (Naylor, 1976a, p. 415)

Of the same claims about amplitude independence, James MacLachlan wrote:

> Now, if anyone swings two equal pendulums through such unequal arcs [Galileo's 80° and 5°] it is easy for him to observe that the more widely swinging one takes a longer time to complete the first oscillation, and after a few more it will have fallen considerably behind the other. (MacLachlan, 1976, p. 178)

Indeed the difference in period between a 90° and a 3° swing is 18%.

On the mass independence claims, MacLachlan wrote:

> As for Galileo's remark that they [cork and lead balls] would not differ even in a thousand oscillations, I have found that a cork ball 10 cm. of diameter is needed just to continue oscillating 500 times. However, for the lead to be 100 times heavier than that, it would have to be more than 10 cm. in diameter, and it would make even fewer oscillations (perhaps only 93) in the time that the cork bob made a hundred. (MacLachlan, 1976, p. 181)

The discrepant results of Naylor, MacLachlan, and others, do not mean that Galileo's work was just "imaginary." Undoubtedly these experiments were conducted, but equally undoubtedly the results were "embellished." Ronald Naylor provided a reasonable summary of the historical evidence:

This paper suggests that while Galileo did undoubtedly devise and use experiments similar to those described in the *Discorsi*, it seems evident that in publication he idealized and simplified the results of these researches. Thus some experimental accounts in the *Discorsi* appear to contain the essential distillate of many experiments rather than the description of any actual experiment. Ultimately, the idealized versions of the experiments seem to fuse with Galileo's theoretical model. Even so, it seems likely that at times Galileo was just as capable of providing a totally imaginary experiment in order to support his case. (Naylor, 1974a, p. 25)

Ignoring discrepant data, and embellishing results, is a methodological two-edged sword: It allows the experimenter to keep their eyes on the main game,[27] but sometimes the discrepant, or outlying, results are not the product of "accidents" and "impediments," but of basic mechanisms in the world. This was the case, as will be seen in the following chapter, with Galileo's continued commitment to the circle being the isochronous curve. There is an epistemological lesson to be learned here about how experimental results relate to theoretical commitments. That is, we need, in the beginning, to distinguish the *theorized* objects of science, and their properties, from the *material* objects of the world, and their behavior. These philosophical matters will be taken up in Chapter 10.

CHAPTER 6

Christiaan Huygens and the Pendulum Clock

Christiaan Huygens (1629–1695) stands out among the great scientific minds of the 17th century who addressed themselves to the improvement of timekeeping and the solution of the longitude problem. Huygens possessed both manual and intellectual skills of the highest order. He was the son of a well-connected and wealthy Dutch diplomat, who, with good reason, called his son *mon Archimède*. Upon Huygens death, Leibniz wrote: "The loss of the illustrious Monsieur Huygens is inestimable; few people knew him as well as I; in my opinion he equaled the reputation of Galileo and Descartes and aided theirs because he surpassed the discoveries that they made; in a word, he was one of the premier ornaments of our time" (Yoder, 1991, p. 1).

Huygens was born in The Hague in 1629. By the age of 13 he had built himself a lathe. By 17, he had independently discovered Galileo's time-squared law of fall and Galileo's parabolic trajectory of a projectile,[1] by 20, he had completed and published a study of hydrostatics. By 24 he formulated the laws of elastic collision. By 25, he was an optical lens grinder of national renown. By 26, and using one of his own telescopes, he discovered the ring of Saturn. When he was in his mid-30s many thought of him as the greatest mathematician in Europe—no slight claim given that his contemporaries included Pascal, Mersenne, Fermat, Descartes, Leibniz, and Newton.

In 1663 he was made a member of England's newly founded (1662) Royal Society. In 1666, at age 37, he was invited by Louis XIV to be founding president of the *Académie Royal des Sciences*, an invitation he accepted, and a position he held, organizing its scientific affairs in a manner inspired by Roger Bacon's view of the scientific commonwealth until 1681. Even during the bitter war between France and Holland that broke out in 1672, and that saw Huygens' father and brother occupying high positions in the opposing court of William III of Orange, Louis XIV nevertheless retained Huygens as president of the *Académie*.[2]

Huygens made fundamental contributions to mathematics (theory of evo-

121

Figure 54. Christiaan Huygens (Gould, 1923, p. 26)

lutes and probability), mechanics (theory of impact), optics (both practical with his own lenses, and theoretical with his wave theory of light), astronomy (discovery of Saturn's rings and determination of the period of Mars), and to philosophy (his elaboration of the mechanical world view, and proposals for a form of hypothetico-deductive methodology in science). In 1695 he died in The Hague, where he was born.[3]

John Locke, in his 1690 *Essay* on the intellectual revolution occurring around him, placed Huygens just one notch behind the "incomparable Mr. Newton." Locke wrote:

> The commonwealth of learning is not at this time without master-builders, whose mighty designs in advancing the sciences will leave lasting monuments to the admiration of posterity; but everyone must not hope to be a Boyle or a Sydenham; and in an age that produces such masters as the great Huvgenius, and the incomparable Mr. Newton, with some others of that strain, it is ambition enough to be employed as an under-labourer in clearing the ground a little, and removing some of the rubbish that lies in the way of knowledge. (Locke, 1690/1924, p. 6)

A modern scholar wrote:

> It is no exaggeration to say that from the 1650s till the 1680s Huygens was Europe's most creative mathematician and natural scientist. He was indeed recognized as such. Leibniz and Newton, the men whose achievements Huygens at the end of his career would come to acknowledge as equal to, if not surpassing his own, admired and praised his work. (Bos, 1986, p. xxv)

Huygens' work falls, both temporally and conceptually, between that of Galileo (*Dialogue*, 1633, *Discourse,* 1638) and Newton (*Principia,* 1687, *Opticks,*

1704). It was a bridge between the announcement and the fulfillment of the new science. It embodied that combination of sophisticated mathematics and refined experimental technique that so characterized the new method of science introduced by Galileo and perfected by Newton.

Huygens played a significant role in the mathematization of physics, something that is the hallmark of modern science. Mathematical physics appeared slowly in Galileo's work and was never brought to completion. Mathematics is absent from Galileo's early writing (ca.1585–1590) where he engaged in discussion of physical questions in the same way that Aristotelians had done for centuries. Galileo's early arguments and analyses are verbal, qualitative, conceptual, and full of appeals to ancient and medieval authorities. In the 250 pages of the *Physical Questions* there is neither a single measurement nor a geometrical figure.[4] Geometrical constructions and measurements begin appearing in Galileo's 1590 *De Motu* (Galileo, 1590/1960). They are more apparent in his 1633 *Dialogue* (Galileo, 1633/1953), and even more so in his final 1638 *Discourse* (Galileo, 1638/1954). Nevertheless in Galileo's entire corpus, there is not an equation to be found, and there is no hint of algebra. When, in the *Assayer*, Galileo spoke of the language of the Book of Nature being mathematical, he said that its characters are "triangles, circles, and other geometric figures without which it is humanly impossible to understand a single word of it; without these, one wanders about in a dark labyrinth" (Drake, 1957, p. 238).

But Galileo's geometrical language was about to be replaced. Coming over the intellectual horizon were changes ushered in by Descartes' creation of coordinate geometry and the work of Fermat and Cardan. Both Galileo's *Dialogue* and the *Discourse* can be read with enjoyment and comprehension by any literate person. Not so Newton's *Principia* (1687). This was published almost exactly 100 years after Galileo's *Juvenalia* and, in the interval, the scientific revolution had occurred. Geometrical demonstrations and proofs begin in the third paragraph of the *Principia*, and continue through its three books, getting progressively more complex.[5]

Huygens belongs to the Newtonian end of this Galileo—Newton mathematical continuum. Huygens pushed classical Euclidean geometry as far as it could go in the solution of physical problems. Although the utilization of geometry in physics led to enormous advances (it has oft been said that the scientific revolution was kick-started by geometry) nevertheless it was limiting and could not deal adequately with bodies in motion. Mathematics is the tool of science. Huygens did as well as he possibly could with the geometrical tool he inherited, but his understanding and analysis was limited by the absence of calculus and differential equations. Huygens' achievements and failures nicely exhibit the dependence of science on the state of mathematics and also the way in which scientific problems act as a spur for the development of more sophisticated mathematics.

However it is in virtue of Huygens' creation of the pendulum clock at Christmas-time 1656, in itself enough to secure his place in the history of science and technology and his persistent efforts to solve the longitude problem that he enters our story.

HUYGENS' REFINEMENT OF GALILEO'S PENDULUM THEORY

Huygens carried forward Galileo's program in mechanics. He wrote that "it has been necessary to corroborate and to extend the doctrine of the great Galileo on the fall of bodies" (Dugas, 1988, p. 181). In particular, he wrote: "The most desired result and, so to speak, the greatest, is precisely this property of the cycloid that we have discovered. . . . To be able to relate this property to the use of pendulums, we have had to study a new theory of curved lines that produce others in their own evolution" (Dugas, 1988, p. 181). The cycloid was such a curve traced out by a point on the circumference of a wheel as it rolls.

Huygens refined Galileo's theory of the pendulum and timekeeping.[6] His first book on the subject was the *Horologium* (1658),[7] his second and major work was the *Horologium Oscillatorium* ("The Pendulum Clock" (1673). He recognized the problem identified by Descartes: "The simple pendulum does not naturally provide an accurate and equal measure of time since its wider motions are observed to be slower than its narrower motions" (Huygens, 1673/1986, p. 11). Huygens changed two central features of Galileo's theory: namely the claims that period varied with length, and that the circle was the tautochronous curve (the curve on which bodies falling freely under the influence of gravity reach its nadir at the same time regardless of where they were released). In contrast Huygens showed mathematically that period varied with the square root ("subduplicate" as Huygens refers to it) of length, and that the cycloid was the tautochronous curve.[8] Huygens thought so highly of the second result that he described it in a 1666 letter to Ismaël Boulliau as "the principal fruit that one could have hoped for from the science of accelerated motion, which Galileo had the honor of being the first to treat" (Blay, 1998, p. 19). No small praise for something so abstruse and arcane; and a reminder, given its utilization in producing truly isochronic pendulum motion in clocks, of the utility of "theoretical" research.

These were both major theoretical breakthroughs. Galileo had argued that the circle was the tautochronous curve, and thus that an oscillating freely suspended pendulum would be isochronous because its bob moved in a circle. Galileo's proposal for a new and accurate clock was to utilize this property of the pendulum and, by his novel escapement, link the swinging pendulum to the drive train of mechanical clocks. His clock's timekeeping quality was thus to be independent of the amplitude of the pendulum's oscillation, with a large oscillation helping to overcome incidental perturbations caused by external bodies, friction and the like.

In the Footsteps of Geometry . . .

Huygens showed that it was not the circle, but the cycloid that was tautochronous and isochronic. He provided the following account of this discovery:

> We have discovered a line whose curvature is marvelously and quite rationally suited to give the required equality to the pendulum. . . . This line is the path traced out in air by a nail which is fixed to the circumference of a rotating wheel which revolves continuously. The geometers of the present age have called this line a cycloid and have carefully investigated its many other properties. Of interest to us is what we have called the power of this line to measure time, which we found not by expecting this but only by following in the footsteps of geometry. (Huygens, 1673/1986, p. 11)

It is informative to follow in the geometrical footsteps of Huygens. Walking a short distance is enough to illustrate the wonderful interaction of mathematics and experiment so characteristic of the scientific revolution, and also the limitations of geometry as a method for the analysis of the pendulum. Scientific analysis is dependent upon good intuitions about what is happening in the physical world, but it is also dependent upon methods of representation. Geometry advanced the science of pendulum motion, but it also placed limits on it.

Huygens mentions "geometers of the present age" having identified the cycloidal curve and having "investigated its many other properties." This was certainly the case. The cycloid (*roulette* as it is known in French) is the curve traced out by a point on the circumference of a circle as the circle rolls along a plane. It was unknown to ancient geometers and seems to have been first identified by Charles Bouvelles in 1501 but he did not examine any of its properties. It was the illustrious group of 17th century mathematicians who, at Mersenne's suggestion (Boyer, 1968, p. 355), first gave serious attention to the cycloid. Galileo knew of the curve, but not much about its properties. Gilles Persone de Roberval demonstrated in 1634 that the area under a cyloid's arc was precisely three times the area of the generating circle. Torricelli (Galileo's pupil) independently arrived at this conclusion sometime prior to 1644 when he published the result. In 1638 Roberval

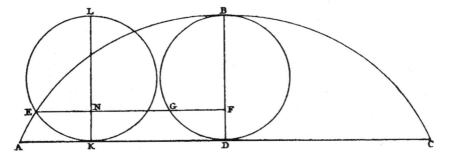

Figure 55. Cycloid curve generated by point on moving circle (Huygens, 1673/1986, p. 50)

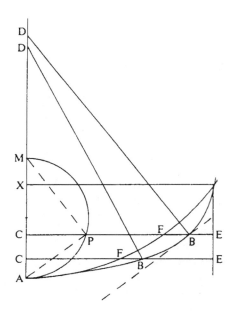

Figure 56. Cycloid curve with defining
reature being parallelism of tangent at B to
chord on generating circle (Yoder, 1988, p. 60)

showed how a tangent could be drawn to any point of the cycloid (Fig. 56). A
procedure that Fermat and Descartes also established. Christopher Wren (1632–
1723), in London in 1658, worked out that the length of the cycloid's arc was four
times the diameter of the generating circle. This was the rich mathematical context of Huygens' investigations of the pendulum and of the cycloid.

By 1659, Huygens (aged 30 years) had recognized the empirical problems
with Galileo's pendulum claims, problems pointed out by del Monte, Mersenne,
Ricardi, Descartes, and others. He posed himself the problem: "It is asked, what
ratio does the time of a very small oscillation of a pendulum have to the time of
perpendicular fall through the height of the pendulum" (Yoder, 1988, p. 50).
Mersenne had tried, unsuccessfully, to solve this problem empirically by letting
the pendulum swing through a quarter circle on to a vertical wall while simultaneously dropping a weight from the point of suspension of the pendulum on to a
horizontal board. Mersenne tried to adjust the horizontal board until the sound of
the two impacts coincided. His results were gross and unsatisfactory. Huygens
attacked the problem mathematically and in terms of infinitesimal displacements.
This was the crucial step toward showing that the tautochronous and isochronic
curve was cycloidal.

Huygens arrived at his celebrated geometrical proof of cycloidal isochronism
during December 1659. Manuscripts from the period clearly show him working
on pendular problems bequeathed by Galileo; with the opening line of one manuscript posing the problem of the comparison of time taken for a physical point to

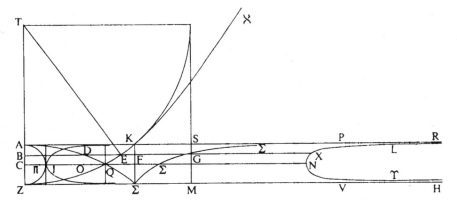

Figure 57. Huygens' 1659 geometric derivation of cycloidal isochronism (Yoder, 1988, p. 51)

describe a very small pendular motion and that same physical point to fall verti-cally from an equivalent height.[9]

Figure 57 is the start of Huygens' proof that the curve of isochronic motion is cycloidal. He examined a circular pendulum suspended at T and describing a small arc KEZ. Huygens compared its fall with the fall of a body from A under uniform motion whose fixed speed is that which a body falling from A under the force of gravity would attain at Z. Huygens already has the mean-speed theorem, so he knew that if v is the speed at point Z of a body falling under the influence of gravity from A, then a similar body moving from A with a uniform velocity of v will reach Z in half the time.

Huygens genius was to consider curvilinear motion as a series of infinitesi-mally small rectilinear motions, namely inclined plane motions. The curve is the summation of a large number of inclined plane motions, each plane having a slightly different gradient. He then began his analysis by comparing the time of fall for the pendulum at E for an infinitesimal period on the arc EQZ, with the time through an infinitesimal period of the freefalling body at B. He established that this proportion is equal to TE/BE. Huygens, following Galileo, knew that at any point such as E on a curve the speed of a body under the action of gravity equaled the speed of a body falling freely from the initial height A to the same vertical height B. Since the vertical distance fallen AB is proportional to the square of the speed, this speed can be represented graphically by a parabola whose vertex is at the initial height A and whose base ZS can be arbitrarily chosen equal to AK (because everything is expressed proportionally). Huygens went on through many steps and many constructions that involved the summation of infinitesimals, until he finally showed that if points K, E, Q, Z are points not on a circle, but on a new curve $cKEQZ$ then the time of descent along the arcs QZ, EZ, KZ are all equal (Yoder, 1988 pp. 50 ff; Blay, 1998, pp. 19–27). It then remained for him to identify the new curve c*KEQZ* with the cycloid. Having done this he wrote that his achievements were "Great matters not investigated by the men of genius among our forefathers" (Yoder, 1988, p. 61).

Having proved, by way of infinitesimals in 1659, that the cycloid was the sought-for isochronic curve,[10] Huygens in his 1673 *Horologium Oscillatorium* (Proposition XXV) stated the matter more directly:

> On a cycloid whose axis is erected on the perpendicular and whose vertex is located at the bottom, the times of descent, in which a body arrives at the lowest point at the vertex after having departed from any point on the cycloid, are equal to each other; and these times are related to the time of a perpendicular fall through the whole axis of the cycloid with the same ratio by which the semicircumference of a circle is related to its diameter. (Huygens, 1673/1986, p. 69)

He made the accompanying construction (Fig. 58) and then went on to prove his isochrony proposition in the following manner.

> Let *ABC* be a cycloid whose vertex A is located at the bottom and whose axis *AD* is erected on the perpendicular. Select any point on the cycloid, for example B, and let a body descend by its natural impetus through the arc *BA*, or through a surface so curved. Now I say that the time of this descent is related to the time of a fall through the axis *DA* as the semicircumference of a circle is related to its diameter. When this has been demonstrated, it will also be established that the times of descent through all arcs of the cycloid terminating at *A* are equal to each other.
>
> On the axis *DA* draw a semicircle whose circumference cuts the straight line *BF*, which is parallel to the base *DC*, at *E*. Draw the line *EA*, and parallel to it draw the line *BG* which is tangent to the cycloid at *B*. *BG* will cut the horizontal straight line drawn through *A* at *G*. Also on *FA* draw the semicircle *FHA*.
>
> Now according to the preceding proposition, the time of descent through the arc *BA* of the cycloid is related to the time of a uniform motion through the line *BG*, with half the velocity from *BG*, as the semicircular arc *FHA* is related to the straight line *FA*. But the time of the indicated uniform motion through *BG* is equal to the time of a naturally accelerated fall through *BG*, or through *EA* which is parallel to it, and thus is equal to the time of an accelerated fall through the axis *DA*. Hence the time through the arc *BA* will be related to the time of a fall through the axis *DA* as the semicircular circumference *FHA* is related to the diameter *FA*. Q.E.D.
>
> Now if it is assumed that the whole cavity of the cycloid has been constructed, it is clear that, after a body had descended through the arc *BA*, it will continue its motion and ascend through another arc equal to *BA* [Proposition IX], and the same amount of time will be consumed in the ascent as in the descent [Proposition XI]. After this, it will return again through *A* to *B*; and the times of each of these oscillations, completed in either large or small arcs of the cycloid, will be related to the time of a perpendicular fall through the axis *DA* as the whole circumference of a circle is related to its diameter. (Huygens, 1673/1986, pp. 69–70)

Once Huygens demonstrated mathematically that the cycloid was the isochronous curve, he then could derive the ratio of one oscillation in a cycloid to the time for a body to fall along the diameter of its generating circle. In effect, as Westfall pointed out, he had a formula for the period of the pendulum in terms of

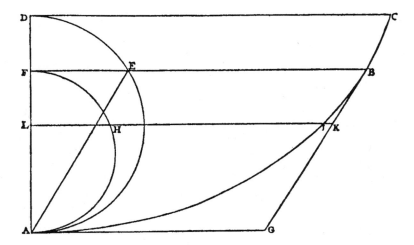

Figure 58. Huygens' isochrony proof (Huygens, 1673/1986, p. 70).

its length and **g** (Westfall, 1990, p. 73). Huygens then calculated that a body falls from rest through approximately 15 (Parisian) feet, 1 inch (or 16 feet, 1½ inches). This is the value still accepted. Huygens thus mathematically by-passed the tedious, and error-prone, observations of Mersenne and Riccioli (Harper and Smith, 1995).

Alexandre Koyré commenting on this theoretical and mathematical achievement of Huygens, said:

> The moral of this history of the determination of the acceleration constant is thus rather curious. We have seen Galileo, Mersenne, and Riccioli endeavouring to construct a timekeeper in order to be able to make an experimental measure of the speed of the fall. We have seen Huygens succeed, where his predecessors had failed, and by his very success, dispense with the making the actual experiment. This, because his timekeeper is, so to say, a measurement in itself; the determination of its exact period is already a much more precise and refined experiment than all those that Mersenne and Riccioli have ever thought of. The meaning and value of the Huygensian circuit—which finally revealed itself as a shortcut—is therefore clear: not only are good experiments based upon theory, but even the means to perform them are nothing else than theory incarnate. (Koyré, 1968, p. 113)

. . . And With the Assistance of Calculus

From a modern perspective, what was so acutely lacking in Huygens' treatment of the pendulum was the formalized mathematical idea of a function, of one variable changing as a result of changes in another variable. Brilliant mathematician though Huygens was, this modern-day commonplace was just beyond his conceptual reach.

Huygens was Leibniz' mathematical mentor, but became an onlooker as his student outstripped him and developed the rudiments of the calculus. Huygens clearly understood Leibniz' achievement, but hung onto the mathematical procedures with which he was familiar and which had returned him such powerful results.[11] Like the pony express triumphing in communications, only to be overtaken by the telegraph, so the calculus was quickly to render obsolete, in physics, the painstaking geometrical analyses of Huygens and Newton. William Whewell (1794–1866) said of the latter's renowned geometrical demonstrations that:

> Newton's successors, in the next generation, abandoned the hope of imitating him in this intense mental effort; they gave the subject over to the operation of algebracical reasoning, in which symbols think for us, without our dwelling constantly upon their meaning. (Whewell, 1958, in Chandler, 1998, p. 37)

The function idea was latent in Galileo's work, especially in his frequent treatment of proportions. In his work on freefall he wrote that "the spaces described by a body falling from rest with a uniformily accelerated motion are to each other as the squares of the time intervals employed in traversing these distances." And for inclined plane motion he explained "The times of descent along inclined planes of the same height, but of different slopes, are to each other as the lengths of these planes." Galileo constantly utilizes this proportional reasoning, translating it into geometric form, and arguing geometrically. Still, one can search all his work and never find an equation. He never used anything like a simple $x = y + A$, much less more complex functions. Galileo had the distinction of being the first to relate distance and speed of fall to *time* elapsed, to treat time as an independent variable as we might say, but he completely lacked, as did Huygens, the modern mathematical idea of a function.

The cycloid occasioned the idea of a geometric curve, or line, being generated by a moving point. This was an important step toward the function concept. Mersenne in 1615 defined the cycloid as the locus of a point on a wheel that rolls along the ground. This gave a dynamic origin to a geometric form. The idea of the curve as the path of a moving point was developed by Roberval and Barrow, and given explicit formulation by Newton in his *Quadrature of Curves* (1676), where he said:

> I consider mathematical quantities in this place not as consisting of very small parts, but as described by a continued motion. Lines [curves] are described, and thereby generated, not by the apposition of parts but by the continued motion of points . . . These geneses really take place in the nature of things, and are daily seen in the motion of bodies. (in Kline, 1972, p. 339)

Newton used the term *fluent* to represent any relationship between variables. Leibniz introduced the term *variable* to refer to the quantities being discussed; that is, one quantity—a variable—changing as another quantity or variable changed. The modern notation, *f(x)* was introduced by Euler in 1734. Add the calculus and

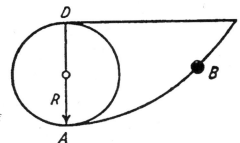

Figure 59. Contemporary demonstration of
cycloidal isochrony (Dugas, 1988, p. 186).

differentiation, and all these developments would eventually render the analysis
of pendulum motion a simple enough matter for introductory physics students. In
1690, James, one of the illustrious Bernoulli brothers, published his solution to
the isochrone problem (the curve where period is independent of amplitude) as a
differential equation. This made Huygens' geometric method look cumbersome
and awkward. Bernoulli's differential equation for the isochrone was:

$$dy\sqrt{(b^2y - a^3)} = dx\sqrt{a^3}$$

The integral of the equation was:

$$\frac{3b^2y - 2a^3}{3b^2}\sqrt{(b^2y - a^3)} = x\sqrt{a^3}$$

This function defined a cycloid, hence the cycloid was the isochrone!

René Dugas provided a proof of Huygens proposition about the isochronism
of the cycloid using the modern calculus. He took a heavy particle B on a cycloid
curve AB (Fig. 59), and said:

> The motion of a heavy particle on a cycloid is defined by the differential equation:
>
> $$d^2s/dt^2 + gs/4R = 0$$
>
> Where s denotes the distance of the particle from A as measured along the arc. If the
> particle starts from rest at the point B, and if the arc AB is equal to s_0, it follows that s
> $= s_0 \cos \sqrt{gt/4R}$. The summit, A, is attained in a time $T = \pi/2\sqrt{4R/g}$. Now if the particle
> were allowed to fall freely along DA, it would reach A after a time T' given by $2R =$
> $\frac{1}{2}gT'^2$, from which it follows that $T' = \sqrt{4R/g}$. Therefore it must be that $T'/T = \pi/2$,
> which is the statement Huygens made. (Dugas, 1988, p. 184)

Access to the calculus meant that isochrony was not the last secret of the
cycloid to be uncovered. In 1696 Jean Bernoulli gave the problem of identifying

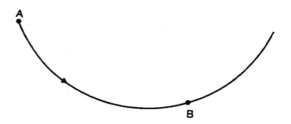

Figure 60. Cycloid as brachistochrone (Gjertsen, 1986, p. 146)

the "curve of fastest descent" to the "sharpest mathematicians in the world." This was the problem of identifying the brachistochrone. That is, the curve of fastest descent for a freefalling body (*brachistos,* meaning fastest, *chronos,* meaning time). The problem was for any two points *A* and *B* such that *B* is below *A*, but not directly underneath it (Fig. 60), find the path of quickest descent from *A* to *B* for a freely falling body influenced only by gravity.

Bernoulli gave his mathematical colleagues six months to find the solution. He underestimated the abilities of some of them. Leibniz solved the problem the day he received it; and he correctly predicted that only five people would solve it: himself, Newton, the two Bernoulli brothers, and Guillaume de l'Hospital. Newton received the problem one day in January 1697 and solved it the same night.[12]

Thus the quickest line of freefall descent between two points is not the short-est line (the straight line or chord), nor the arc of a circle, but the arc of the longer cycloid containing the two points. This is one of the numerous counterintuitive results that indicate the gulf between common sense intuitions or judgemnts, and those of science. Even Huygens, who had extensively investigated the cycloid curve, did not recognize that its inverse was the line of quickest descent, the brachistochrone.

It is also counter intuitive that a body released from further away on a curve, should get to the nadir of the curve at the same time as one released from near to the nadir—the tautochronos property of a cycloid. By looking at the cycloid, one can see why this is so. The further away a body is, the steeper, and hence faster, is its initial descent; the closer the body, the less steep its initial descent and thus the slower is its movement. The distant body moves further but faster, the closer body moves less but slower. The tautochronous curve will thus also be an isochronic curve—the time taken for a freely falling body to reach the bottom will be the same, no matter from where on the curve the body is released. Thus it is that a cycloidal, not a circular, pendulum is isochronic.

J.E. Hofmann commented on Huygens' mathematical conservatism and his failure to embrace the calculus of Leibniz:

> The reason probably lies in the fact that the new tendency did not entirely please him. Leibniz strove for a technique of representation that is simplified and formalized down to the last [detail] by means of appropriate symbols, yet that cannot be immediately grasped but must be learned. Whoever is able to acquire this has an unimaginable advantage over the unschooled, even when he has no particularly deep insights into the structure of the problem: *the formalism thinks for him*. It was precisely this possibility that Huygens saw as undesirable. He was the last and most important representative of the old school, which arrived at its results on the basis of truly ingenious, but particular, arguments that stood in isolation. (in Mahoney, 1990, p. 480)

Recall Galileo's famous statement, made in his 1623 *The Assayer*, of the marriage of physics (natural philosophy) and mathematics:

> Philosophy is written in this grand book, the universe, which stands continually open to our gaze. But the book cannot be understood unless one first learns to comprehend the language and read the letters in which it is composed. It is written in the language of mathematics, and its characters are triangles, circles, and other geometric figures without which it is humanly impossible to understand a single word of it; without these, one wanders about in a dark labyrinth. (Galileo, 1623/1957, p. 237)

For Galileo the language of nature was geometrical (triangles, circles, etc.), not algebraic. After Leibniz and Newton there was a new language which could be learned by natural philosophers. Both geometry and the calculus *represented* nature, provided the "inscriptions" (as some would like to say) whereby nature was theorized, and constituted the theoretical object of Galilean and then Newtonian science. This historical change in language can be usefully examined to tease out the difference between physics and mathematics and to raise questions about how one judges the adequacy or otherwise of *particular* systems of mathematical representation. The emergence of functions and differential equations in late seventeenth-century mathematics transformed early modern physics, and gave impetus to the program of rational mechanics (Bos, 1980c).

HUYGEN'S CLOCK

After the isochronic properties of the cycloidal curve were established, Huygens, in order to construct a pendulum timekeeper, had to devise a way of making Galileo's pendulum swing not in a circle, but in an arc of a cycloid. He did this by suspending the pendulum by a flexible chord (unlike Galileo's rigid rod), and placing cycloidal laminates at the fulcrum or point of suspension (Fig. 62). Once this was done, and once he found an effective measure for the length of a cycloid pendulum and its center of oscillation, then the period became independent of the amplitude and the way was cleared for the construction of reliable pendulum clocks and possibly solving the longitude problem.

Huygens' timekeeping achievements are brought together in his *Horologium*

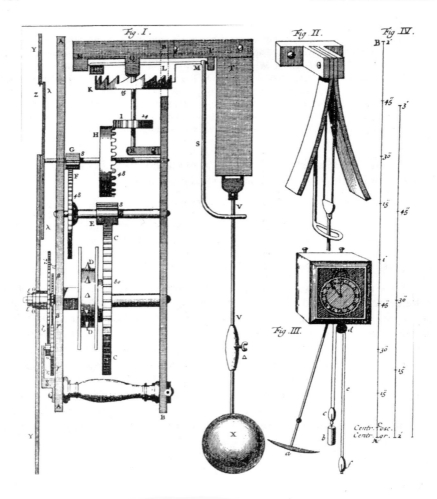

Figure 61. Huygens' pendulum clock (Huygens, 1673/1986, p. 14).

Oscillatorium of 1673, whose dedication is to "Louis XIV, Renowned King of France and of Navarre" (who the year before had commenced a bitter war with Holland).[13] Huygens, in an understatement, admitted, "I believe I have done my best work when I have dealt with things whose usefulness is connected with some subtlety of thought and difficulty of discovery." He added, "I confess that I have pursued this double goal with greater success in the invention of my clock than anything else." (Huygens, 1673/1986, p. 7)

Huygens is aware of the clock's role in the problem of longitude: "They are especially well suited for celestial observations and for measuring the longitudes

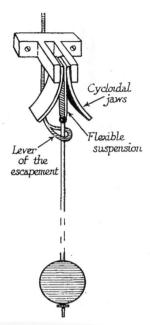

Figure 62. Huygens cycloidal suspension (Hogben, 1940, p. 280).

of various locations by navigators" (Huygens, 1673/1986, p. 8). Indeed, one scholar claims that "it could be argued that the longitude problem was at the basis of the invention of the pendulum clock" (Leopold, 1998, p. 103)

Huygens made his own pendulum clock (Fig. 61) and then illustrated it on the opening page of his 1673 book. He elaborated on the crucial role played by the escapement:

> It is quite evident that the motion of the pendulum *VX*, after it has first been started by hand, is sustained by the force of the wheels which are pulled by the weight, and also that the periodic swings of the pendulum lay down a law and a norm for the motion of all the wheels and of the clock as a whole. For the small rod *S*, which is moved very slightly by the force of the wheels, not only follows the pendulum which moves it, but also helps its motion for a short time during each swing of the pendulum. It thus perpetuates the motion of the pendulum which otherwise would gradually slow down and come to a stop on its own, or more properly because of air resistance. (Huygens, 1673/1986, p. 16)

Huygens moved the verge to a horizontal position, away from its customary vertical alignment, and of course substituted, as Galileo recommended, a pendulum for the medieval foliot that, unlike the pendulum, had no natural period. Figure 63 shows Huygens' crown wheel (*C*) which is turned by falling weights (not shown), the pallets (*D*) on the horizontal verge, the pendulum rod (*B*) and the pendulum weight (*E*), whose position can be altered to control the time of oscilla-

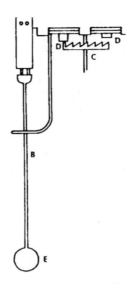

Figure 63. Huygens' verge escapement (Macey, 1980, p. 27).

tion. The pendulum *both* regulates the escape of the crown wheel, and is actuated by the crown wheel.

Huygens mentioned the problem of circular motion not being isochronic, a problem first identified by Mersenne, Descartes, and Riccioli, and said that "The oscillations of clocks carried at sea are especially unequal because of the continual heaving of the ship" (Huygens, 1673/1986, p. 20). And hence "corrections must be made in all these clocks, in general, and especially in marine clocks, so that the times of the wider or narrower oscillations of the pendulum can turn out to be equal" (ibid.). These corrections are achieved by means of cycloid laminates, between which the pendulum is made to swing.

Huygen's persisted with the verge escapement (it can be seen in Figure 61 in Huygen's own clock diagram, and is labeled as *L* and *K*). For 300 years this had been the standard escapement of mechanical clocks. A significant drawback of the verge escapement was that the pallets required large amplitude, and hence nonisochronic, swings to engage the crownwheel. Galileo's pinwheel escapement was a more accurate device, but it was not utilized by Huygens. With all his technical virtuosity, Huygens still was not able to make this decisive break with horological tradition.

SEAGOING TESTS FOR LONGITUDE

Huygens' cycloidal pendulum clock was immediately called upon to solve the problem of longitude at sea.[14] The first such use was by the Scottish nobleman

Alexander Bruce, the earl of Kincardine.[15] Bruce's politics forced him to leave Scotland during the Commonwealth period, and take up residence in Holland where his aristocratic connections brought him into contact with Christiaan and the Huygens' family. After the Restoration, Bruce returned to Scotland and had benefit of the large revenues from his land and mine holdings. Huygens met him during his London visit of 1661, and the following year Bruce had two clocks made to be taken on a voyage to Denmark to test their reliability (Leopold 1993). The clocks were spring-driven, rather than driven by a falling weight. Their pendulums (which regulated the spring-driven wheels) were 9¾ inches in length and thus they beat half-seconds. Robert Hooke, in a paper read to the Royal Society, gave the following account of these trials:

> The Lord Kincardine did resolve to make some trial of what might be done by carrying a pendulum clock to sea, for which end he contrived to make the watch to be moved by a spring instead of a weight, and then, making the case of the clock very heavy with lead, he suspended it underneath the deck of the ship by a ball and socket of brass, making the pendulum but short, namely, to vibrate half seconds; and that he might be the better enabled to judge of the effect of it, he caused two of the same kind of pendulum clocks to be made, and suspended them both pretty near the middle of the vessel underneath the decks. This done, having first adjusted them to go equal to one another, and pretty near to the true time, he caused them first to move parallel to one another, that is to the plane of the length of the ship, and afterwards he turned one to move in a plane at right angles with the former; and in both cases it was found by trials made at sea (at which I was present) that they would vary from one another, though not by very much. (Gould, 1923, p. 29)

In 1663 improved clocks were tested on a voyage to Lisbon during which time the larger of the two worked reasonably well, while the smaller failed. Huygens wrote that he was pleased that his clocks could withstand the worst storms without stopping (Robertson, 1931, p. 149).

Captain Robert Holmes, in 1664, gave the clocks their most extensive sea tests, tests which vindicated Huygens' faith in their ability to solve the longitude problem. Ironically Holmes was on a trans-Atlantic voyage of pillage against Dutch possessions in Africa, New Amsterdam (New York), and South America. Huygens, in his *Horologium Oscillatorium*, repeated Holmes' account of the scientific part of this voyage which had been published in the first volume of the *Philosophical Transactions* (Holmes, 1664):

> After leaving the coast of Guinea, he proceeded to the island called Sao Tomé, which lies under the circle of the equinox. There he adjusted his clocks in relation to the sun. He then followed a westward course and sailed without interruption for about seven hundred miles. Then he turned and sailed towards the coast of Africa with the help of a south-west wind. After holding this course for about two or three hundred miles, he was advised by the captains of the other ships, who feared that they would run out of drinkable water before reaching Africa, to change course towards the American islands called Barbados to obtain water. He then conferred with the captains and ordered them to produce their charts and individual computations, from which he found that

their calculations differed from his, one by 80 miles, another by 100, and the third by still more. From the evidence of his clocks the Commander concluded that they were no more than about thirty miles from the island of Fuego, which is one of the group of islands not far from Africa which is called the Cape Verde Islands, and that they could reach it on the next day. Having thus consulted his pendula, he ordered that this new course be followed, and at noon the next day that island came into view, and a few hours later provided a stopping place for the ships. This was reported by the Commander himself. (Huygens, 1673/1986, p. 28)

Holmes pinned his faith on Huygens' clocks and gambled against his crew dying of thirst. His faith was rewarded. Huygens wrote to a friend:

Major Holmes at his return has made a relation concerning the usefulness of pendulums, which surpasses my expectations. I did not imagine that the watches of this first structure would succeed so well, and I had reserved my main hopes for the new ones. But seeing that those have already served so successfully, and that the others are yet more just and exact, I have the more reason to believe that the finding of the longitude will be brought to perfection. (Gould, 1923, p. 29)

Huygens reported another successful trial of his clock—by the Duc de Beaufort, in the Mediterranean Sea. On that trial "between the port of Toulon and the city of Candia [in Crete] a difference of one hour, 22 minutes was found, that is, a difference of longitude of 20°30′ and almost exactly the same difference was found on the return voyage. This agreement is a most certain indication of truth" (Huygens, 1673/1986, p. 29). Unfortunately for Huygens the difference in longitude between these two ports is only 18½ degrees, which amounts to an error of 120 miles. The astronomer Jean Richer took clocks on a voyage to Canada in 1670. The results were unimpressive: Both clocks stopped during a storm. There were other trials of the seconds pendulum clock when Varin took one to Cape Verde and St. Thomas, and Fontenay took one to China. During the 1660s the Italian brothers, Giuseppe and Matteo Campani experimented with seagoing pendulum clocks, and in 1662 had built one that was encased in a glass bell jar to protect it against atmospheric pressure changes and had a double compensatory system to nullify the effects of ocean pitching (Proverbio 1985, p. 100). There were also trials by Dutch astronomers and navigators. Huygens remarked on these trials in a letter of 1679:

The last trials with pendulum clocks at sea have not been wholly unsuccessful; but, as they necessarily suffer from the motion (*agitation*) of a vessel, there is more hope of success with balances with a spiral spring, but made on a large scale, because accuracy is proportionately increased, and it would be well worth while to make this test. (in Robertson, 1931, p. 162)

Huygens strove to make the clocks better for navigation purposes by eliminating the effect of irregular external motion being transferred to the clock. This was done by the use of gimbals, heavy weights, and by drastically reducing the length of the beating pendulum. It was no longer to be a seconds pendulum, but a fraction-of-a-second pendulum. That is, six inches instead of 39. He constructed it in triangular form to neutralize the effect of impressed force.

Huygens spoke of these improved designs (see Figures 64 and 65):

> Instruments which have been modified in this way have yet to be taken to sea and put to the test. But they offer an almost certain hope of success because, in experiments conducted so far, they have been found to hold up under every kind of motion much better than prior clocks. (Huygens, 1673/1986, p. 32)

The first sea trials of these improved clocks took place in 1686 when Thomas Helder took them in the *Alcmaer* from Texel in Holland to the Cape of Good

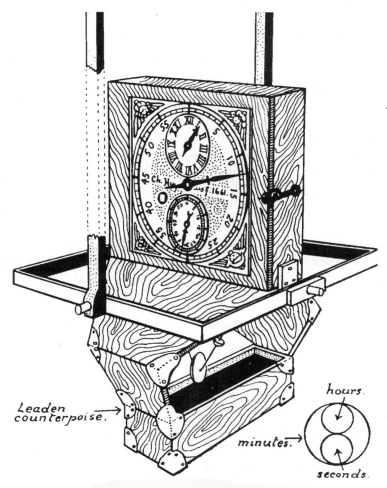

Figure 64. Huygens' marine clock (Wolf, 1950, p. 115).

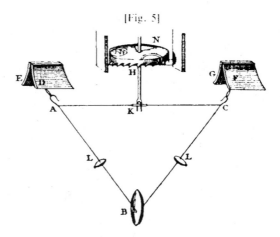

Figure 65. Huygens' improved pendulum suspension (Huygens, 1673/1986, p. 31).

Hope and back (Schliesser and Smith, 1996; Leopold, 1998, pp. 109–110). Huygens reported to the East India Company, which was sponsoring the voyage, that there was a small problem with longitude as determined by his clocks: The ship was logged as sailing through the *middle* of England and Ireland! He attributed this to a systematic error resulting from the rotation of the earth—the "spinning-off" effect on bodies, and hence their loss of weight, was greater at the equator and diminished toward the poles.

This trial took place *after* Jean Richer's 1672 voyage to Cayenne, and hence Huygens was alerted to the need to "adjust" for the slowing of the pendulum with decrease in latitude. With these adjustments, Huygens calculated that the ship did not sail through England and Ireland, and that at the end of a return journey of 118 days, Texel was, on the basis of the ship's adjusted clock reading, only 17" of arc, or 19 km, east of its known longitude. For the period, this was a high degree of accuracy.

Another trial was undertaken, again to the Cape, in 1690. Huygens reported that "the clocks have not proved such a success as we had hoped for" (Robertson, 1931, p. 165). He did not press the Company for further trials, but said:

> I have on this occasion invented something quite different and incomparably better, which I have in hand at the present moment, whereby any little difficulty in the use of this invention will once for all be removed, of which in due course I hope to give Your Excellencies further particulars, and remain, etc. (in Robertson, 1931, p. 165)

Drummond Robertson, who extensively documented Huygens' attempts to solve the longitude problem (Roberston, 1931, Chap. 9), suggested this "new invention" could have been a perfected spring-balance timekeeper.

HUYGEN'S PROPOSAL FOR AN INTERNATIONAL STANDARD OF LENGTH

As he elaborated his theory of pendulum motion and clockwork design, Huygens argued that the seconds pendulum could provide a new international standard of length. Undoubtedly this would have been a major contribution to simplifying the chaotic state of measurement existing in science and everyday life. Within France, as in other countries, the unit of length varied from city to city, and even within cities. A not insignificant problem for commerce, trade, construction, and technology, to say nothing of science. Many attempts had been made to simplify and unify the chaotic French system. Emperor Charlemagne in 789 was among the earliest to issue edicts calling for a uniform system of weights and measures in France. Henry II repeated these calls, issuing this decree in 1557:

> Weights and measures shall be reduced to clearly defined forms and shall bear the appellation of royal weights and measures. Since in all duchies, marquisates, counties, viscountcies, baronies, castellanies, cities and lands observing the laws of our kingdom, weights and measures are of diverse names and dimensions, wherefore many of them do not correspond to their designations; and often, indeed, there coexist two weights and two measures of different sizes, the smaller one being used in selling and the other in buying, whence innumerable dishonest deals arise. . . . All shall be required to regulate their measures according to ours. (Kula, 1986, pp. 168–169)

This decree was about as effective as King Canute's admonition to the rising Baltic tide. One traveller to France in 1789 was infuriated by a country "where infinite perplexity of the measures exceeds all comprehension. They differ not only in every province, but in every district and almost in every town" (Alder, 1995, p. 43). There were 700 or 800 different units or measures, with towns and localities having their own versions of the measure or unit. As Heilbron noted, "In Paris a pint held a little less than a liter; in Saint-Denis, a liter and a half; in Seine-en-Montagne, two liters; and in Précy-sous-Thil, three and a third" (Heilbron, 1989, p. 989). One estimate is that in France alone there were 250,000 different local measures of length, weight, and volume (Alder, 1995, p. 43).

The situation was little better in the German states. Although the "Common German Mile" was widely used, and taken to be 1/15 of a degree at the equator, in Vienna it was subdivided into 23,524 "work shoes," while in Innsbruch it was subdivided into 32,000 "work shoes." Other cities had their own idiocyncratic divisions. Some Italian States used the "Italian Mile" which was 1/60 of an equatorial degree, and contained 5,881 Vienna workshoes. The multiplicity of measures facilitated widespread fraud, or just smart business practice: Merchants routinely bought according to "long" measures and sold according to "short" measures.

The English situation was slightly better. In 1305 Edward I established the standard yard as the length of the iron ulna kept in the Royal Palace, and 1/36 part of it was to be an inch. Away from the palace, in places without access to a copy of the Royal Ulna it was ordained that "three grains of barley, dry or round, make an inch; 12 inches make a foot; three feet make an ulna; 5½ ulna(s) make a perch." A slightly shorter yard standard was decreed by Henry VII in 1497, and a slightly shorter one again by Elizabeth I in 1588. These English standards were arbitrary and artificial; the yard was not a natural unit. The Royal Society at its inception was asked to investigate the reform of the length standards, and Christopher Wren proposed, as did Huygens, the length of a pendulum beating seconds as an English, and also international, length standard.[16] Both Huygens and Wren assumed a seconds pendulum would beat seconds no matter where in the world it was taken. The technical problems were (1) how to get a pendulum to beat seconds, and (2) how to measure its effective length.

Huygens, in the early pages of his *Horologium*, after giving the length of his seconds pendulum, said:

> When I say "three feet," I am not speaking of "feet" as this term is used in various European countries, but rather in the sense of that exact and eternal measure of a foot as taken from the very length of this pendulum. In what follows I will call this an 'hour foot', and the measurement of all other feet must be referred to it if we wish to treat matters exactly in what follows. (Huygens, 1673/1986, p. 17)

This topic is taken up later in the book when he discussed the center of oscillation of bodies:

> Another result, which I think will be helpful to many, is that by this means I can offer a most accurate definition of length which is certain and which will last for all ages. (Huygens, 1673/1986, p. 106)

He elaborated on this "useful result," saying:

> A certain and permanent measure of magnitudes, which is not subject to chance modifications and which cannot be abolished, corrupted, or damaged by the passage of time, is a most useful thing which many have sought for a long time. If this had been found in ancient times, we would not now be so perplexed by disputes over the measurement of the old Roman and Hebrew foot. However, this measure is easily established by means of our clock, without which this either could not be done or else could be done only with great difficulty. (Huygens, 1673/1986, p. 167)

In brief, Huygens said that if a seconds clock is built and tested against the rotation of the fixed stars (as described on pp. 23–25 of his 1673 book, and as initially done by Riccioli and his colleagues), and if a pendulum is set swinging with a small amplitude and its length adjusted until it swings in time with the seconds clock, then:

> . . . measure the distance from the point of suspension to the center of the simple pendulum. For the case in which each oscillation marks off one second, divide this distance into three parts. Each of these parts is the length of an hour foot . . . By doing all this, the hour foot can be established not only in all nations, but can also be reestablished for all ages to come. Also, all other measurements of a foot can be

expressed once and for all by their proportion to the hour foot, and can thus be known
with certainty for posterity. (Huygens, 1673/1986, p. 168)

Thus his basic unit of length was to be three horological feet (0.9935m, 39 1/8
English inches), less than a millimeter short of the original meter adopted a century later. The astronomer Picard concurred in this recommendation, saying a
"universal foot" should be 1/3 the length of a seconds pendulum.

Huygens was enough of a craftsman to realize that spheres may not be uniform in their composition nor perfect in their sphericity, and that when taken
down from suspension, the suspending wire will contract. It was for these reasons
that he concentrated on "centers of oscillation" rather than the more obvious "center
of the sphere," and why he devoted Part IV of his book to the determination of
centers of oscillation. In this final part, he showed that "in all isochronous pendula
the distance from the point of suspension to the center of oscillation is exact and
equal" (Huygens, 1673/1986, p. 169). He defined the center of oscillation of a
compound pendulum to be a point at distance *l* from the point of suspension on a
line through the point of suspension and the center of gravity, and where *l* is the
length of a simple pendulum that beats in time with the oscillations of the compound pendulum. Centers of oscillation and suspension are convertible.

The effect can be seen (see Fig. 66) if an irregularly shaped piece of board is
suspended by a nail *O* and set swinging, and a string with a small but heavy
weight *C*, a simple pendulum, is suspended from the same nail and set swinging
though small arcs (angle a is small, so that sine a approximates a), and the length
of the string adjusted until the simple pendulum beats in synchrony with the irregular board. Both can then be stopped and a mark made on the board where the
center of the small weight lies *G*. This is the center of oscillation, and of gravity,
of the board. If it is suspended through this point it will swing with the same

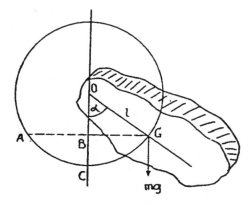

Figure 66. A compound pendulum (Booth and Nicol, 1962, p. 75).

frequency as it did when suspended through O. The distance l between the two points (center of suspension and point of oscillation) is called the *equivalent simple pendulum*.

Jean Richer's Cayenne Voyage of 1672–1673

Huygens thought his universal length standard, the seconds pendulum, was dependent only upon the force of gravity, which he took to be constant all over the earth, and thus the length standard would not change with change of location. The standard was to be portable over space and time. He did recognize that the centrifugal force exerted by the rotating earth (the force that tends to throw bodies off the earth's surface) varied from the equator (largest) to the pole (smallest), but he did not think that its effect on the pendulum would be measurable.[17] In this he was wrong.

When Jean-Dominique Cassini (1625–1712) became director of the *Académie Royale des Sciences* in 1669, its secretary observed that it "began to discuss sending observers under the patronage of our most munificent King into the different parts of the world to observe the longitudes of localities for the perfection of geography and navigation" (Olmsted, 1942, p. 120). The *Académie* instituted the modern tradition of scientific voyages of discovery, and Jean Richer's voyage to Cayenne in 1672–1673 was the second such voyage, the first being Jean Picard's voyage to Uraniborg in Denmark in 1671. Even by this second voyage scientific interests were being mixed with political ones.[18] Cayenne was in French Guiana, at latitude approximately 5° N. It was chosen as a site for astronomical observations because equatorial observations were minimally affected by refraction of light passing through the earth's atmosphere—observer, the sun, and planets were all in the same plane.

The primary purpose of Richer's voyage was to ascertain the value of solar parallax and to correct the tables of refraction used by navigators and astronomers. A secondary consideration was to check the reliability of marine pendulum clocks, which were being carried for the purpose of establishing Cayenne's exact longitude.

The voyage was spectacularly successful in its primary purposes: The obliquity of the ecliptic was determined; the timing of solstices and equinoxes was refined, and, most important, a new and far more accurate value for the parallax of the sun was ascertained: 9.5" of arc. This figure—equivalent to the angular size of the earth's radius when viewed from the sun—provided the only known way to measure the distance of the sun from the earth, and hence the dimensions of the solar system. Using Richer's parallax figure, Cassini calculated the sun to be 21,600 semi earth-diameters (87,000,000 miles) from the earth.[19] The enormity of the solar system was staggering to contemporaries, especially to nonastronomers. Voltaire thought the earth and a human's place on it, had been shrunk to insignificance. For believers, swallowing Copernicus' displacement of God's chosen and redeemed people from the center of the universe had been difficult enough; many choked when asked to accept a 90-odd million mile displacement from the center.

But it was the unexpected consequences of Richer's voyage which destroyed Huygens' vision of a universal standard of length "for all nations" and "all ages." Richer found that a pendulum set to swing in seconds at Paris had to be shortened for it to swing in seconds at Cayenne. Not much—2.8 mm (0.28%)—but nevertheless shortened. Although, with good reason, many doubted his experimental ability,[20] Richer found that a Paris seconds-clock lost 2½ minutes daily at Cayenne.

However, Newton acknowledged the veracity of Richer's claims, when he wrote in his *Waste Book* of 1682:

> Monsr. Richer sent by ye French King to make observations in the Isle of Cayenne (North Lat 5^{gr}) having before he went thither set his clocke exactly at Paris found there in Cayenne that it went too slow as every day to loose two minutes and a half for many days together and after his clock had stood & went again it lost 2½ minutes as before. Whence Mr Halley concludes that ye pendulum was to be shortened in proportion of—to—to make ye clock true at Cayenne. In Gorea ye observation was less exact. (Cook, 1998, p. 116)

In his *Principia* (Bk.III, Prop.XX, Prob.IV), Newton utilized Richer's, and Halley's comparable observations from St. Helena, to develop his oblate account of the earth's shape.

Richer's demonstration raised the problem of an independent measure of time. He did not have a second timepiece (a digital watch, for instance) against which to measure the speeding up or slowing down of his pendulum clock. The only independent clock he had was the clock of the heavens. He probably measured the number of pendulum swings against the number of seconds in a solar day (noon to noon) or a sidereal day, or in portions of those days. This was a difficult enough technical exercise, and it was compounded by the fact that the solar day actually varies in length by plus or minus 15 minutes through the year. But the yearly variation, the equation of time, was known, and the technical problems of timing the sun's transit were overcome. Richer was adamant that Huygens' clocks slowed, that they had to be shortened. This was tantamount to saying the force of gravity, and hence the weight of bodies, diminished from Paris to the equator—an astonishing conclusion.

The reality that the weight of a body changed from place to place, as was manifest in the variation of the pendulum's period, sowed the seed for the conceptual distinction between weight and mass. The intuition was that although weight changed with change in gravity, nevertheless something about the "massiness" of the body remained the same. Jean Bernoulli first introduced the distinction between mass and weight, and Newton, as will shortly been seen, clarified it by introducing the idea of inertial mass.

Some Methodological Matters

Richer's claim that the pendulum clock slows in equatorial regions nicely illustrates some key methodological matters about science and about theory testing
(matters that will be further developed in Chapter 10). The entrenched belief since
Erastosthenes in the second century BC was that the earth was spherical (theory T),
and on the assumption that gravity alone affects the period of a constant length
pendulum, the observational implication was that period at Paris and the period at
Cayenne of Huygens' seconds-pendulum would be the same (O). Thus

$$T \rightarrow O$$

But Richer seemingly found that the period at Cayenne was longer ($\sim O$).
Thus, on simple, Popperian, falsificationist views of theory testing:

$$T \rightarrow O$$
$$\sim O$$
$$\therefore \sim T$$

But theory testing is never so simple, a matter that was recognized by Popper
and articulated by Kuhn (1970), Lakatos (1970), Feyerabend (1975), and a host
of other contemporary philosophers of science. In the 17th century, many upholders of T just denied the second premise, $\sim O$. The astronomer Jean Picard, for
instance, did not accept Richer's findings. Rather than accept the message of varying
gravitation, he doubted the messenger. Similarly Huygens was not favorably dis-
posed toward Richer. In 1670, on one longitude testing voyage to the West Indies
and Canada, Richer behaved irresponsibly with regard to Huygens' clocks: He
did not immediately restart them when they stopped in a storm, and he allowed
them to crash to the deck (Mahoney, 1980, p. 253). Huygens did not require much
convincing that it was Richer's ability, not gravity, that was weak at Cayenne.

Others saw that theories did not confront evidence on their own. There was
always an "other things being equal" assumption made in theory test; there were
ceretis paribus clauses (C) that accompanied the theory into the experiment. These
clauses characteristically included statements about the reliability of the instruments, the competence of the observer, the assumed empirical state of affairs,
theoretical and mathematical devices used in deriving O, and so on. Thus

$$T + C \rightarrow O$$
$$\sim O$$
$$\therefore \sim T \text{ or } \sim C$$

Some people maintained belief in T and said the assumption that other things
were equal was mistaken: Humidity had interfered with the swings, heat had length-

ened the pendulum, friction increased in the tropics, and so on. These, in principle, were legitimate concerns. But more and more evidence came in, and from
other experimentors including Sir Edmund Halley, confirming Richer's observations. Thus ~ O became established as a scientific fact, as Fleck (1935/1979)
would have said, and upholders of T, the spherical earth hypothesis, had to adjust
to it. This was not easy.

In 1738 Voltaire, a champion of Newtonian science, wrote on the Richer
episode, drawing attention to the problems of adjustment that scientists experienced:

> At last in 1672, Mr Richer, in a Voyage to Cayenna, near the Line, undertaken by
> Order of Lewis XIV under the protection of Colbert, the Father of all Arts; Richer, I
> say, among many Observations, found that the Pendulum of his Clock no longer made
> its Vibrations so frequently as in the Latitude of Paris, and that it was absolutely
> necessary to shorten it by a Line, that is, eleventh Part of our Inch, and about a Quarter
> more.
>
> Natural Philosophy and Geometry were not then, by far, so much cultivated as at
> present. Who could have believed that from this Remark, so trifling in Appearance,
> that from the Difference of the eleventh of our Inch, or thereabouts, could have sprung
> the greatest of physical Truths? It was found, at first, that Gravity must needs be less
> under the Equator, than in the Latitude of France, since Gravity alone occasions the
> Vibration of a Pendulum.
>
> In Consequence of this it was discovered, that, whereas the Gravity of Bodies is by so
> much the less powerful, as these Bodies are farther removed from the Center of the
> Earth, the Region of the Equator must absolutely be much more elevated than that of
> France; and so must be farther removed from the Center; and therefore, that the Earth
> could not be a Sphere. (Fauvel and Gray, 1987, p. 420)

He commented further:

> Many Philosophers, on occasion of these Discoveries, did what Men usually do, in
> Points concerning which it is requisite to change their Opinion; they opposed the
> new-discovered Truth. (Fauvel and Gray, 1987, p. 420)

From the Seconds Pendulum to the Meter: Some Political Matters

Richer's findings did in an age of precision rule out the seconds pendulum as an
invariant, universal, and portable standard of length verifiable for all nations. But
once a location or latitude was specified, then the length of the seconds pendulum
would be invariant, and it could still be a universal standard. Moreover such a
standard would be natural, not arbitrary. It would of course be a matter of some
national pride as to what location was chosen: Would it be, for instance, the length
of the seconds pendulum in Paris, London, Madrid, Berlin, Rome, St. Petersburg,
or Washington? La Condamine and others in an expedition to Peru in 1735 ascertained the length of the equatorial seconds pendulum to be 439.15 lines. In 1739
the length of a Paris seconds pendulum was determined to be 440.5597 lines. La
Condamine proposed, perhaps as a way of rising above nationalist rivalries in

Europe, that the equatorial seconds pendulum be the universal unit of length (all major European powers having equatorial colonies where they could do their own measurements).

The French revolutionaries of 1789 not only proposed to do away with the arcane feudal system of the *ancien régime*, but also with the cacophony of measures that people saw as propping it up, robbing them in every transaction and as requiring elitist knowledge to judge and manipulate. The people wanted a simple, rational, democratic, and universal system. They wanted one where, as Sir John Riggs Miller the English champion of measure reform would say, "the meanest intellect is on a par with the most dextereous" (Heilbron, 1989, p. 990). One of the first decisions of the Estates General was to direct the *Académie to* establish a Committee on Weights and Measures to recommend reform of French measurement. This committee was duly established and included Lagrange and Laplace. Talleyrand, the bishop of Autun, was not too worried about nationalist bias and, in 1790, suggested to the new, post-revolution, National Assembly that the unit of length be the seconds pendulum at the 45° latitude, a latitude conveniently running through France. Talleyrand wrote to Riggs Miller in the English parliament and urged him to use his influence to have a common system adopted between France and England, saying "Too long have Great Britain and France been at variance with each other, for empy honour or for guilty interests. It is time that two free Nations should unite their exertions for the promotion of a discovery that must be useful to mankind" (Berriman, 1953, p. 141). Alas this entreaty fell upon deaf ears, even when followed by Louis XVI being asked by the Assembly to write to George III seeking a unified standard.

The committee's second report, in 1791, rejected the pendulum measure and instead revised a version of Abbe Gabriel Mouton's 1670 suggestion for using the length of a geodetic minute of arc decimally divided. Cassini had also advocated this geodetic foot measure in 1720, wanting it to be 1/6,000 of a terrestrial minute of arc. The committee recommended that the length standard be 1/10,000,000 of the distance of the quadrant of the arc of meridian from the north pole to the equator passing through Dunkirk, Paris, and Barcelona. Dunkirk and Barcelona "anchored" the segment of meridian at sea level, and the Paris meridian was the obvious choice for the French standard. This was painstakingly measured by the astronomer–geodesists Delambre and Méchain during the years 1792–1799 using classical surveying techniques and the Toise (approximately 2m) as their unit of length (Chapin, 1994, p. 1094).

The committee's third report, 1793, named the new geodesic unit a *metre* (from the Greek word *metron*, meaning "measure"). The convention, which had replaced the assembly, accepted this, and the new standard was enacted in the law of the 18th Germinal, year III (7 April 1795), 5th Article. These new metric measures were officially termed "republican," indicating their political dimension. As the minister of finance in year II of the republic said, "the introduction of new weights and measures was extremely important on account of its association with the Revolution, and for the enlightenment and in the interest of the people" (Kula, 1986, p. 239). This association also meant that Lavoisier, Laplace, and Coulombe were dismissed from the Agency of Weights and Measures for unrepublican associations. The brass standard meter was put in place in 1799.

It is interesting, but problematic, why the academy's committee opposed Talleyrand's proposal of the seconds pendulum, which Huygens' suggested a hundred years earlier. The length of the seconds pendulum at a given latitude was constant, public, recoverable, natural, and portable. It seemed to fit all the criteria for a good standard. Ostensively the committee's reason was that it introduced temporal considerations into a length standard. Some have suggested the real reason is the academy wanted to preserve its intellectual and technical territory and boost its funding. The determination of *its* standard was a highly complex and elitist matter that required not only the most sophisticated technology, but also an agreement on the degree of oblateness of the earth. On August 8, 1791, it received 100,000 livres, about twice its normal annual budget, as a downpayment for the geodesic survey of the meridian sector. Estimates of the total cost of determining the length of the 10 millionth part of the quarter meridian through Paris vary from 300,000 livres to millions of livres (Heilbron, 1989, p. 991). There was, in the committee's recommendation, a certain element of fundamental self interest hiding behind noble academic ideals. Early in the following century, the surveyor Jean Baptiste Biot, wrote: "If the reasons that the Academy presented to the Assembly [to obtain support for the project] were not altogether the true ones, that is because the sciences also have their politics" (Heilbron, 1989, p. 992). It is salutory to recognize that politics was involved in the determination of the length standard upon which all measurement in modern science is predicated, and at the same time to recognize that the intrusion of politics does not necessarily mean epistemological compromise.

There was more than just impersonal scientific interest involved in adopting a unified system of measurement. As one early 19th century French commentator remarked:

> The conquerors of our days, peoples or princes, want their empire to possess a unified surface over which the superb eye of power can wander without encountering any inequality which hurts or limits its view. The same code of law, the same measures, the same rules, and if we could gradually get there, the same language; that is what is proclaimed as the perfection of the social organization . . . The great slogan of the day is uniformity. (Alder, 1995, p. 62)

Huygens' seconds pendulum length standard did survive its rejection by the academy. After years of patient measurement of the meridian sector, and the expenditure of a great deal of state money, the academy chose a fraction of the meridian distance that coincided with Huygens "three horological feet," and accepted the seconds pendulum as a *secondary* reference for its new length standard. Their meter differed from that of Huygens by only 0.3 of a millimeter, or 0.0003 m. This is a remarkable cosmic coincidence. The number of seconds in a day (86,400) is purely conventional, yet the length of a pendulum beating seconds turns out to be exactly 1/40 million part of the circumference of the earth.

The academy's argument about not mixing temporal and length considerations failed at the time to convince sceptics; two hundred years later in 1983, it did not convince delegates to the General Conference on Weights and Measures meeting in Paris who defined the standard universal meter as "the length of path travelled by light in vacuum during a time interval of 1/299,792,458 of a second." Thus time and space become inextricably linked. Again, pleasingly, this seemingly arbitrary figure, that is among the first things to confront students opening modern textbooks, is within a millimeter of Huygen's original and entirely natural length standard.[21]

PENDULUM MOTION, THEORIES OF GRAVITY AND THE EARTH'S SHAPE

Huygens, when it was confirmed by others, applied himself to explaining Richer's effect, and in so doing brought into relief the scientific and methodological differences between himself and Newton.[22] Huygens was a neo-Cartesian. He followed Descartes in embracing the mechanical world view and the thesis that the movement of bodies had to be explained in terms of impact. Newton's claim for attraction at a distance struck Huygens (as well as Descartes, Bernoulli, and Leibniz) as occult.[23] In its place Huygens embraced a version of Descartes' "subtle matter," or aether, theory, a theory that postulated subtle or minute matter completely filling all space, whirling around and so moving bodies to the center. The rotation of the subtle matter gives it a centrifugal *conatus*, or tendency to fly outwards. As it moves outwards, it pushes other matter (ordinary objects) towards the center. That is, the centrifugal tendency of the rotating subtle matter creates a centripetal tendency in objects. It is this effect that we call gravity. Huygens demonstrated this effect to the *Académie des Science* using powder on the surface of a rotating cylinder of water: The whirling water "pushes" the powder to the center.[24]

In his *Discourse on the Cause of Gravity* (1690), which was published shortly after reading Newton's *Principia* (1687), and in the same year as Locke's *Essay*, Huygens wrote:

> I have supposed that gravity is the same both inside the Earth and at its surface; . . . Mr. Newton . . . makes use of a completely different assumption . . . I will not examine it here, because I do not agree with a Principle that he assumes in this and other computations. This is: that all small parts that one may imagine in two or more different bodies attract or tend to mutual approach. I could not admit this, since I clearly saw that the cause of such an attraction cannot be explained by any principle of Mechanics or by the rules of motion. I am also not convinced of the necessity of the mutual

attraction of whole bodies; for I have shown that, even if the Earth did not exist, the bodies would not cease to tend to a center by the so-called gravity. (in Martins, 1993, p. 207)

To use Rom Harré's terminology (Harré, 1964), Huygens' general conceptual scheme (the mechanistic world view) did not allow him to accept Newton's force concept as an explanation. For a mechanist, it explained nothing. Rather, attractive force at a distance was in need of explanation. Both parties could agree on the phenomena, but Huygens believed that only the mechanists were offering respectable scientific explanations of phenomena. He thought that Newtonians were reintroducing "occult'" explanations. Huygens thought "gravity is the same both inside the Earth and at its surface," a claim denied by believers in the universal applicability of Newton's inverse square law of attraction. Huygen's was prepared to recognize, on the basis of the *Principia's* arguments, the inverse square law for celestial gravitation, but not "occultists" attraction at a distance explanation of the law. He had fundamental philosophical objections to terrestrial attraction at a distance, and thus the inverse square law.

Huygens believed Richer's findings could be fully explained by the difference in centrifugal force that rotating bodies on the earth's surface experience as they move from the equator to the poles. As bodies move upwards (or downwards) in latitude their radius of revolution diminishes from $r = 4,000$ miles at the equator to $r = 0$ at the poles. They all rotate through 360° in 24 hours, or 86,400 seconds. Bodies at the equator will rotate faster than those at the pole (zero circular movement). Consequently those at the equator will have a greater propensity to "fly off." This propensity (centrifugal force) counteracts gravitational attraction (weight) and so the effective downwards force on the body (its effective weight, mg) is less at the equator. Huygens' calculated the centrifugal effect of the earth's rotation at the equator to be 1/289 the force of gravity at the equator. That is, the "throwing off" effect is only about 1/300 the weight of the body. This can be expressed as either **g** being less at the equator, or **g** being constant, but effective weight being less. Huygens initially believed the latter.[25]

The foregoing was explained on the assumption of a spherical earth. The mathematics did not give the shortening that was required to have Richer's pendulum beat seconds.[26] That is, the centrifugal force, when calculated, counteracted gravitational attraction, but not enough to produce the longer period found at the equator. Huygens surmised that the shortfall could be explained by a flattening of the earth at the poles, so that bodies at the equator were further away from the earth's center, and so the force of gravitational attraction (which he explained mechanically) was also lessened on this account. Between increased centrifugal force throwing rotating bodies off at the equator, and lessened gravitational attractive force pushing them down, Richer's readings could be explained.

That is, from the formula, $T = 2\pi\sqrt{l/g}$, it can be seen that if T is increased (the clock runs slower), then **g** has to be decreased.

And from the formula for the force of gravitational attraction:

$$F = Gm_1m_2/r^2$$

we have, if m_1 is the mass of the earth, and m_2 the mass of the pendulum,

$$m_2g = Gm_1m_2/r^2, \text{ or } g = Gm_1/r^2$$

Thus if **g** has to be lessened in order to account for the slowing of the pendulum, then r (the distance from the pendulum to the earth's center), has to be increased. In other words, the earth is oblate shaped, a view first propounded by Newton and then shared, for different reasons, by Huygens.[27] In this Newton and Huygens were united against the bulk of continental atomists who, following Cassini, thought that the earth was not a flattened sphere, but an elongated one.

SPRING BALANCE CLOCKS

Huygens response to Richer's findings contributed to knowledge of the varying effects of gravity on bodies, and pushed the Cartesian mechanical world view to its limits, but it also spelled the end of Huygens' and anybody else's commitment to the pendulum as a solution to the longitude problem. In 1675 Huygens gave up on the pendulum as a regulator for marine clocks. He acknowledged the insoluble problem of variation in period caused by variation in gravitational attraction and the lesser, but still critical, problem of periodic variation with temperature variation. Instead he turned to spring-driven and balance-wheel regulated clocks. Timekeeping as the key to the longitude problem was not abandoned, only the particular timekeeper. These spring balance proposals precipitated a bitter controversy with Robert Hooke, who thought Huygens had appropriated his own designs and inventions.

Since the 16th century, spring-driven mechanisms had, been used in chamber clocks in place of falling weights. The uncoiling spring moved the wheels which were then regulated by the verge and foliot, or some other such regulator. Because the wound spring, unlike the falling weight, lost motive power as it unwound, there needed to be some means of compensating for this. The fusee gear, which was tapered such that less and less had to rotate as more and more of the spring unwound, successfully did this. Huygens' great innovation was to use a spring to also move a balance wheel, which was set up in such a way that it rotated back and forth with isochronic motion and so it could replace the pendulum as a regulator of clockwork.

Huygens in 1675, while he was receiving a pension from Louis XIV, applied successfully to minister Colbert for a *privilège* (patent) for his new balance-wheel

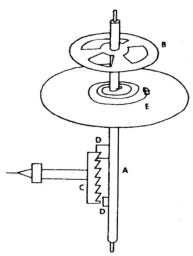

Figure 67. Huygens' spring-balance mechanism (Macey, 1980, p. 29).

regulator. Writing of a rival claimant's invention, Huygens said: "It will be seen how much they differ; since, besides a totally different form and application of the spring, I employ a balance that turns on its pivots, and my invention consists in the combination of these two things" (*Works*, vol.vii, p. 457, in Robertson, 1931, p. 177). On February 15, 1675, Huygens wrote to Oldenburg, the secretary of the Royal Society in London, inquiring whether an English patent for his device would be of any value. This letter, as will be seen in the next chapter, got caught up in the Huygens–Hooke controversy over priority of invention of the balance-wheel regulator.

Huygens, for his spring-balance timekeeper (see figures 67 and 68), attached a wound spring (*E*) and an oscillating balance wheel (*B*) to the verge (*A*), which he was persisting with. The pallets (*D*) on the verge alternately engaged the teeth of the crown wheel (*C*), which then led off to the gears and ultimately the clock face.[28]

Once again the idea was right, but its implementation was beyond the limits of technology. As one historian of horology wrote:

> Huygens' method of applying the spring was original, for he geared up the balance, so that instead of describing, like Hooke's, an arc of 120° or so, it revolved several turns at each beat. This device is termed a "pirouette." It is theoretically objectionable on account of the friction in the gearing, and never came into general use. He found himself, however, baffled by the effect of temperature on the strength of the spring. (Gould, 1923, p. 30)

Huygens made an enormous contribution to timekeeping, even if, ultimately, he failed in his quest for a solution to the longitude problem. The effect of tem-

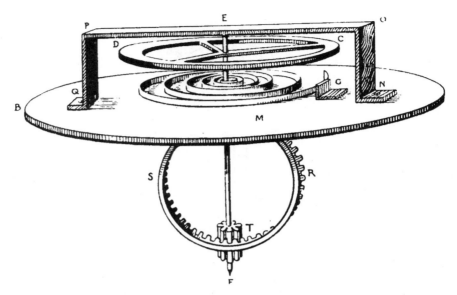

Figure 68. Huygens' spring-nalance mechanism (Wolf, 1950, p. 114).

perature change on metals was simply not understood in the late 17th century. It baffled Huygens, who was unable to make compensations in either his pendulum rod or clock spring.

Huygens' *Horologium Oscillatorium* (1673) was the standard horological reference work at the beginning of the era of precision timekeeping, being re-printed in 1724 and again in 1751 (Leopold, 1998, p. 113). Huygens had an impact on all European clockmakers of the late seventeenth and eighteenth centuries, including John 'Longitude' Harrison. Huygens unlocked the secret of isochronic motion when he recognized that the restoring force on his cycloid pendulum was proportional to its displacement from the nadir. He, and others, were then able to manufacture other types of isochronic motion that could be used as clock regulators.

Huygens, in his mathematics, science and technology, exemplified the new era of *precision* that typified modern European science. Jean Richer's Cayenne measurements destroyed the seconds pendulum as a universal measure of length, yet the fatal adjustment was a mere 0.28 cm. At other times, and in other places, this adjustment would count for nothing. Huygens could not live with it; nor could the emerging world of modern science.

The Dutch scientist 's Gravensande, in his 1724 essay on 'The Life of Huygens,' sums up Huygens' horological contribution:

He was persuaded that his clocks could be used at sea, and that nothing more was requisite on board for determining longitudes. For it is well known that the solution of this useful problem of the longitude—a need long felt, and maybe destined still to be—depends on the exact measurement of time.

It was not, however, sufficient to determine the motions of clocks by fixed laws; but, as he himself admitted—to preserve an equal motion when the ship was labouring in a storm—*hic opus, hic labor erat.* But he always hoped to be able to overcome the difficulty; many were the trials he made; and, in order to reach the goal he perservered with fresh attempts almost to the day of his death. Nevertheless, albeit he was unable to attain it, whosoever peruses the first volume of these works [of Huygens] may judge with what genius, with what penetration he pursued the subject. Not all his attempts have, however, been made public. (in Robertson, 1931, p. 174)

CHAPTER 7

Perfecting Mechanical Timekeeping and Solving the Longitude Problem

The development of physics and astronomy in 17th century Europe required greater and greater precision in timekeeping. Investigation of the law of fall, the action of gravity, and of the shape and periodicity of planetary orbits, plus a host of other problems, required clocks accurate to a few minutes, if not seconds, per day. Scientists worked with clockmakers to realize this accuracy. This theoretical and technological endeavour went hand in hand with the unfolding scientific revolution, a revolution that subsequently transformed scientific, and most other, modes of thought and action. Contributors to the new science labored intellectually and technically, to lay bare the laws of nature, particularly the laws of motion. Newton, who spoke of himself as "standing on the shoulders of giants," thought that these efforts were so successful that heathens reading his *Principia* would come to believe in the divine lawgiver responsible for the terrestrial and celestial laws that his book presented.[1] The achievements and brilliance of Galileo, Huygens, Newton, Leibniz, and Boyle were such that many lesser contributors were overshadowed at the time and, their reputation has been eclipsed with the passage of centuries.

ROBERT HOOKE

One of the lesser figures of the Scientific Revolution was Robert Hooke (1635–1703) whose familiarity to science students is largely due to the law of deformation that bears his name (*for an elastic deformation, stress is proportional to strain*). But as one writer reminds us, "Hooke was one of the outstanding figures of his age. His mind ranged over the entire scientific studies of his time, and there was hardly any branch of science which he did not consider, and to whose advancement he did not contribute" (Gould, 1923, p. 24). He was, for instance, the author of one of the earliest books on microscopy—*Micrographia* (1665), and the author of the earliest systematic work on springs and deformation—*De Potentia Restitutiva or of Spring* (1678). Hooke was the first to state clearly that the motion of heavenly bodies must be regarded as a mechanical problem, and he approached in a remarkable manner the discovery of universal gravitation. In addition he was an architect of some consequence, who worked with his cousin Christopher Wren on many projects after the 1666 Great Fire of London. He was a member of the small group of Oxford and London "philosophers" that met informally at Gresham College (others included Wren and Boyle) which would become in 1661 the Royal Society.[2] Hooke was elected a Fellow of the Society in 1663, and he held the position of Curator of Experiments for the duration of his life. It is said that: "In personal appearance Hooke made but a poor show. His figure was crooked and his limbs shrunken; his hair hung in dishevelled locks over his haggard countenance. His temper was irritable, his habits penurious and solitary. He was blameless in morals, and reverent in religion" (Jourdain, 1913, p. 357).[3]

The longitude problem was a mainstay of the early years of the Royal Society, and articles on the topic appeared regularly from the first issue of the society's *Philosophical Transactions* (Holmes, 1664; Rooke, 1665; Rooke, Moray and Hooke, 1667; Huygens, 1669; Oldenburg, 1676).[4] Hooke was a regular participant in these discussions.

One of the branches of science to which Hooke made a major contribution was horology. On the technical side, he invented the balance-wheel regulator and the anchor escapement; on the theoretical side, he identified a vibrating spring and a swinging pendulum as comparable motions (simple harmonic) each with isochronic periods. The technical advances overcame the problems that bedeviled Huygens' pendulum clocks and opened the way for John Harrison's subsequent creation of a successful marine chronometer.

Around 1659 Hooke first turned his attention to the improvement of portable timekeepers for solving the longitude problem at sea. His early efforts were not encouraged by his friends. He wrote that:

> All I could obtain was a Catalogue of Difficulties, *first*, in the doing of it, *secondly* in the bringing of it into publick use, *thirdly*, in making advantage of it. Difficulties were proposed from the alteration of *Climates, Airs, heats* and *colds*, temperature of *Springs*, the nature of *Vibrations*, the wearing of *Materials*, the motion of the *Ship*, and divers others. Next, it would be difficult to bring it to use, for Sea-men knew their way already to any Port, and Men would not be at the unnecessary charge of the *Apparatus*, and Observations of the Time could not be well made at Sea, and they would nowhere be of use but in the East and West *India* voyages, which were so perfectly understood that every common Sea-man almost knew how to pilot a ship thither. And as for making *benefits*, all People lost by such undertaking; much has been talked about the *Praemium* for the *Longitude*, but there has never been any such thing, no King or State would ever give a farthing for it, and the like; all which I let pass . . . (Gould, 1923, p. 24)

Notwithstanding the advice of this circle of Jeremiahs, Hooke pushed on with his attempts to solve the longitude problem (Wright, 1989), and lectured on the question at Gresham College in 1685 and incorporated his work into a course of astronomy (Hooke, 1705/1969, pp. 511–512). Hooke's comfort and competence in both the worlds of theory and of practice is illustrated in a diary entry for Saturday May 2, 1674 where he wrote of visiting the celebrated London clockmaker Thomas Tompion:

> Told him the way of making an engine for finishing wheels, and a way how to make a dividing plate; about the forme of an arch; about another way of Teeth work; about pocket watches and many other things. (in Hall, 1963, p. 31)

Given the eminence of Tompion as a craftsman, technician and clockmaker, it says a lot about Hooke that he could be giving such advice.

The Balance-Wheel Regulator

Hooke's invention of the wheel regulator was a major advance in timekeeping, and thus in solving longitude. Its principles can be explained by reference to Figure 69.

The weighted bar is attached to an axle, which is joined to the inner end of a spring. The other end of the spring is anchored to a stationary body. When the bar is initially moved out of its equilibrium position (to the etched position) this tensions the spring which then causes the bar to swing back. The inertia of the masses takes the bar past its equilibrium point, thus tensioning the spring again. This causes the bar to swing back the other way, past its equilibrium point again, and the cycle repeats with the arc of swing (amplitude) slowly diminishing due to friction.

Whereas the motion of the pendulum was dependent upon gravity, and thus changeable with location, the motion of the balance was independent of gravity and of location. Further, from his law relating tension to deformation, Hooke

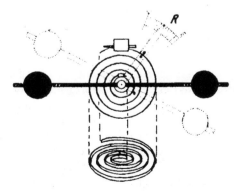

Figure 69. The balance spring (Gould, 1923, p. 22).

established that the period of oscillation of the balance, at least in the ideal case, was isochronic. Its isochrony resulted from the fact that its restoring force was proportional to its displacement—something which Huygens had identified as the *sine qua non* of isochronic motion. Thus the balance could replace the pendulum as a regulator of clockwork. Actual springs and watches, however, were far from ideal cases: Friction at the axle pivots, variation in the quality of springs, and most important, changes in temperature all affected the oscillations. Jewelled pivots minimized pivot friction; developments in metallurgy improved springs, and a "regulator" to change the effective length of the spring assisted in compensating for temperature change. Hooke lectured on the longitude problem at Gresham College in 1685, and systematized his understanding of it.

The Anchor Escapement

There was at the time, and still remains, dispute about whether Huygens or Hooke deserve credit for the invention of the balance spring regulator—"the most important single improvement ever applied to portable timekeepers" (Gould, 1923, p. 26). There is no such dispute about Hooke's claim to be the inventor of the anchor escapement, another significant advance in the perfection of timekeeping (Fig. 70). Four centuries had passed since the last significant (verge and foliot) development in escapement technology (Galileo's pinwheel escapement having not been sufficiently publicized or tested).

Many clockmakers recognized the inherent limitations of the verge and foliot escapement. It required the pendulum to swing through a wide arc, hence departing from the isochronic cycloidal path, and it checked and interfered with the movement of the pendulum. Huygens nevertheless persisted with the verge, as can be seen in the foregoing diagram (Fig. 61) of his 1673 pendulum clock where the verge pallets are marked by the letter "L." The anchor escapement was proposed by Hooke around 1676, and the following year was incorporated into commercial clock making by William Clement. Apart from designing the escapement, Hooke made a machine for the automatic cutting of gear teeth—a necessity for large-scale, accurate clock production.

The anchor escapement required a pendulum swing of only 3° to be activated, whereas the older verge escapement required one of 40°. The small amplitude allowed longer and heavier pendula, with an attendant increase in accuracy. The English used 39.1 inches as their standard pendulum length. This was the pendulum length required to beat seconds at London (at the equator it was 39.00 inches, and in Edinburgh it was 39.11 inches). The anchor was in the same plane as the driving wheels, and so considerably reduced the checking effect which bedevilled the verge. In Figure 70, it can be seen that as the pendulum swings to the left, it forces fluke A of the anchor between teeth of the escape wheel; as the pendulum swings to the right, it raises A and the wheel driven by the weight (not shown) revolves until checked by the descending fluke B.

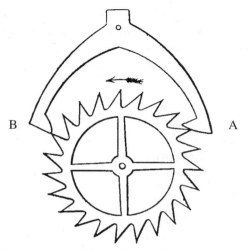

Figure 70. Anchor escapement (Dale, 1992, p. 38).

In 1719, shortly after Hooke's anchor escapement was made, John Flamsteed, the astronomer royal, in conjunction with the renowned clockmaker, Thomas Tompion, made a still better escapement, the "dead-beat" escapement (Fig. 71). This operated without recoil, minimized friction between the escapement teeth and the escape-wheel teeth and, in addition, imparted a measured impulse to the pendulum rod, so as to keep it oscillating. This escapement, perfected by George Graham (1673–1751), was relatively simple to make (once teeth cutting gears had

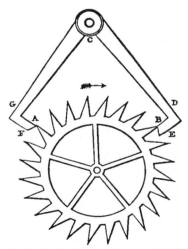

Figure 71. Graham's dead-beat escapement (Dale, 1992, p. 38).

been mechanized and perfected) and it became standard equipment on subsequent pendulum clocks.[5] Graham was also the first to use mercury as a compensator for expansion of the pendulum rod with rising temperature—the rod finishing in a jar of mercury so that as the rod expanded, the mercury rose, thus keeping the center of gravity fixed!

Theoretical Achievements

In an era blessed with so many great natural philosophers, Hooke, understandably, did not stand out as a theoretician. Nevertheless he was no mediocrity.[6] Hooke was not a mere technician, or curator of demonstrations. In 1674, 15 years before the *Principia*, Hooke came tantalizingly close to stating Newton's theory of universal gravitation, when in the *Philosophical Transactions* of the Royal Society, he wrote:

> I shall hereafter explain a system of the world, differing in many particulars from any yet known but answering in all thing to the common rules of mechanical motions. This depends upon three suppositions. First, that all celestial bodies whatsoever have an attraction or gravitating power towards their own centers whereby they attract not only their own parts and keep them from flying from them, as we may observe the earth to do, but that they also do attract all the other celestial bodies that are within the sphere of their activity . . . The second supposition is this, that all bodies whatsoever that are put into a direct and simple motion will so continue to move forward in a straight line till they are, by some other effectual powers, deflected and bent into a motion describing a circle, ellipsis, or some other more compounded curve line. The third supposition is that these attractive powers are so much the more powerful in operating, by how much nearer the body wrought upon is to their own centers. (in Jourdain, 1913, pp. 367–368)

Mathematics is the stumbling block for Hooke. He proceeded to say:

> Now what these several degrees are I have not yet experimentally verified; but it is a notion which, if fully prosecuted as it ought to be, will mightily assist the astronomer to reduce all the celestial motions to a certain rule, which I doubt will never be done true without it . . . This I only hint at present to such as have ability and opportunity of prosecuting this inquiry . . . (Ibid)

And then he turned to the pendulum to explain this system of the world:

> He that understands the nature of the circular pendulum and of circular motion will easily understand the whole ground of this principle, and will know where to find direction in nature for the true stating thereof. (Ibid)

Hooke played an important part in the Baconian experimentalist tradition that characterized of the scientific revolution. Hooke's work that perhaps best shows this interplay of scientific theory, mathematics and practical technique, so characteristic of the experimental method introduced in the 17th century, is *De Potentia Restitutiva or of Spring* (1678). In this work he developed his theory of springs and of vibrations, both of which bore upon timekeeping.

In the *Potentia Restitutiva* (Fig. 72) Hooke said that he had discovered his law (extension is proportional to force applied) 18 years earlier.

He expressed the law as "the power of all Springs is proportionate to the degree of flexure." He proposed it as an experimental law and demonstrated it using wire helix, spiral watch springs, straight wire suspended perpendicularly, and air. He suggested it as a general law for all springy material, but one can note that the law is actually then built into the definition of springy material: If a material does not obey his law, then it is not springy. He described his "philosophical scales," a spring balance, with which he demonstrated, among other things, the variation in gravitational attraction at different points above and below the earth's surface.

Hooke believed that the "work" (perhaps "effort" is more correct) done in

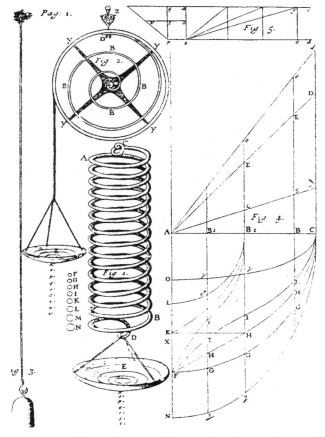

Figure 72. Frontispiece of 'De Potentia Restitutiva (Patterson, 1952, p. 311).

stretching a spring is proportional to the square of the extension. As he put it, a spring bent 2 spaces receives on its return 4 degrees of impulse, 3 in the first space returned and 1 in the second; bent 3 spaces, it receives 9 degrees of impulse; bent 10 spaces, it receives 100 degrees of impulse (Patterson, 1952, p. 309). He asserted, without proof, the law of conservation of mechanical energy, and then inferred that the oscillations of a released spring must be isochronous. The physical intuition was that the further a spring was stretched the greater "impulse" it had when released, and thus the greater velocity. So, although it covered more distance than a mildly stretched spring, it did so at a greater speed. When a stretched spring reached its equilibrium point, its stored "impetus," "force," or "impulse" carried it beyond equilibrium to a compressed state, and then the process started over again. This is, in modern terms, simple harmonic motion: that is, vibratory motion in which acceleration is proportional to displacement, and is always directed towards the equilibrium point. Figures 73 and 74, from a contemporary physics text, illustrate this motion.

Hooke's Dispute with Huygens

It is not unexpected that questions about priority of invention should arise. Huygens made his first pendulum clock in 1656. His first book on the subject, the *Horologium*, was published in 1658. He worked with the clockmaker Samuel Coster on the provision of pendulums for church clocks in a number of Dutch towns during 1657 and 1658. His major work, *Horologium Oscillatorium* which demonstrated the isochronic properties of the cycloid, and suggested the spring balance, was published in 1673. Thus his work on the theory and practice of timekeeping stretched over almost 20 years (during which time of course he also made

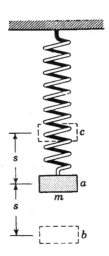

Figure 73. Coiled spring in simple harmonic motion (Weber, White, and Manning, 1959, p. 134).

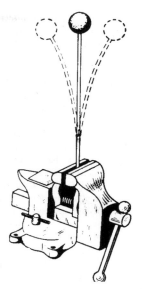

Figure 74. Straight spring in simple harmonic motion (Weber, White, and Manning, 1959, p. 134).

significant contributions to other branches of science). As we have seen, he was not the only person in western Europe trying to perfect timekeeping. In London, Christopher Wren, Robert Hooke, and others in the Royal Society were similarly occupied, with Hooke beginning his horological investigations at least as early as 1658.

Not satisfied with waging a long and acrimonious war against Isaac Newton over claims for priority of inventing the refracting telescope and for discovering the universal law of gravitational attraction, Hooke opened up a second front when he entered into a comparable dispute with Christiaan Huygens. The horologist, J. Drummond Robertson, discovered in the 1920s an unpublished Royal Society lecture of Hooke's, in which he reviews Huygens' just published *Horologium Oscillatorium* (1673). The lecture is noteworthy for the subject matters that it canvases, and the light that it sheds on contemporary disputes about priority of clockwork inventions.[7] The full text is in Robertson (1931, pp. 167–173), extracts are reproduced below.

> I have lately [30th May 1673] Received from the Inquisitive Hugenius van Zulichem a book written by himself containing a Description of severall mechanicall & mathematicall Inventions Intitled *Christiani Hugenii Zulichemii Const .f.² Horologium Oscillatorium sive de motu pendulorum ad Horologia aptato demonstrationes geometricae.*
>
> There are in it indeed many things very ingenious and very usefull but there are not wanting also severall things that are of a quite contrary nature as I shall shew you by some few observations which I have made in the cursory reading of it having not yet had time to examine every particular thereof more strictly.

And in the first Place the Author gives us an account how about 15 years since he first published his Invention of applying a pendulum to a clock and thinks thereby he hath sufficiently securred and warranted himself to have been the first inventor thereof because there was noe body before him that had made publication thereof to the world, and is very unwilling to Allow Galileo any share in the honour of the Invention. Whether Galileo or his Son did find out a way of applying it to a clock I can not affirme but sure I am that the Greatest excellency of the Invention is to be ascribed to Galileo who first found out the Vibrations of a pendulum were very near of equal Duration. Nor is Mersennus or Ricciolus to be deprived of their share in further examining & promoting the Doctrine of pendulous motions nor that Frence author who writt animadversions upon Galileos mechaniques, who does not only speak of the applycation of a pendulum to clocks, but also enigmatically Describes a way of using it at Sea for the finding out the Longitudes, nor Indeed after the knowledge of the aequality of pendulous motion was it Difficult to find out away of applying it to clocks. Dr. Wren, Mr. Rook, Mr. Ball & others made use of an Invention of Dr. Wrens for numbring the vibrations of a pendulum a good while before Mon^r Zulichem publisht his and yet did not cry Euraka, and, I my self had an other way of continuing and aequalling the vibrations of a pendulum by clock work long before I heard of Mr. Zulichem. And yet though I think I doe deservedly much preferr it before Mr. Zulichems way, nay though aequaled wth. A Cycloeid yet I have not either cryd Euraka or publisht it and yet I think I can produce a sufficient number of credible witnesses that can testify for it about these 12 years. Soe that the argument that he soe much Relys upon to secure to him the Invention is not of soe great force as to perswaid all the world that he was the first & sole inventor of that first particular of applying a pendulum to a clock.

The next thing wch he mentions is his invention that all the vibrations of a pendulum moved in a cycloeid are of aequall Duration. This for ought I know he is the first Inventor of for I never heard of any one that claimed the honour from him of it. It is indeed an Invention very extraordinary and truly excellent and had been honour enough for him Justly to have gloried in the happinesse thereof, and I believe there is none that would have gone to have deprived him of his Due praise, But he should also have Rememberd that Golden Rule to doe to others as he would have others doe to him & not to have Vain-gloriously & most Disingenuously Indeavoured to Deprive others of their Inventions that he might magnify himself, and wth the Jack Daw pride himself in the plumes of others, which how much and often he hath Done in the rest of th book I shall Indeavour to explaine.

But before I come to these particulars which are indeed noe ways pleasant to me were there not a necessity and duty incumbent on me to Doe it Give me Leave a little to animadvert upon those two Inventions wch for ought I know may Justly be his own. That is the way of applying a pendulum to a clock and the aequation of the motion of a pendulum by a cycloeid. For the first I say the Invention is very simple and plaine and therefore soe much the more to be preferred, but yet if thereby the pendulum becomes affected wth every inaequality that the Clock work which it really doth is subject to, and that inaequality be not removed by the aequation of the cycloeid as certainly & experimentally it is not, then be the Geometricall subtilty and demonstration thereof never soe excellent yet it is in it self but a lame invention, and he hath come short by a point if he hath made it dulce & not utile. (Robertson, 1931, pp. 167–173)

Some version of this lecture was forwarded to Huygens' father, Sir Constantijn Huygens, who forwarded it to his son. Christiaan Huygens, in August 1673, wrote a most civil reply:

> I am obliged to the civilitie of Mr. Hooke for what he writes to you concerning my Book . . . I do not wonder at what he sayth to have observed touching the insufficience of the Pendolo's to find the longitude, because he hath seen the experience of those which the Earl of Kinkardin had caused to be made and such like, for they had yet very considerable defaults. The last form I have reported in my Book is a great deale better and I am still in good hope of it, expecting that a tryal be made of it at sea, which without these unhappy warres had been done by this time . . . I beseech you to communicate all this to Mr. Hooke. (Robertson, 1931, p. 173)

Huygens had earlier visited Oldenburg, the first secretary of the Royal Society and had subsequently written to him offering him some interest, or share, of English patent rights for Huygens' own pendulum clock and balance spring watch. In October of 1675 Oldenburg wrote to Huygens:

> Since my last an incident has occurred which obliges me to write this letter. Mr. Hook, having learnt that you had given me permission to avail myself of any interest you might claim from a patent in this country for your watch, has been so bold and impudent as to say publicly that you gave me this permission as a reward for having revealed to you his invention, adding with the utmost effrontery that I am your spy here for communicating to you what ever of moment is discovered here, and that I wanted to rob him of the profit of his invention . . . Such accusations are as atrocious as false . . . I have never communicated to you anything about this invention, nor of any other until after you had revealed your own to us. Sir, I beg you to tell the whole truth as fully as you possibly can...Sir, after having made use of this letter, put in the fire, I beg of you. (Robertson, 1931, p. 184)

England and Holland were now at war, and Oldenburg was imprisoned in the tower for two months as a spy. Upon receipt of Oldenburg's letter, Huygens wrote to Lord Brouncker (October 31, 1675):

> I take the liberty of writing you these lines, being compelled thereto by the interest that I feel in Mr. Oldenburg's honour and my own . . . I had invented a new clock of which I sent him the secret . . . Mr. Hook said that he had invented a similar thing, as he thought, some years ago...It is his customary vanity to claim the invention of everything. . . . As to the manner of his construction I am still in ignorance. It is therefore false that, as a reward for having revealed something to me, I wanted to confer a favour on Mr. Oldenburg; and my sole motive in this is that I thought that he certainly deserved to derive some profit from new inventions which he has taken so much trouble in publishing in his Transactions to the great benefit of science, and moreover because he might have more need of it than any other of my friends in England . . . But no one will be so unjust as to condemn him on the simple accusation of such a man as Mr. Hook who is rather in the present case impelled by his own interest. (Robertson, 1931, p. 185)

The controversy over priority for invention of the balance-spring regulator is difficult to resolve. It is perhaps best not to attribute malice or deceit to either Huygens or Hooke. Despite Holland and England being rivals and at war, the scientific barriers between the two countries were tissue thin. Lord Bruce was a political refugee in Holland during the Cromwellian period and had close dealings with Huygens, bringing his early clocks back to England to test them in seagoing longitude trials. Oldenburg, the secretary of the Royal Society, also went back and forth to Holland, visited Huygens, and became the English agent for Huygens clocks. Huygens, the first President of the *Académie Royale des Sciences* visited London in 1661, 1663, and 1689. He attended meetings at Gresham College of the embryonic Royal Society, became personal friends with a number of its members—Newton called him *summa Huygenius*—and became a member of the Royal Society when it was officially formed in 1663. Both Huygens and Hooke had close dealings with clockmakers in their respective countries, who in turn had dealings with each other. Moray and Brouncker who initially provided financial and other support for Hooke's timekeeping project, then switched their support to Huygens and sought to market Huygens clocks in Holland, France, Spain, Sweden, and Denmark. Both Huygens and Hooke learnt from Christopher Wren's analysis of pendulums and his investigations of clockwork. With all these exchanges going on, and with so many records no longer extant, just who learned what, from whom, when, is probably indeterminate.[8]

On the Shape of the Earth and Variation in Gravity

Hooke examined, and rejected, Huygens' proposal for a universal standard of length based upon the length of a seconds pendulum. This rejection flowed from Hooke's advanced theory of gravitation. Hooke asked whether

> . . . if a pendulum be made of such a length as to make 86,400 single vibrations [24 × 60 × 60] in the time that any fixt Star passes from the Meridan, 'till it returns to the same the next Night, that length may be taken for a perpetual measure of length or a standard Yard. (Hooke, 1705/1969, p. 458)

And then, presciently, commented that:

> But against this way of finding a natural universal, and perpetual Standard measure of length there may be divers Objections not inconsiderable; as, First, That if the Gravitating Power of the Earth be greater in one place than in another, as towards the Poles more than towards the Aequator, then the length of a Pendulum to vibrate Seconds, must also be considerably longer towards the Poles than towards the Aequinoctial, otherwise the Pendulum of the same length with what is determined in France, which is about the middle between the North Pole and the Aequinoctial, will go too quick near the Poles, and too slow near the Aequinocitial. Now, what I many years since discover'd in this place reading about Penduls for Longitude, and what I have now before mention'd concerning a probability, if not a necessity of such an inequality of Gravitation, and consequently of a Boul-like form of the Earth, does at least hint that some Experiment of that Kind ought to be try'd at some place near the Aequator, to see whether it be so or not, and till that be done there can be no certain Conclusion made thereupon. (ibid)

This was precisely the experiment that, by accident, Jean Richer conducted in 1672 at Cayenne. Hooke then listed various "imperfections" that might interfere with the supposed constant oscillation of the pendulum. If the earth's rate of daily (diurnal) rotation varies from summer (aphelion) to winter (perihelion), then even a constant pendulum will not beat seconds; the expansion and contraction of the pendulum occasioned by change of temperature will affect rate of oscillation; and differing air densities will affect the rate of oscillation. For all these reasons, Hooke ruled out Huygens' proposed standard of length.

In elaborating these imperfections, Hooke also (see Chapter 10) undermined Aristotelian-reductionist accounts of time, and prepares the way for Newton's nonreductionist, absolutist conception of time. Aristotle identified time with the unchanging, completely regular, motion of the heavenly bodies. Copernicus, Galileo, and Kepler introduced disorder and irregularity into planetary motions. This ruled out their motions as the standard of time, at least if time were assumed to be steady, unchanging, and regular in its rate of passage. Reductionists looked then to terrestial movements to instantiate time, and thought that Huygens' pendulum gave them just such a regular motion. Hooke dashed their hopes for this. Newton then moved, as will be seen, to give up the whole reductionist program by

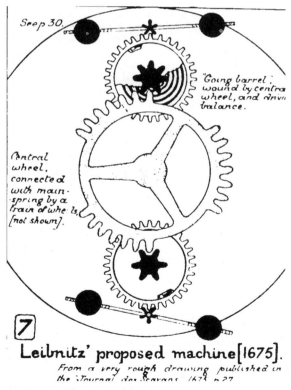

Figure 75. Leibniz's 1675 clock design (Gould, 1923, p. 26).

claiming that time was a reality in itself, of which the irregular planetary and terrestial motions were only pale copies or measures.

Leibniz' Clockwork Proposal

In the same year that Hooke proposed his successful anchor escapement, Gottfried Wilhelm von Leibniz, one of the towering intellects not just of the 17th century but of all time, also proposed, in the *Journal des Scavans* (the journal of the French Académie), an escapement mechanism and new form of clock that he thought would solve the longitude problem (see Fig. 75). He proposed using two balances and two spiral springs. Each spring was wound in turn by the train and then allowed to run down freely, controlled only by the inertia of the balance, the other spring being meanwhile wound. When the first spring had run down, it unlocked the second and so the cycle was to be repeated. He was oblivious to the fact, well known to Hooke, that as the spring unwound its motive power decreased. Dismissing these "practical problems," the great Leibniz said:

> . . . all these defects, that proceed from the imperfection of the matter, may be surmounted by a general remedy, without examining them here in particular. And that is, that for executing it in great, we may make use of massy springs, as are those of cross-bows, we being the masters of them, not wanting force or place in a ship to govern a great weight that may serve to bend them continually again . . . And it is easy to demonstrate, that by augmenting the size of the engine, and the force of the massy springs, we may make the error as small as we please . . . ; which answer is so clear and so universal, that all those who have considered it have expressed their satisfaction therein. (*Journal des Scavans*, 1675, p. 130. In Gould, 1923, p. 31)

Hooke might well have commented that this was "as clear as mud," and that although Leibniz had made a famous calculating machine, and was no technological innocent, nevertheless "as a clockmaker, Leibniz made a great mathematician." Hooke clearly possessed gifts that bordered on genius, it was his personal misfortune to be surrounded by individuals whose gifts more than just bordered. More particularly, Hooke lacked the mathematical brilliance of Newton, Huygens, and Leibniz. And from their time, major advances in physical science were dependent upon extraordinary mathematical competence: Intuition alone did not suffice.

JOHN HARRISON

Gemma Frisius in 1530 identified accurate timekeeping as the way to solve the longitude problem. This idea was ahead of 16th-century clockmaker's technology. Despite the efforts of Galileo, Huygens, Leibniz, Hooke, and countless other theoreticians and technicians of the highest calibre, and notwithstanding the offers of kings' ransoms in rewards, Frisius' idea continued to outrun horological reality for the next 234 years. The longitude problem was not solved until 1764 and then not by a theoretical genius, but by a dogged, tenacious, mechanical ge- nius, who devoted the entire 70 years of his adult life to nothing other than per- fecting a clock that would do as Frisius required, namely "records with the great- est exactness the time of the place of departure, and that it is not allowed to run down during the voyage…in this manner it is possible to find the longitude, even in a distance of a thousand miles, even without knowing where we have passed, and without knowing the distance travelled" (Pogo, 1935, p. 470).

The craftsman was John Harrison, known as "Longitude Harrison," was born in 1693, who the son of a Yorkshire carpenter, and died in London in 1776.[9] Harrison died three years after receiving the British Longitude Board's reluctant final payment of the £20,000 reward established by the Longitude Act of 1714. The prize required the then unheard of accuracy of 2 minutes during a 8–10 week voyage to the West Indies: Harrison's clock would keep within 30 seconds of correct time on a voyage out and back to the Indies!

At age 22 years, and with no formal education, Harrison built his first grand- father clock, which is still preserved in the Museum of the Clockmakers Com- pany, London. It is noteworthy for having all of its wheels made of lignum vitae, a remarkably hard and naturally oily wood. A short while later he invented the "gridiron" pendulum to overcome the deleterious effects of temperature change on the length, and hence period, of clock pendulums.[10] This was achieved by al- ternating steel (low expansion) and brass (high expansion) rods in a grid in such a way that the differential expansion rates compensated each other. He also made a new, highly accurate escapement, called the "grasshopper" escapement (King, 1998, p. 178; Burgess, 1998, p. 266; Hastings, 1993); and also a "going ratchet" to keep his clocks going while they were being wound up. At the age of 30 years he was making wooden clocks that did not vary more than a second a month, and that only needed cleaning each 30 years (King, 1998). This was the greatest pre- cision ever achieved by a pendulum clock. But it would not go to sea, and Harrison knew he would need a new design if he were to gain the longitude prize.

Isaac Newton was a member of the Longitude Board. In 1714 he advised the Board that the timekeeping method was unlikely to be successful, saying: "one method is by a Watch to keep time exactly. But, by reason of the motion of the Ship, the Variation of Heat and Cold, Wet and Dry, and the Difference of Gravity in different Latitudes, such a watch hath not yet been made."[11] Newton firmly believed that only astronomical methods—lunar distances or eclipses of Jupiter's moons—would successfully solve the longitude problem; he dismissed mechani- cal timekeeping means as utopian (Andrewes, 1998b).

Harrison, from the age of about seventeen years, single-mindedly set out to make what Newton doubted could be made. Having been stunningly successful, and having all his life been frustrated and doubted by professors and government bureaucrats, Harrison wrote, at age 82 years:

... if it so please Almighty God, to continue my life and health a little longer, they the Professors (or Priests) shall not hinder me of my pleasure, as from my last drawing, viz. Of bringing my watch to a second in a fortnight, I say I am resolved of this, though quite unsuitable to the usage I have had, or was ever to expect from them; and when as Dr. Bradley once said to me . . . if timekeeping could be to 10 seconds in a week, it would, as with respect to the longitude, be much preferable to any other way or method. And so, as I do not now mind the money . . . the Devil may take the Priests. (Gould, 1923, p. 68)

Harrison's first longitude clock, known as H1, now in the Royal Observatory, took six years to make (see Fig. 76). It was regulated not by a pendulum, but by two 5 lb balances connected by wires in such a way that the action of one would be counteracted by that of the other. Much of the mechanism was wood so as to obviate the need for oiling and hence possible clogging of parts. It was the first balance-regulated clock to contain a compensation device for temperature change. The total weight of the machine was 72 lbs (33 kg).[12] In 1736 it was taken on a sea trial to Lisbon. The Captain of the *Orford* reported of the homeward voyage that:

When we made the land, the said land, according to my reckoning (and others), ought to have been the Start; but, before we knew what land it was, John Harrison declared to me and the rest of the ship's company that, according to his observations with his machine [H1], it ought to be the Lizard—the which, indeed, it was found to be, his observation showing the ship to be more west than my reckoning, above one degree and twenty-six miles [90 miles]. (Gould, 1923, p. 46)

This was a gratifying result, surpassing any previous sea trial. But it did not meet the stringent conditions of the Longitude Act, which required a journey not to Lisbon, but to the West Indies and return. Nevertheless the Longitude Board gave him £500 as encouragement. Rupert Gould spoke of this first longitude clock:

Crude though it may seem, it is hard to praise this machine too highly. In constructing it Harrison, working single-handed and self-taught, grappled successfully with several problems which had . . . defied all previous attempts at solution. Even if he had gone no further, this machine would have proved him a mechanical genius, and possessed of 'an infinite capacity for taking pains'. (Gould, 1923, p. 45)

His second clock, H2, weighing 103 lbs, was completed in 1739 and, although it was never tested at sea, the Royal Society reported testing it most vigorously by "heating cooling and agitating it for many hours together, with greater violence than what it could receive from the motion of a ship in a storm" (Sobel, 1995, p. 85). Despite it passing these on-land tests, Harrison was not satisfied with what he had made (Andrewes, 1998b, p. 214).

He began work on H3 in 1740 and worked almost without interruption until its completion in 1757. Harrison was known as a nonstop and tireless worker, yet fully 17 years were spent on the construction of this one clock. During this time, the Longitude Board forwarded him £2,500 in five installments. When the 66 lb clock was finally finished, its timekeeping ability was, by Harrison's demanding standards, less than acceptable. He had calculated the moment of inertia of the balance weights to be given by the formula mv, when he should have used mv^2, where m was the mass of the weights, and v their radial velocity (there was after all some place for professors). But upon completion Harrison did suggest to the Longitude Board that he build two small pocket-size timekeepers. These were to be H4 and H5. Gould says of H4 that: "by reason alike of its beauty, its accuracy, and its historical interest, [it] must take pride of place as the most famous chronometer that ever has been or ever will be made" (Gould, 1923, p. 49). The watch beats five to a second, the pivot holes are ruby, the ends of the axles are diamond, the internal plates are brass and the case is silver.

Harrison's H4 (see Fig. 77) was completed in 1759, when Harrison was 67 years old. He immediately asked for, and was granted, a sea trial. This trial was to bring him, after unconscionable prevarication from the Longitude Board, his long sought for Longitude Prize.

Harrison's son, William, left for the West Indies on November 18, 1761, with the H4 on the *Deptford*. Its future contribution to navigation was glimpsed when the boat was only nine days sail out of Plymouth. The captain, by dead reckoning, put their longitude as 13°50' west of Greenwich, but H4 indicated that they were 15°19' west of Greenwich. William Harrison, standing by his father's timepiece, said that they would sight the Madeira Islands the next morning. At 6 AM they did so. On arrival in Jamaica, H4 was only five seconds slow, corresponding to a longitude error of only 1¼ seconds of arc (Randall, 1998, pp. 241–243). This entitled John Harrison to the Longitude Board's prize of £20,000 provided the timekeeper "was a method generally practicable and useful."

The Board did not pay the prize. They maintained that they could not determine whether the clock was "generally practicable and useful" until they took it apart, examined the mechanism, and had copies built. Harrison said that he would not open the watch until he was paid his money—he had, after all, been working non-stop for 35 years on a chronological solution to the longitude problem, and his H4 had achieved an incomparable degree of accuracy and reliability.

During this period Harrison finetuned the isochronism of the clock's balance. He had ascertained that the balance took slightly longer to move through a long arc than it did through a smaller arc. To correct for this he changed slightly the slope of the pallet faces, and fitted a "cycloid-pin" inside the curve of the spring—reminiscent of Huygen's cycloidal plates. Harrison wrote of H4:

> My Time-keeper's Balance is more than three times the Weight of a large sized common Watch-balance, and three times its diameter; and a common Watch-balance goes through about six Inches of Space in a Second, but mine goes through about twenty-four Inches in that Time: So that had my Time-keeper only these Advantages over a common Watch, a good Performance might be expected from it. But my Time-keeper is not affected by the different Degrees of Heat and Cold, nor Agitation of the Ship; and the Force from the Wheels is applied to the Balance in such a Manner, together

with the Shape of the Balance-spring, and (if I may be allowed the Term) an artificial
Cycloid, which acts at this Spring; so that from these Contrivances, let the Balance
vibrate more or less, all its Vibrations are performed in the same Time; and therefore,
if it go at all, it must go *true.* So it is plain from this, that such a Time-keeper goes
entirely from Principle, and not from Chance. (Gould, 1923, p. 62)

Eventually a second trial of H4 was agreed to, and again William Harrison,
on March 28, 1764, sailed with the watch to Jamaica (Randall, 1998, pp. 246–
247). On return to Portsmouth, after a return voyage of five months, the watch
was found to have slowed just 15 seconds, an error of *less than a tenth of a second
per day.* This was a staggering result; it was the culmination of centuries of effort
to perfect timekeeping. Harrison was, understandably, pleased. He wrote:

I think I may make bold to say, that there is neither any other Mechanical or
Mathematical thing in the World that is more beautiful or curious in texture than this
my watch or Time-keeper for the Longitude . . . and I heartily thank Almighty God
that I have lived so long, as in some measure to complete it. (Gould, 1923, p. 63)

The Longitude Board was less pleased. Under the strong influence of the
Astronomer Royal, Nevil Maskelyne (1732–1811), who was a partisan of the astronomical lunar distance method, the Board kept delaying payment of the prize,
maintaining that only public display of the mechanism, and commercial produc-

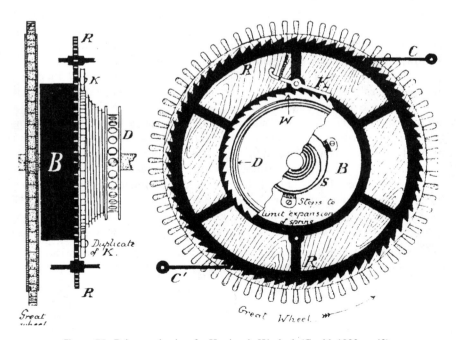

Figure 76. Drive mechanism for Harrison's H1 clock (Gould, 1923, p. 43).

Figure 77. Harrison's H4 longitude watch (5 inch diameter) (Gould, 1923, p. 53).

tion of the clock would establish that it was "a method generally practicable and useful." Harrison was going blind, when finally King George III intervened on his behalf, and the balance of the reward money was paid in 1773.[13]

Under instructions from the board, the watchmaker Larcum Kendal (1728–1779) made a duplicate of H4 (called K1) which was taken in 1773 by Captain James Cook on his second voyage of discovery to the South Seas. Using K1 Cook produced extensive and, for the first time, accurate maps of the Pacific islands. He said that: "It would not be doing justice to Mr Harrison and Mr Kendall if I did not own that we have received very great assistance from this useful and valuable

Table 1.

Port	Halley's Longitude	Modern Longitude (w)
Barbados	59° 11'	59°
Madera	16° 45'	17°
Martinique	60° 20'	60° 30'
Recife	35° 30'	34° 52'
Rio de Janeiro	44° 45'	43° 17'
St Helena	6° 30'	5° 43'
San Iago	22° 40'	22° 40'
Carlisle Bay	59° 05'	59° 37'
Saba	62° 55'	63° 26'

timepiece" (Sobel, 1995, p. 150). Cook, described as the greatest navigator of his age, was so impressed with K1 that he took the timepiece on his ill-fated third voyage.[14]

The accuracy and reliability of Harrison's chronometers, along with rapid improvement in the sophistication and efficiency of clockmaking, meant that by the 1790s most major British shipping companies equipped their ships with marine chronometers for the determination of longitude.[15] Edmund Halley, for instance, during that period was able to draw up a table of the longitudes of Atlantic ports with the values remarkably close to modern ones (Cook, 1998, p. 257). Such tables were of no small value to the conduct of British trade and naval affairs.

DETERMINING LOCAL TIME

All the methods for finding longitude, astronomical and mechanical, required that *local time* be accurately known. No matter how precisely lunar distances were measured, or how reliably a clock kept time during a voyage, the determination of longitude was only as accurate as the observer's knowledge of local time. Harrison's H4 clock reliably kept London time during its voyage to the West Indies; but for him to know the longitude of Port Royal, Jamaica, the time shown on H4 had to be compared to local time. The situation was nicely expressed by William Maitland in the middle of the 18th century who, in his *An Essay Towards the Improvement of Navigation, Chiefly with Respect to the Instruments used at Sea*, wrote:

> Mr. Harrison's Clocks, undoubtedly bid fairest for measuring Time truly at Sea, of all that ever were invented for that Purpose; and his unwearied Pains to bring his curious Machines to still greater Degrees of Exactness, will in Process of Time (we hope) have the desired Effect. But granting we had a just Measure of Time at Sea, there is still requisite a good Method of finding the apparent Time, in that Place where the Ship is; as it is from comparing the true and apparent Time that the Difference of

> Longitude must be had; and I must frankly acknowledge, that I know of no Method by which to determine apparent Time at Sea, with less Error than that of some Minutes, which if we take to be only five, this Error, when reduced to its proportional Part of the Equator, will amount to no less than 75 Sea-Miles (Maitland 1750, in Andrewes 1998c, p. 394)

The basic theory was straight forward: local noon occurs when the sun is directly overhead, and so at noon the time on the travelling clock (for mechanical methods of determining longitude) can be read, and the time difference converted to degrees of longitude east or west of one's origin. However theory is not practice: technology and technique are needed to bridge the gap. Andrewes reports that four methods were commonly used in the 18th century for determining local time: first, recognizing the moment of noon; second, measuring the exact altitude of the sun; third, measuring the exact altitude of a star when the horizon was visible; fourth, measuring the meridian angle of the sun or a star and using spherical trigonometry and logarithms to find the true time (Andrewes, 1998c, p. 395). Instruments were needed for all of these measurements, and they were progressively refined by craftsmen and technicians—the cross-staff, Jacob's staff, astrolabe, backstaff, quadrant, octant and sextant (Andrewes, 1998c). The first sextant was made by John Bird (1709–1776) in the mid-1750s. Subsequently countless thousands of exquisite examples were made, and it became a necessary part of all ships' navigational equipment up to the introduction of the satellite-based Global Positioning System in the 1960s.

These instruments enabled local noon to be determined; but once that was done some form of clock was usually needed to record the passage of local time—not all observations could be done at noon! Sundials, hour-glasses and, from the 18th century, mechanical clocks and watches were the usual local timekeepers. These could be checked and adjusted each 24 hours, or as their accuracy demanded.

THE HOROLOGICAL REVOLUTION

Huygens' pendulum clocks, first made in 1657, instituted a revolution in horology. They did not solve the problem of longitude. They were constrained by variations in gravity and by technological considerations such as friction, uneven impetus given to the oscillations, length variation induced by temperature, and so on, but on land and in stable environments, they achieved standards of timekeeping that hitherto had only been dreamed of (Bertele, 1953). In 1760, John Harrison's marine chronometers proved capable of solving the longitude problem and were to put their technological stamp on all mechanical timekeepers up to the 20th century. The one hundred year period between Huygens' pendulum clocks and Harrison's marine chronometer (1660–1760) is justly termed "the horological revolution." The successful pursuit of precision and reliability in timekeeping, was instrumental in allowing the fruits of the Galilean–Newtonian scientific revolution to flower. The horological revolution also had far-reaching political, commercial, cultural, philosophical, and religious consequences.

Before Huygens' pendulum clock (1657) clocks could not keep time more accurately than within 15 minutes per day (99% accuracy), and they had to be checked and set daily against sundials. Within twenty years, they were accurate to within ten seconds per day (99.99% accuracy). By 1761, John Harrison's marine chronometer would keep time accurately to within 15 seconds on a half-year voyage (99.9999% accuracy). This progression in accuracy of timekeeping is well portrayed in the accompanying graph (Fig. 78) where the vertical axis measures accuracy in seconds per day.

Pendulum clocks became so accurate—within one second in one hundred days (Burgess, 1998, p. 259)—that the question of how one checked their accuracy became pressing. What was the standard that they would be checked against?

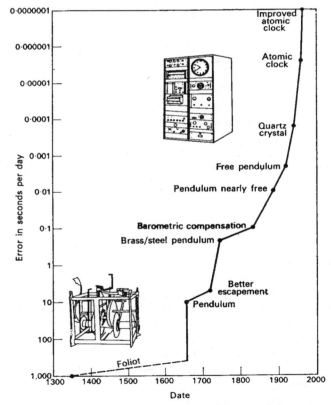

Figure 78. Development in accuracy of time measurement (Whitrow, 1972, p. 87).

And what was time that the standard supposedly measured? The old standards, for instance the sundial, were no longer good enough; they did not differentiate seconds, and they were calibrated to the solar day, not the mean solar day.

For instance, Charles Bellair, in July 1659, wrote to Huygens and said:

> Allow me to ask you whether in Holland, in those places where there are several pendulum clocks, they continue very long to sound the hours together; because I have had two of them converted and put a seconds pendulum in each. I have not yet been able to keep them going together four days in a row. Not that they're very far apart, and when one checks them with sundials, one cannot see a difference even after a week; but the precision of hearing is much more sensitive than that of sight. (Landes, 1983, p. 121)

Until 1967, when the atomic clock standard was adopted, all clocks had to be regulated against heavenly motion—the Heavenly Clock. This is what John Harrison, and Christiaan Huygens before him, had done. By sighting successive transits of a star across the meridian, a sidereal day could be ascertained and clock time could be compared to it. Rather than the meridian, Harrison used a sight line from the edge of a window in his house to the edge of a chimney on a neighbor's house. The period between successive transits of a selected star across this sight line was a sidereal day (3 minutes 56.4 seconds less than a solar day), and Harrison used this period to correct his clocks.

Watchmakers and clockmakers embodied the very finest craft skills, and they moved easily in the highest social circles. Tompion, Graham, Coster, Berthoud, and the Le Roy family enjoyed immense social prestige and large incomes. Their work was prized by kings and nobility; their clocks combined utility with exquisite artistry.[16] The kings of Poland, Prussia, and China maintained clock factories. Louis XVI and George III both took an active interest in the state of their nation's timekeeping industry and their leading clockmakers. The latter, as we have seen, personally intervened for John Harrison in his long-running battle with the British Longitude Board.

The horological revolution resulted in immensely greater availability of timekeepers. It is estimated that in 1703, 170,000 watches were made in just one English center—Clerkenwell. This could be an overestimate. But there are reasonably firm figures suggesting that in 1798, 50,000 watches were being produced in England for the home market alone, excluding those made for export (Macey, 1980, p. 35). Watches and clocks were no longer the preserve of the nobility or the extremely wealthy: Most of the middle class could aspire to ownership.

The watchmaker's craft was also revolutionized. Timekeepers went from being the outcome of individual craftsmen's labor, to being the product of large scale division of labor. According to Charles Babbage (1791–1871), the early 19th-century mathematics professor at Cambridge, there were in his time 102 distinct branches of the watchmaking trade, to each of which a boy might be apprenticed (Macey, 1980, p. 36).[17] Babbage himself was dependent upon the clockmaker's craft for the refined and complex machinery that drove his first "analytical engine," the forerunner of the modern computer.[18]

The spread of technical competence occasioned by clock- and watchmaking was the social foundation for the blossoming of scientific instrument making, which made possible the "world of precision" required for advancing the scientific revolution. It was a craft tradition whose motto was "perfect is near enough." This attention to exactitude and precision was exactly what was needed for the widespread manufacture, and hence democratic use, of reliable scientific instruments. One writer on scientific instruments of the 17th and 18th centuries commented:

> . . . precision mechanics is in [clockmakers'] debt for its first successful constructions and its basic progress until the present time. For five centuries clockmakers in every generation were responsible for the most exact mechanisms known . . . Thus, quite apart from specialized inventions, thanks to which clocks and watches continually increased in precision, the clockmakers were also responsible for putting at the service of mechanics greatly improved tools which instrument makers could adopt. (Daumas, 1972, p. 116)

CHAPTER 8

The Pendulum
in Newton's Physics

Isaac Newton (1642–1727), born the year Galileo died, was the towering figure of the scientific revolution.[1] In a period rich with outstanding thinkers, Newton was simply the most outstanding. Goethe, for instance, labeled 1642 the "Christmas of the modern age." Pope's epitaph, written for Newton's Westminster Abbey grave, captured the esteem in which Newton was held:

> *Nature and Nature's Laws lay hid in night:*
> *God said, Let Newton be! and all was light.*

This esteem has not diminished much with the passage of years. Even the most revisionist histories—beginning with the Soviet historian Boris Hessen's thunder-clap presentation to the International History of Science Congress in 1931[2]—have barely touched it.

Newton dominated late 17th century mechanics and astronomy (with his *Principia,* 1687) and optics (with his *Opticks,* 1704). He wrote on chemistry, and he also wrote 50,000 words on alchemy. In addition, he wrote over a million words on theological and biblical issues. He was a member of parliament, president of the Royal Society, and Master of the Mint. And, of importance for the story being told in this book, he chaired the government's Longitude Board, the board set up by the 1714 Longitude Act to encourage and reward the solving of the longstanding longitude problem.

Newton set his methodological stamp upon the course of modern science. The Newtonian method quickly became the norm in all branches of science and was aspired to in fields of inquiry beyond science, including ethics, politics, history, theology, and philosophy.[3] In philosophy, no less a figure than Immanuel Kant wrote that the "true method of metaphysics is basically the same as that introduced by Newton into natural science and which had such useful conse-

Figure 79. Isaac Newton, aged 50 years (Newton, 1729/1934, frontispiece).

quences" (Kant, 1968, p. 17). In the scholarly world of the 18th and 19th centuries, *Principia* Envy was, if not universal, nevertheless widespread.[4]

The centrality of the pendulum in Newton's work is often overlooked. The
pendulum is pervasive in Newton's mechanics and philosophy. Pendulum experiments are mentioned in his early *Notebooks*, and in each of the three books of his
monumental *Principia*. Richard Westfall has done something to rectify the
pendulum's neglect, saying in one publication on Newton, that "It is not too much

to assert that without the pendulum there would have been no *Principia* (Westfall, 1990, p. 82). This is no small praise for the pendulum.

Newton used the pendulum to establish the gravitational constant (which at the time was regarded as the distance all bodies fall in the first second of freefall). He also used it to improve timekeeping, to disprove the mechanical philosophers' aether assumption, to show the proportionality of mass to weight, to determine the coefficient of elasticity of bodies, to determine the speed of sound, to demonstrate the conservation laws, and much more.

THE HOROLOGICAL BACKGROUND

Measurement of time was an acute problem in the 17th century. As previously noted, accurate and reliable timekeeping was necessary for solving the longitude problem and for setting the foundation for work in mechanics and astronomy. Newton had met Huygens a number of times and was familiar with Huygens' horological deliberations. Huygens published his first *Horologium* in 1658,[5] and then his masterful *Horologium Oscillatorium* in 1673 (Huygens, 1673/1986). Newton was engaged not only by the subject matter, but also by the rigor of Huygens' geometrical argument, demonstrations, and deductions, and by the prospect that it held for accurate clock making.

In Aristotle's scheme of things, the perfectly steady, circular, regular and eternal motion of the unchanging heavenly bodies provided the perfect clock. Aristotelian time was identified with, and reduced to, the motion of the cosmic clock. Terrestrial clocks were just approximations, and crude ones at that, to this cosmic clock. The Copernican revolution (1543 and after) made only a slight dent in this horological picture. Copernicus still had the heavenly bodies moving regularly in circles, and the earth's annual and diurnal motions were thought to be perfect and regular.[6] Although the earth was no longer at the center of solar system, horological adjustments could easily be made for this: The solar day was no longer defined in terms of the orbit of the sun, but in terms of the rotation of the earth. These were different definitions, but they resulted in the same phenomenological experience and the same measurements.

After Johanne Kepler's (1571–1630) *Harmonices Mundi* was published in 1630, the Aristotelian cosmic clock began to appear less regular. Kepler established that the solar system was elliptical, that the speed of the earth varied with its orbital position, and that the axis of rotation of the earth itself rotated. The cosmic clock was beginning to wobble. But if the standard is irregular, what could replace it? For the most part, scientists (natural philosophers) gave up hope of any exact, unchanging, celestial measure of time. But the laws of motion, and dynamics more generally, involved time as an independent variable. If there were no reliable standard for time, then the enterprise of mechanics and natural philosophy would be based on unstable foundations. This was not something which the age's quest for certainty could live comfortably with.

Newton was well aware of the terminal condition of Aristotle's cosmic re-ductionist account of time. If time were regular, if it "flowed equally" as Newton wrote, then it could not be reduced to, or identified with, a process that was ir-regular. Reductionists could give up the cosmic clock and retreat to Huygens pen-dulum clock as the instantiation of time. This haven was short-lived, being de-molished by Jean Richer's 1672 experiments in French Guiana and Halley's experiments in St. Helena, both of which established the "unevenness" of a given pendulum's motion as it moved up and down in latitude. Eventually, in the *Principia*, Newton cut the Aristotelian nexus between time and its measure:

> Absolute, true, and mathematical time, of itself, and from its own nature flows equably without relation to anything external, and by another name is called duration: relative, apparent, and common time, is some sensible and external (whether accurate or unequable) measure of duration by the means of motion, which is commonly used instead of true time; such as an hour, a day, a month, a year. (Newton, 1729/1934, p. 6)

This does not mean Newton was indifferent to the practical measure of time: He admired Huygens' pendulum clock and thought it the best available. He saw it, to use Westfall's terms, as the key to the "world of precision," and accurate measure-ment that modern natural philosophy required.

THE METAPHYSICAL BACKGROUND

One of the first uses to which Newton put the pendulum was the testing of a metaphysical claim, namely the aether hypothesis required by Cartesian mechanical philosophy. This was an important hypothesis for Newton to test. He was an early adherent of the new, mechanistic philosophy, although he began to waver on some core aspects of the mechanical world view as he realized its incompatibility with his notion of attraction at a distance (Westfall, 1980, p. 374).

Cartesian mechanism was a philosophy that resuscitated views, or at least orientations, of the ancient atomistic opponents of Aristotle, especially Democritus and Epicurus.[7] Descartes, Gassendi, Huygens, Boyle, and all others who embraced the new, anti-Aristotelian, mechanical world view, or "corpuscularian philoso-phy" as it is sometimes called, were committed to showing that all celestial and terrestrial phenomena could be explained on the assumption of moving atoms and their collisions. It was an archtypal reductionist program.

Mechanists rejected outright the Aristotelian metaphysical categories of sub-stance, nature, form, potency, and especially the idea of teleological self-move-ment. In the Aristotelian world view, the world consisted of matter that was con-stituted into specific species, or kinds, by virtue of their form. The form of a human was different from the form of a dog even though, conceivably, the matter was the same. It was the form that differentiated kinds and that controlled their development and natural motion. Development, or growth, was the unfolding of the form within; it was the process of "realizing one's potential," as medievals used to say and as educators frequently still do say. Aristotle defined motion itself as "the fulfillment of what is potentially" (Barnes, 1984, p. 343). Change was then a teleological process. It was directed towards the fulfillment of some natural order: Oak seeds grew into oak trees, not into horses. Stones fell down toward their natural place; they did not move up and float about (except when subject to violent motions). Failure to achieve potential was a real loss. The Aristotelian world was ordered, purposeful, and self-directing.

While medieval scholastic philosophers, such as Albert the Great and Aquinas, in part embraced Aristotelianism because of its seeming compatibility with Christian faith,[8] the mechanical philosophers regarded this world picture, and its constituents, as occult, and as an impediment to true understanding (science). Even "tendencies" were banished by the mechanists. All change was the result of collison or rearrangement at the macro or micro (atomistic) level. Bodies could only affect each other when they were in direct or indirect contact with each other. Matter was inert, there could be no action at a distance.[9] Descartes' world was cold, passive, and colorless.[10] Descartes followed Galileo in banishing all secondary qualities (heat, color, pain, pleasure, scent, sound, etc.) to the subjective, mental, realm.

The mechanical philosophers—despite certain differences between themselves, particularly about the degree of divisibility of atoms,[11] and the reality of forces[12]—set out to explain everyday and experimental phenomena, in terms of the motion and collision of inert atoms. Huygens work on light, and Boyle's in chemistry are perhaps the crowning achievements of the mechanical world view.[13] Mechanical philosophers needed something to explain the regular rotation of the heavenly bodies. Aristotle's crystalline spheres, which carried the planets, would have been a nice mechanical explanation, but Galileo and Kepler's identification of comets as heavenly bodies shattered those spheres. They also needed a mechanical explanation for freefall, and here neither Aristotle's natural motion or Newton's action at a distance was available to them.

They postulated an invisible, all-pervasive, microscopic aether to explain both sets of phenomena. The aether was supposed to consist of the tiniest particles set in motion. For hardline mechanists, such as Descartes, there was no void because the aether filled all space. By postulating certain motions and properties for the aether, explanations of planetary motion and of free fall could be generated. The aether saved the mechanical world view, but did it exist?

Newton devised an ingenious pendulum experiment to test the aether hypothesis. This is described at the end of his *Principia* chapter where "we may find the resistance of mediums by pendulums oscillating therein" (Newton, 1729/1934, p. 316). After determining the relative resistances of air, water, mercury, and other media, he asked:

> Lastly, since it is the opinion of some [the mechanical philosophers] that there is a certain ethereal medium extremely rare and subtle, which freely pervades the pores of all bodies; and from such a medium, so pervading the pores of bodies, some resistance must needs arise; in order to try whether the resistance, which we experience in bodies in motion, be made upon their outward surface only, or whether their internal parts meet with any considerable resistance upon their surfaces, I thought of the following experiment. (Newton, 1729/1934, p. 325)

Newton identified the crucial point that the aether offered not just resistance to external surfaces, but because it pervaded the insides of all bodies, it also offered internal resistance to bodies moving through it. Thus formulated, the metaphysical aether hypothesis has empirical consequences. Although metaphysical, it became a scientifically testable hypothesis. It is worth describing Newton's experiment at length as it illustrated his attention to detail and his commitment to the experimental method in philosophy.

> I thought of the following experiment. I suspended a round deal box by a thread 11 feet long, on a steel hook, by means of a ring of the same metal, so as to make a pendulum of the aforesaid length. The hook had a sharp hollow edge on its upper part, so that the upper arc of the ring pressing on the edge might move the more freely; and the thread was fastened to the lower arc of the ring. The pendulum thus being prepared, I drew it aside from the perpendicular to the distance of about 6 feet, and that in a plane perpendicular to the edge of the hook, lest the ring, while the pendulum oscillated, should slide to and fro on the edge of the hook; for the point of suspension, in which the ring touches the hook, ought to remain immovable. I therefore accurately noted the place to which the pendulum was brought, and letting it go, I marked three other places, to which it returned at the end of the 1st, 2nd, and 3rd oscillation. This I often repeated, that I might find those places as accurately as possible.

Having set up his pendulum, and established benchmarks (literally), he organized the experimental test of the aether assumption. The test was based on the point that the aether offers resistance to internal parts, as well as external surfaces. Air affected the surface of bodies; aether was supposed to pervade all bodies and so was extra-resistive to the movement of bodies.

> Then I filled the box with lead and other heavy metals that were near at hand. But, first, I weighed the box when empty, and that part of the thread that went round it, and half the remaining part, extended between the hook and the suspended box; for the thread so extended always acts upon the pendulum, when drawn aside from the perpendicular, with half its weight. To this weight I added the weight of the air contained in the box. And this whole weight was about 1/78 of the weight of the box when filled with the metals. Then because the box when full of the metals, by extending the thread with its weight, increased the length of the pendulum, I shortened the thread so as to make the length of the pendulum, when oscillating, the same as before.

After this preparation, the test proceeded.

> Then drawing aside the pendulum to the place first marked, and letting it go, I reckoned about 77 oscillations before the box returned to the second mark, and as many afterwards before it came to the third mark, and as many after that before it came to the fourth mark.

Newton then drew on the reasoning he had used in his earlier experiments on resistance to pendular oscillation:

> From this I conclude the the whole resistance of the box, when full, had not a greater proportion to the resistance of the box, when empty, than 78 to 77.

Newton inserted a crucial premise linking dampening effects on a pendulum to its inertia:

> For if their resistances were equal, the box, when full, by reason of its inertia, which was 78 times greater than the inertia of the same when empty, ought to have continued its oscillating motion so much the longer, and therefore to have returned to those marks at the end of 78 oscillations.

But as he observed,

> . . . it returned to them at the end of 77 oscillations.

Newton was clearly careful with this experiment. He said the first time that he did the experiment, some years earlier,

> . . . the hook being weak, the full box was retarded sooner. The cause I found to be, that the hook was not strong enough to bear the weight of the box; so that, as it oscillated to and fro, the hook was sometimes bent this and sometimes that way. I therefore procured a hook of sufficient strength, so that the point of suspension might remain unmoved, and then all things happened as is above described.

The result, as with so much in science, is capable of divergent interpretation. The two main ones being:

(i) The filled box does dampen one oscillation (1/78th) prior to what is expected from just differences in inertia. Thus the difference is to be accounted for from the added resistance of the aether affecting the internal parts of the metal fill. Thus the experiment supports the aether hypothesis.

(ii) The results are roughly what is expected from just considerations of inertia. The slight discrepancy (1/78th) can reasonably be attributed to extraneous factors, and thus there is no additional dampening due to the aether. Thus the experiment does not confirm the aether hypothesis.

One can imagine Galileo, who was so impatient with impediments and disturbances that got in the way of theoretically predicted experimental results, reasoning in the manner of (ii). What did Newton do?

We might expect that Newton would choose (ii), because in the opening words of the *Principia* (the first definition dealing with quantity of matter) has already put his anti-aether cards on the table: "I have no regard in this place to a medium, if any such there is, that freely pervades the interstices between the parts of bodies" (Newton, 1729/1934, p. 1). However, in the first edition of the *Principia* he followed the reasoning of (i), but said:

> This reasoning depends upon the supposition that the greater resistance of the full box arises from the action of some subtle fluid upon the included metal. But I believe the cause is quite another. For the periods of the oscillations of the full box are less than the periods of the oscillations of the empty box, and therefore the resistance on the external surface of the full box is greater than that of the empty box in proportion to its velocity and the length of the spaces described in oscillating. (Westfall, 1980, p. 376)

He concluded:

> Hence, since it is so, the resistance on the internal parts of the box will be either nil or wholly insensible. (Westfall, 1980, p. 377)

Thus he drew the *anti*-aether conclusion from the result. Curiously, in the second edition of the *Principia*, he left out the first edition's intriguing mention of the heavier box having a lesser period than the empty box (something contrary to one of the basic Galilean pendulum laws), and left out the strong anti-aether conclusion. He in effect left the reader to choose between (i) and (ii) above. On this question, even the great Newton is ambivalent.

SYNTHESIS OF CELESTIAL AND TERRESTRIAL MECHANICS

To the mechanical philosophers' basic explanatory elements of inert matter and motion, Newton added attractive force between bodies. He was not the first to propose some form of gravitational force as the explanation for why apples fell out of trees. Many before Newton had also rejected Aristotle's "self-moving" (natural motion) explanation for falling bodies and had speculated on some kind of force being responsible for the fall.[14] To avoid charges of occultism, many people interpreted the force in mechanical, specifically contact, terms. Forty years before Newton, Gassendi, a leading atomist and proponent of the mechanical worldview (although holding a slightly different world view from Descartes) thought that bodies fell to earth because they were attached by very fine, invisible strings. Newton's theory of attraction that (bodies experience an attractive force varying as the product of their masses and inversely as the square of the distance between them) postulated action at a distance, but to the chagrin of mechanical philosophers in England and on the continent, he avoided specifying any mechanism for its operation:

> I here use the word *attraction* in general for any endeavour whatever, made by bodies to approach to each other, whether that endeavour arise from the action of the bodies themselves, as tending to each other or agitating each other by spirits emitted; or whether it arises from the action of the ether or of the air, or of any medium whatever, whether corporeal or incorporeal, in any manner impelling bodies placed therein towards each other. (Newton, 1729/1934, p. 192)

The big issue then was whether this *attractive force* was truly *universal;* that is, did it apply not only to bodies on earth, but also between bodies in the solar system? Aristotle, as with all ancient philosophers, made a clear distinction between the heavenly and terrestrial (sublunar) realms, the former being eternal, unchanging, perfect, etc., the latter being changeable, imperfect, etc. It was thus "natural" that the science of both realms would be different; and, to speak anachronistically, laws applying to the terrestrial realm would not apply to the celestial realm. This cosmic divide lasted for 2,000 years. Copernicus, by relegating the earth to a position alongside the other planets, and Galileo, with his telescope observations showing the lunar surface to be similar to the earth's, and his observations on comets as true celestial bodies, had begun the process of breaking down the distinction. If Newton's law of attraction were to be shown to hold not just on earth, but also between celestial bodies, then the fundamental Aristotelian cosmic division would collapse.

It was the analysis of pendulum motion that rendered untenable the celestial–terrestrial distinction, and enabled the move from "the closed world to the infinite universe" (Koyré, 1957). The same laws governing the pendulum were extended to the moon, and then to the planets. The longstanding celestial–terrestrial distinction in physics was dissolved. The same laws were seen to apply in the heavens as on earth: There was just one world, a unitary cosmos.

At the beginning of Book III of the *Principia*, Newton in Proposition IV stated:

> The moon gravitates towards the earth, and by the force of gravity is continually drawn off from a rectilinear motion, and retained in its orbit. (Newton, 1729/1934, p. 407)

At 22 years of age, while ensconced in Lincolnshire to avoid London's Great Plague, Newton began to speculate that the moon's orbit and an apple's fall might have a common cause (Herivel, 1965, pp. 65–69). He was able to calculate that in one second, while traveling about one kilometer in its orbit, the moon deviates from a straight-line path by about a 1/20 of an inch. In the same period of time an object projected horizontally on the earth would fall about 16 feet. The ratio of the moon's "fall" to the apple's fall is then about 1:3,700. This was very close to the ratio of the square of the apple's distance from the earth's center (the earth's radius), to the square of the moon's distance from the earth's center (1:3,600). Was this a cosmic coincidence? Or did the earth's gravitational attraction apply equally to the apple and the moon? That is, was there one gravitational force acting on both the apple and the moon, whose effect varied as the square of the distance of each object from the earth's center?

Twenty years later, the above conjecture appeared in the first edition of the *Principia*. The pendulum tied the two together. In the *Principia*, he said:

> The mean distance of the moon from the earth in the syzygies [new and full moon] in semi-diameters of the earth, is, according to *Ptolemy* and most astronomers, 59; according to *Vendelin* and *Huygens*, 60; to *Copernicus*, $60\frac{1}{3}$; to *Street*, $60\frac{2}{5}$; and to

Tycho, 56½. . . . Let us assume the mean distance of 60 diameters in the syzygies; and suppose one revolution of the moon, in respect of the fixed stars, to be completed in 27 days, 7 hours, 43 minutes, as astronomers have determined; and the circumference of the earth to amount to 123,249,600 *Paris* feet, as the *French* have found by mensuration. And now if we imagine the moon, deprived of all motion, to be let go, so as to descend towards the earth with the impulse of all that force by which it is retained in its orb [gravitational attraction to the earth], it will in the space of one minute of time, describe in its fall 15 1/12 *Paris* feet. . . . For the versed sine of that arc, which the moon, in the space of one minute of time, would by its mean motion describe at the distance of 60 semidiameters of the earth, is nearly $15^1/_{12}$ *Paris* feet, or more accurately 15 feet, 1 inch, and 1 line 4/9. Wherefore, since that force, in approaching to the earth, increases in the proportion of the inverse square of the distance [Newton's supposition], and, upon that account, at the surface of the earth, is 60 × 60 times greater than at the moon, a body in our regions, falling with that force, ought in the space of one minute of time, to describe 60.60.15 1/12 *Paris* feet; and, in the space of one second of time, to describe 15 1/12 of those feet; or more accurately 15 feet, 1 inch, and 1 line 4/9. (Newton, 1729/1934, pp. 407–408)

Having conjectured that the inverse square law of attraction applied equally to the moon falling towards the earth and to apples falling to the ground, Newton then drew measurable consequences. He knew the distance fallen by the apple in one second (this was the then gravitational constant, about 16 feet) and, using the inverse square law and knowing the distance to the moon in terms of the radius of the earth, he could ascertain the distance the moon should fall in one second.[15] It must then be determined what velocity is required to keep the moon in its observed orbit. Alternatively, if the moon is 60 times further from the earth's center than an object, such as an apple, at the surface of the earth, then the earth exerts a gravitational force on the moon equal to 1/60 × 60 that which it exerts on the apple (from the inverse square law).

Following the dictates of his own method, Newton then experimentally investigated whether the derived consequences are seen in reality. He deferred to Huygens' experimental measurement:

And with this very force we actually find that bodies here upon earth do really descend; for a pendulum oscillating seconds in the latitude of *Paris* will be 3 *Paris* feet, and 8 lines ½ in length, as Mr *Huygens* has observed. And the space which a heavy body describes by falling in one second of time is to half the length of this pendulum as the square of the ratio of the circumference of a circle to its diameter (as Mr *Huygens* has also shown), and is therefore 15 *Paris* feet, 1 inch, 1 line 7/9 And therefore the force by which the moon is retained in its orbit becomes, at the very surface of the earth, equal to the force of gravity which we observe in heavy bodies there. (Newton, 1729/1934, p. 408)

Newton then drew his conclusion:

And therefore the force by which by which the moon is retained in its orbit is that very same force which we commonly call gravity.

Later in the *Principia* Newton rendered more intelligible, or more commonplace, this moon–apple similarity, by referring to a projectile shot from a cannon on a mountain top (Fig. 80). With low velocity it falls close by, with increasing muzzle velocity it falls further away, until at last if it is shot with sufficient velocity, it does not fall until it gets back to the same mountain after circling the earth. With slightly more muzzle velocity, it just keeps circling the earth; it becomes a satellite.

Given that the force (centripetal) keeping an orbiting body in motion is given by mv^2/r, and that after Newton, we know that this is equivalent to the force of its attraction to the earth (gravitational attraction), given by

$$\frac{Gm_1 m_2}{r^2},$$

then for different altitudes, we can work out velocities required for satellites to have whatever period we wish, including a period of 24 hours which makes them "stationary" above the earth. These stationary satellites need to orbit at 22,000 miles above the earth's surface—a long way from Huygen's simple pendulum, but historically dependent upon it.

Newton's argument can be explained in terms of a body of mass m rotating on the end of a string of length r (Fig. 81). If there were no string, or if the string were cut when the body is at P, then it would move straight ahead, along the tangent at P, with the speed v it had at P.

In the accompanying figure, assume that the body moves through a distance c and reaches Q, during the time t that it would otherwise have reached R on its

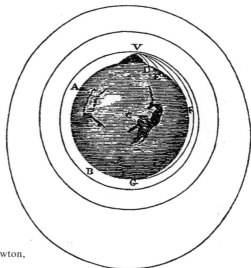

Figure 80. Newton's "satellites" (Newton, 1729/1934, p. 551).

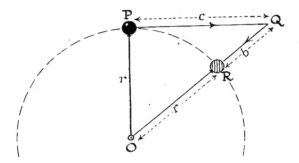

Figure 81. Forces in circular motion (Hogben, 1940, p. 273).

circular path. Thus $c = vt$. Hence for maintenance of its circular movement, the body would have to be pulled or pushed a distance b along the radial direction QRO. If its acceleration along this radial path is a, then the length $b = \frac{1}{2}at^2$. The speed of the regularly moving body at P is equal to the circle's circumference divided by the time it takes to travel the circumference. That is $v = 2\pi r/T$. Hence along the tangent PQ, the distance c, which is given by vt, equals $2\pi rt/T$. Given that the body has to accelerate, a, along the radial path QRO, in time t, in order to "regain" its circular orbit, we then know that $b = \frac{1}{2}at^2$.

Considering the right-angled triangle PQO, and using Pythagoras' theorem, the distance relationships are given by:

$$(r + b)^2 = c^2 + r^2$$
$$\therefore 2br + b^2 = c^2$$
$$\therefore 2b(r + \tfrac{1}{2}b) = c^2$$
$$\therefore at^2 (r + \tfrac{1}{2}b) = c^2$$
$$\therefore at^2 (r + \tfrac{1}{2}b) = 4\pi^2r^2t^2/T^2$$
$$\therefore a(r + \tfrac{1}{2}b) = (4\pi^2/T^2).r^2$$

When the weight has not moved appreciably beyond P, the distance b does not differ appreciably from zero and the direction QO does not differ appreciably from PO. So, if we neglect b, a becomes the instantaneous acceleration along the radius when the weight is at P, and:

$$ar = (4\pi^2/T^2).r^2$$
$$\therefore a = (2\pi/T)^2.r = v^2/r$$

If the mass of the weight is m, the force it exerts on the string is thus:

$$F = ma = m(2\pi/T)^2 = mv^2/r$$

The inward force is called the *centripetal* force, and hence *centripetal* acceleration is directed towards the centre. As the body does not actually move toward the centre, there must be an opposite force balancing it, this is called the *centrifugal* force.

If the circular speed is n revolutions per second, it takes $1/n$ seconds to rotate once, and $T = 1/n$. Thus the force, F, exerted on a mass of m rotating at a distance r from the centre of wheel making n revolutions per second, is given by:

$$F = (2\pi\, n)^2 mr$$

Newton was able to show the forces that had to be assumed to keep the moon in regular obit around the earth, satisfied the equation for forces required to keep a body orbiting on a string. And that the centripetal acceleration of the moon was the same as the downward (centripetal) acceleration of a falling apple. This was an enormous triumph for the Newtonian method, and a vindication of the role of idealization and counterfactual argument in scientific theorizing. Recall that Newton's argument, related above, in the *Principia* pp. 407–408, is that: *If* (a) the earth is stationary, *and* (b) the earth to moon distance is 60 earth radii, *and* (c) the moon's period is 27 days, 7 hours, 43 minutes, *then* assuming inverse square gravitational attraction, (d) bodies should fall to the surface of the earth at 15½ feet per second. All of the premises (a), (b) and (c) are false, and the conclusion (d) is also false. But Newton rightly saw in his demonstration the unification of terrestrial and celestial mechanics, the vindication of the universality of the law of gravitational attraction. The argument is a counterfactual one. The key points are that the premises and conclusion are *approximately* correct, and further, as the premises are refined, then the prediction (d) approaches more closely to the observed rate of acceleration of earthly bodies (Laymon, 1982, p. 115).

Newton's argument also demonstrated how interlocked and interdependent were strands of the new science. The proof of Newton's conjecture about a single gravitational force operative on the earth's surface and also in space, depended upon French cartographers' measurement of the earth's radius, the determination by Huygens and others of the distance of the moon from the earth, ascertaining the distance of freefall in one second as demonstrated by a seconds pendulum, and so on. Newton was a genius, but as he said of himself, he "stood on the shoulders of giants." Newton's ability to see assumed certain features about the scientific endeavor: It was communal, democratic, and concerned with truthful and candid reporting of results.

MATTER, MASS, AND WEIGHT

A few pages after Proposition IV, Book III, of the *Principia* (the proposition that broke the conceptual distinction between the heavens and the earth) Newton used the pendulum to flesh out the troublesome concept of "mass," and its relation to "matter" and "weight." Jean Richer's previously mentioned 1671 Cayenne voyage for the first time made the relationship of mass and weight problematic. When his Paris seconds pendulum clock lost about 2½ minutes per day at Cayenne, Huygens, Newton, and others realized that a fundamental conceptual problem had been unearthed. Hitherto mass and weight (the tendency and power of falling) had been thought of as the same thing. Indeed weight was frequently regarded as an essential and unchanging property of bodies. Richer's results, and those of Halley, Varin, Hayes, Feuillé and others are discussed by Newton in Book III, Proposition XX, Problem IV of the *Principia*. They showed that weight changed as a body moved up and down in latitude (others showed that it also changed as one moved up and down in altitude). But everyone reasonably thought that something about the body, the quantity of its matter, its mass, did not change. But how to specify the unchanging? And how to measure it?

Ernan McMullin remarked:

> The notion of a "matter" that underlies can reasonably be said to be the oldest conceptual tool in the Western speculative tradition. Scarcely a single major philosopher in the short but incredibly fertile period that separates Thales from Whitehead has omitted it from the handful of basic ideas with which he has set out to make Nature more intelligible to man. In many instances, ancient and modern, an initial judgement about the role to be attributed to matter has been decisive in orienting a philosophic system as a whole. (McMullin 1963a, p. 1)

Newton's system is no exception. His attribution of "inertia" to matter is one of the major landmarks in the history of the concept. As with much else, the way for Newton's discussion was prepared by Galileo who addressed the question of matter and its properties in *The Assayer* (Galileo, 1623/1957) where he said:

> Now I say that whenever I conceive any material or corporeal substance, I immediately feel the need to think of it as bounded, and as having this or that shape; as being large or small in relation to other things, and in some specific place at any given time; as being in motion or at rest; as touching or not touching some other body; as being one in number, or few, or many. From these considerations I cannot separate such a substance by any stretch of my imagination. (Galileo, 1623/1957, p. 274)

For Galileo, shape, size, location, contiguity, number, and motion are the primary and essential qualities of matter. Other qualities are secondary or subjective:

> But that it [material substance] must be white or red, bitter or sweet, noisy or silent, and of sweet or foul odor, my mind does not feel compelled to bring in as necessary accompaniments . . . these are no more than mere names so far as the object in which we place them is concerned, and that they reside only in the consciousness. Hence if the living creature were removed, all these qualities would be wiped away and annihilated. (Galileo, 1623/1957, p. 274)

Interestingly, "heaviness," or "massiness," is not listed amongst Galileo's essential properties of matter. For Aristotle, heaviness ("gravitas") or, lightness ("levitas") was such an essential property. Buridan's impetus was transferred to a body in accord to the amount of matter the body contained: Hence a stone could be thrown further than a feather. Matter was a measure, or cause of a body's power to resist or prolong motion. In getting rid of gravitas, Galileo seems to have eliminated all its cohorts. For Newton, Galileo's matter was too anemic.[16] The first definition of his *Principia* addressed the question of the *quantity of matter*.

> It is this quantity that I mean hereafter everywhere under the name of body or mass. And the same is known by the weight of each body, for it is proportional to the weight, as I have found by experiments on pendulums, very accurately made, which shall be shown hereafter. (Newton, 1729/1934, p. 1)

In his third definition (p. 2), Newton said:

> The *vis insita*, or innate force of matter, is a power of resisting, by which every body, as much as in it lies, continues in its present state, whether it be of rest, or of moving uniformly forwards in a right line.

In elaborating the definition, he added:

> . . . this *vis insita* may, by a most significant name, be called inertia (*vis inertiae*) or force of inactivity. But a body only exerts this force when another force, impressed upon it, endeavours to change its condition; and the exercise of this force may be considered as both resistance and impulse. (Newton, 1729/1934, p. 2)

So inertia is added to the essential properties of matter. But is gravity an essential property? Certainly Aristotelian *gravitas* was not, but what of the power to attract other masses? The Newtonian tradition, and most school texts, answer affirmatively to this,[17] but Newton himself was far more diffident. In his well known letter of February 25, 1692, to Richard Bentley, he said:

> It is inconceivable, that inanimate brute matter should, without the mediation of something else, which is not material, operate upon, and affect other matter without mutual contact, as it must be, if gravitation in the Sense of *Epicurus*, be essential and inherent in it. And this is one reason why I desired you would not ascribe innate Gravity to me. . . . that one body may act upon another at a distance . . . is to me so great an absurdity, that I believe no man who has in philosophical matters a competent faculty of thinking, can ever fall into it. (Thayer and Randall, 1953, p. 54)

It is in Book III of the *Principia* that he gave an account of the pendulum experiments establishing a correlation, but not identity, between mass (the quantity of matter) and weight:

> It has been, now for a long time, observed by others, that all sorts of heavy bodies . . . descend to the earth *from equal heights* in equal times; and that the equality of times we may distinguish to a great accuracy, by the help of pendulums. I tried experiments with gold, silver, lead, glass, sand, common salt, wood, water, and wheat. I provided two wooden boxes, round and equal: I filled the one with wood, and suspended an equal weight of gold . . . in the centre of oscillation of the other. The boxes, hanging by equal threads of 11 feet, made a couple of pendulums perfectly equal in weight and figure, and equally receiving the resistance of the air. And, placing

the one by the other, I observed them to play together forwards and backwards, for a long time, with equal vibrations. And therefore the quanity of matter in the gold . . . was to the quantity of matter in the wood as the action of the motive force . . . upon all the gold to the action of the same [motive force] upon all the wood; that is, as the weight of the one to the weight of the other: and the like happened in the other bodies. (Newton, 1729/1934, p. 411)

This reasoning can be reconstructed as follows:

The period T, of a pendulum whose length is l, swinging where the acceleration of gravity is $\mathbf{g}$ is given by the formula:

$$T = 2\pi \sqrt{l/\mathbf{g}}$$

The weight W of the body, and its quantity of matter m, are related by the formula:

$$W = m\mathbf{g}$$

Thus,

$$T = 2\pi \sqrt{ml/W}$$

And,

$$W/m = 4\pi^2 \, l/T^2$$

Since the experiment showed that the measurable ratio l/T^2 is independent of whether wood, gold, lead, glass, wheat is swinging, then W/m (the proportion of weight to mass) is a constant. Newton, reflecting the accuracy to which pendulum experiments had been brought, concluded his discussion of the experiment by saying:

By these experiments, in bodies of the same weight, I could manifestly have discovered a difference of matter less than the thousandth part of the whole, had any such been. (Newton, 1729/1934, p. 411)

Jean Bernoulli (1667–1748), shortly after publication of the second edition of the *Principia*, clarified Newton's concept of mass and gave it its modern form when he wrote in 1742 that the weight of a body is its mass multiplied by the acceleration of gravity.

THE PENDULUM AND CONSERVATION LAWS

The pendulum played an important role in establishing the foundational laws of classical mechanics: the conservation of momentum and the conservation of energy laws. Again Galileo's pendulum experiments paved the way for Newton's work.

In the *Two New Sciences,* Galileo demonstrated that balls rolling down an incline rose up another incline to the height at which they had started, irrespective of the gradient of the second incline—on the assumption, of course, of zero friction and no air resistance (Galileo, 1638/1954). The "motion" gained at the bottom was sufficient to impel them to the height from which they came. This was independent of time taken or distance traveled. Height dropped was the only determinant of height regained and "motion" acquired.

By making the second incline's gradient smaller and smaller, Galileo had the ball traveling further and further (on the assumption of ideal conditions) before gaining its initial height. It was then a small conceptual leap to having the second incline horizontal and "seeing" that the ball, under ideal conditions, would continue indefinitely. Actually, Galileo realized that it was not a horizontal, but a circular board, one having the same radius as the earth's surface that was required for this "ideal" perpetual motion to be realized. This was Galileo's statement of the law of *circular* inertia, in contrast to Newton's *linear* inertia.

The real, inclined-plane, situation is difficult and messy. It abounds with "accidents" and "impediments." Galileo used a simple but brilliant pendulum experiment to minimize friction, to minimize the impact problem at the join of the planes, and to approach further to the ideal case. The pendulum more easily allows the mental transition from the ideal to the real; it makes more believable Galileo's claims about the behavior of objects under ideal situations.

Galileo suspended a pendulum, as illustrated (Fig. 82), from point *A*, pulled it aside from the vertical *AB* and released it from *C*. It oscillated back and forth between *C* and *D*, with *D* being on the same level as *C*. He then interrupted the swing by placing a nail at point *E*. He found that the bob swung upwards to *G*, on the same level as *DC*, and then back again to *C*.

When the nail was placed below the level of *DC*, at *F*, the bob swung upwards to I, again at the level of *DC*. He concluded that in dropping from any

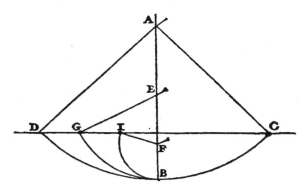

Figure 82. Galileo's "conservation" demonstration (Galileo, 1638/1954, p. 171).

height *h* a body acquires just sufficient "motion" to enable it to regain exactly the same height, provided its path is free from "impediments." This experiment was simple. It was not quantitative. Any imperfections in the result were overlooked as coming from Galileo's usual suspects, namely "impediments" and "accidents."

However identifying and understanding conservation during impact defied all of Galileo's intellectual and experimental efforts. He spoke of the "force of percussion" and attempted to measure it by noting the distance that poles were driven against resistance. Pile-drivers were his model. He not only lacked experimental apparatus to investigate collision phenomena, but he also lacked conceptual apparatus for the investigation: "Force" and "percussion" were gross concepts. It was the refinement of concepts that would allow Newtonian mechanics to make spectacular strides in understanding impacts and their associated conservation laws.

Johann Marcus Marci, a contemporary of Galileo and rector of the University of Prague, was perhaps the first to give scientific recognition to an intriguing phenomenon that was to puzzle impact theorists and cause refinement of the conservation laws. Marci's phenomenon was a commonplace of children's marble games: When a marble was flicked into a string of marbles only the last marble in the string was ejected. No matter how many marbles in the string, and no matter how fast the hitting marble traveled, only one marble was ejected at the other end. The last marble moves as if it were immediately struck without any intermediaries. This was a puzzle. Marci wrote of it in his 1639 *De Proportione Motus*, where he fancifully illustrated it by a cannon ball firing on to a string of cannon balls on a bench. However, having identified the phenomena, Marci was unable to shed any light on its explanation. This would come later. Although Marci's diagram seems fanciful, the use of cannons to illustrate findings of the new science was not. The cannon gave impetus, so to speak, to the science of projectiles in the late medieval period. The cannon regularly appears in mechanics texts of the time. See, for instance, Figure 84, which is from a 1607 book on scientific instruments.

Fifty years later Newton performed his own conservation experiment, the

Figure 83. Marci's 1639 impact "experiment" (Mach, 1883, p. 397).

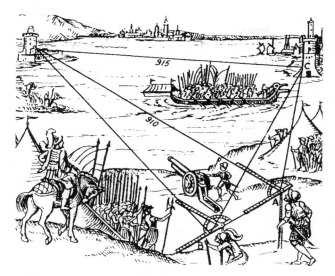

Figure 84. New science, new warfare (Hogben, 1940, p. 237).

difference between them demonstrating the tremendous increase in accuracy and sophistication with which, after Huygens, Wren, Wallis, and Hooke, pendular experiments were being conducted in the seventeenth century. The context of Newton's experiments was the investigation of collision phenomena.[18] Descartes, and others resorted to theological considerations when they proposed their principle of "conservation of motion": God initially created all the matter in the world, then he put it in motion. As he said in his *Principles of Philosophy,*

> As far as the general and first cause is concerned, it seems obvious to me that this is none other than God Himself, who being all-powerful, in the beginning created matter with both movement and rest; and now maintains in the sum total of matter, by His normal participation, the same quantity of motion and rest as He placed in it at the time. . . . there is a fixed and determined quantity of [motion]. (Descartes, 1644/1983, pt.II, #36)

If this totality was not conserved, then it would reflect poorly on the Creator. As Leibniz later said: Do we think that God is like a poor watchmaker who has to keep returning to wind up His creation? Obviously not. But just what was conserved was unclear. The amount of "motion" in the world was surely a constant, at least for those sharing Descartes' religious tradition, but how was this to be defined or measured?

Three properties made the pendulum an ideal vehicle for the investigation of collisions: Pendulums of the same length reach their nadir at the same time irrespective of where they are released from. They reach their nadir simultaneously regardless of their mass, and their velocity at the nadir is proportional to the length of the chord joining the nadir to the point of release. Additionally, the paths of the colliding bodies could be constrained.

Newton set up his conservation experiment (Fig. 86):

> Let the spherical bodies A, B be suspended by the parallel and equal strings AC, BD, from the centres C, D. About these centres, with those lengths as radii, describe the semicircles EAF, GBH, bisected respectively by the radii CA, DB. Bring the body A to any point R of the arc EAF, and (withdrawing the body B) let it go from thence, and after one oscillation suppose it to return to the point V: then RV will be the retardation arising from the resistance of the air. Of this RV let ST be a fourth part, situated in the middle, namely, so that
>
> $RS = TV$, and $RS:ST = 3:2$,
>
> then will ST represent very nearly the retardation during the descent from S to A. Restore the body B to its place: and, supposing the body A to be let fall from the point S, the velocity thereof in the place of reflection [collision] A, without sensible error, will be the same as if it had descended *in vacuo* from the point T. Upon which account this velocity may be represented by the chord of the arc TA. For it is a proposition well known to geometers, that the velocity of a pendulous body in the lowest point is as the chord of the arc which it has described in its descent. After reflection, suppose the body A comes to the place s, and the body B to the place k. Withdraw the body B, and find the place v, from which if the body A, being let go, should after one oscillation return to the place r, st may be a fourth part of rv, so placed in the middle thereof as to leave rs equal to tv, and let the chord of the arc tA represent the velocity which the body A had in the place A immediately after reflection. For t will be the true and correct place to which the body A should have ascended, if the resistance of the air had been taken off. In the same way we are to correct the place k to which the body B ascends, by finding the place l to which it should have ascended *in vacuo*. And thus everything may be subjected to experiment, in the same manner as if we were really placed *in vacuo*.

Having painstakingly set up his experiment, Newton proceeded:

> These things being done, we are to take the product (if I may so say) of the body A, by the chord of the arc TA (which represents its velocity), that we may have its motion in the place A immediately before reflection; and then by the chord of the arc tA, that we may have its motion in the place A immediately after reflection. And so we are to take the product of the body B by the chord of the arc Bl, that we may have the motion of the same immediately after reflection. And in like manner, when two bodies are let go together from different places, we are to find the motion of each, as well before as after reflection; and then we may compare the motions between themselves, and collect the effects of the reflection. Thus trying the thing with pendulums of 10 feet, in unequal as well as equal bodies, and making the bodies to concur after a descent through large spaces, as of 8, 12, or 16 feet, I found always, without an error of 3 inches, that when the bodies concurred together directly, equal changes towards the contrary parts were produced in their motions, and, of consequence, that the action and reaction were always equal.

And then he concluded:

> By the meeting and collision of bodies, the quantity of motion, obtained from the sum of the motions directed towards the same way, or from the difference of those that

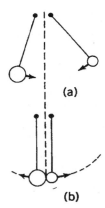

Figure 85. Collision on the vertical plane between unequal mass and amplitude pendulums (French, 1965, p. 440).

(a)

(b)

were directed towards contrary ways, was never changed. . . . (Newton, 1729/1934, pp. 22–24)

Newton does not label this the conservation of momentum. He speaks ill-definedly of conservation of "motion," but it is our momentum, mv, a vector quality, that he described. In modern terminology, his conclusion is given by the formula:

$$m_1 u_1 + m_2 u_2 = m_1 v_2 + m_2 v_2$$

The pendulum has here made a significant contribution to the foundation of classical mechanics. The law of conservation of momentum was true for both elastic (where there is no energy absorbed in the collision itself) and inelastic collisions (where energy is absorbed). For instance, if the above pendulums were made of putty, then when they collided, they would deform and simply come to a halt; there would be no motion after the collision. In this situation one could

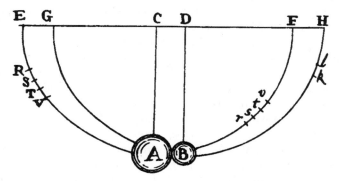

Figure 86. Newton's collision experiment (Newton, 1729/1934, p. 22).

hardly talk of conservation of "motion"—although, Descartes for instance, reso-
lutely maintained that the quantity *mv* was the basic measure of "motion," and *mv*
was the basic entity that was conserved in the world. The disciples of Newton,
who followed Descartes on this matter, when faced with the undeniable fact that
the pendulums of putty just come to a halt when they collide, preserved their
theory by saying that the atoms or corpuscles of putty were put into motion by the
collision, and hence the amount of motion was conserved. That is, an increase in
"internal," or corpuscular, motion made up for the obvious loss of external, or
gross, motion.[19]

The principles of conservation of momentum and conservation of energy
were, understandably, confused in the 17th century. All significant scientific notions are initially "seen through a glass darkly," even though science teachers
frequently expect students to immediately see them through an overhead slide
clearly. The conceptual confusion fueled acrimonious debate,[20] as can be judged
by the title of one of Leibniz' central contributions:

> A Short Demonstration of a Remarkable Error of Descartes and Others, Concerning
> the Natural Law by which They Think the Creator always Preserves the same Quantity
> of Motion; by which, however the Science of Mechanics is totally Perverted. Leibniz,
> 1686/1969[21]

One basic problem was that Descartes' interpreted *mv* as a *scalar*, not a *vec-
torial*, quantity. Thus, for him, the sum of motion of two equal bodies travelling at
equal speed in *opposite* directions is double that of any of the pair. When *mv* is
considered as a vector quantity (the modern *momentum*), the sum of motion in
this case is zero. So although Descartes uses the formulae *mv* for his quantity of
motion, it is not the modern *momentum* he is talking about.

Newton's conservation experiments have a methodological interest. They
illustrate the problematic nature of falsification and theory testing in science, a
matter that has previously been mentioned when discussing Huygens' responses
to Richer's observations about the slowing of the seconds pendulum at French
Guiana. When a theory is falsified by experimental results, do we abandon the
theory? Do we adjust the theory? Do we adjust the background assumptions? Do
we blame the apparatus? Do we blame the experimentor?

Newton's experiments also illustrate how metaphysics can be a heuristic for
science, how metaphysics can suggest avenues of experimental inquiry (Wartofsky,
1968). Finally, Newton's experiments illustrate the complexity involved in distinguishing metaphysical claims from claims about high-level, scientific, theoretical
entities. Is the aether hypothesis about a metaphysical entity or a theoretical entity? Newton of course did not make this distinction, but the later logical positivists made it a central one. The positivists wanted to demarcate metaphysics from
science, and did so, initially, using the verificationist criterion. For them, if any
claim could not be immediately verified by sensory evidence, it was metaphysical
and thus outside the bounds of rational discussion. But, using Newton's conserva-

tion experiment as an example, can the distinction between metaphysical concepts and high-level scientific concepts be clearly drawn? Newton's experimental treatment of the aether hypothesis seems to give support to Kuhn's contention that metaphysics (and methodology) is an intrinsic part of science, and that metaphysics "told scientists what many of their research problems should be" (Kuhn, 1970, p. 41). It could not be excluded from science. Indeed even low-level theoretical terms (magnetic field, electron spin, force, atom) seemed to be rendered unscientific by the positivist criterion. These kinds of problems eventually led to the abandonment of the verificationist criterion.[22]

For over 30 years Newton and his followers disputed with Leibniz and his followers about this conservation law. It was finally shown by Jean d'Alembert in 1758 to be a conceptual, not an empirical, disagreement.

Newton's thoughts, when expressed in modern terms, meant he regarded the rate of change of momentum of a body to be a measure of the force being applied to it. Thus the ratio of two forces f and F acting on two bodies m and M each for a period of t seconds is given by the equation:

$$f/F = ma/MA = mat/Mat = mv/MV$$

That is the ratio of forces is the same as the ratio of momenta produced.

Leibniz took a different tack. In his 1686 *Discourse on Metaphysics*—written because, as Leibniz said he was "somewhere having nothing to do for a few days, I have lately composed a short discourse on metaphysics" (Leibniz, 1686/ 1989, p. 35)—Leibniz proposed a telling thought experiment. He asked just for two propositions to be assumed: (1) at its terminus a falling body acquires an amount of motion (or force of motion) sufficient to return it to its starting point; and (2) the force of motion required to raise a unit body four units in height is the same as that required to raise a four-unit body one unit in height (Leibniz 1686/ 1969). Both propositions were assented to by Cartesians.

Leibniz went on to point out that a body of one unit mass falling four units of distance will have the same force of motion as a body of four units mass falling one unit of distance (Fig. 87). But Galileo had proved that a body falling four units has twice the velocity as a body falling one unit. Thus on the Cartesian assumption that force of motion is measured by the product of the mass and speed, the force of the first body is 1×2, or 2; the force of the second is 4×1, or 4. But by assumption, the forces of motion in the two cases are equal, thus its measure cannot be mass times velocity (mv, or momentum). The measure of the force that is conserved must be mass times velocity squared (mv^2, or double kinetic energy).

Thus a battle between intellectual giants was vigorously joined. At least it was joined until 1743 when Jean d'Alembert showed that *both* Newton and Leibniz were right: It was just a case of them talking about different things.[23] Descartes and Newton's adoption of momentum (mv) as the measure of motion conserved

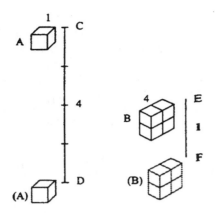

Figure 87. Leibniz's thought experiment (Leibniz, 1686/1989, p. 50).

was correct for forces acting for a given *time*; Leibniz' use of *vis viva* (mv^2, or kinetic energy) was correct for forces acting through a certain *distance*.[24]

d'Alembert used Euler's algebraic expression of Newton's first law ($F = ma$, the expressions used today) to formulate the Newton/Leibniz disagreements:

$$Ft = mat = [m(v-u)/t] \times t = mv - mu = \text{change of momentum}$$

Alternatively,

$$Fs = mas = [m(v-u)/t] \times s = \tfrac{1}{2}m(v^2 - u^2) = \text{change of kinetic energy}$$

The first principle, the conservation of momentum considered as a vector (direction dependent) quantity, was thought to be conserved under all collision conditions: that is, it was conserved in both elastic and inelastic collisions. The second principle, the conservation of *vis viva* (living force or kinetic energy), was understood to be conserved just in elastic collisions. The term "energy," but not yet "kinetic energy" seems first to have been used for this concept by Thomas Young in his 1807 *Lectures on Natural Philosophy*, where he wrote:

> The term energy may be applied with great propriety to the product of the mass or weight of the body, into the square of the number expressing its velocity . . . This product has been denominated the living or ascending force . . . some have considered it to be a true measure of the quantity of motion; but although it has been universally rejected, yet the force thus estimated well deserves a distinct denomination. (in Elkana, 1974b, p. 25)

Young, following in the footsteps of Descartes and Newton, and oblivious to d'Alembert's conceptual refinement, said this energy was not conserved because it manifestly was not conserved in nonelastic collisions.[25] But what happened to kinetic energy in these inelastic collisions was a matter of speculation and debate.

Was it also conserved, but in some different form? And of course there was an element of convention and definition here: Was an elastic collision one between perfectly hard bodies, or one in which momentum was conserved? Could the former be defined independently of the latter?[26] What constituted the limits of the *system* in which energy was or was not conserved? Was it possible to speak about, and then examine, an isolated system? All of these qualifications left plenty of conceptual space for retention of a theory regardless of what the specific experimental outcome was.

This was not so much an experimental but a philosophical matter. Just as on *theological* grounds Descartes, Newton, and Leibniz all maintained that the amount of *motion* in the world was constant (they just differed, as has been seen, on how this quantity was defined and measured), so too it was on *philosophical* grounds that initial debate about conservation of energy was conducted. Was the world constituted in such a way that the overall amount of energy was preserved? There was more faith involved here than experiment—which is not to say that the faith is not justified or vindicated. The usual formulation of the principle of conservation of energy is that in an isolated system, and in some finite time boundary:

$$PE + KE + OE = k$$

Where *PE* is potential energy, *KE* is kinetic energy, *OE* is other energy (electrical, chemical, thermal, nuclear, etc.). The principle need only be stated in order to see that it is immune from experimental disproof, or inductive justification.[27] It is one example of the operation of a priori elements in scientific theory—a standard subject of debate between empiricists, who wish to minimize or eliminate these elements, and rationalists, who wish to maximize them.[28] Energy occupies the borderland between science and philosophy: Both disciplines, and their respective histories, are required for an adequate understanding of the concept.

In 1760, Henry Cavendish (1731–1810) wrote his paper, "Remarks on the Theory of Motion," it well illustrated the conceptual and experimental difficulties that attended the conservation laws almost a century after Newton's *Principia*. Yet again the pendulum is set swinging in order to elucidate the matter. Cavendish distinguishes between a momentum with direction (a vector quality) which he called "ordinary momentum," and that element of it which does not depend upon direction, which he calls "mechanical momentum." Cavendish then proceeded to say:

> From hence appears the nature of the dispute concerning the force of bodies in motion; for if you measure this force by the pressure multiplied into the time during which it acts the quantity of force which a moving body will overcome or the force requisite to put a body in motion or in other words the force of a body in motion will be as the velocity multiplied into the quantity of matter, but if you measure it by the pressure multiplied into the space through which it acts upon the body the force of a body in motion will be as the square of velocity multiplied into the quantity of matter.

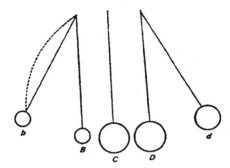

Figure 88. Cavendish's Pendulum Illustration (Elkana, 1974b, p. 45).

> The first way of computing the force of bodies in motion is most convenient in most philosophical enquiries but the other is also very often of use, as the total effect which a body in motion will have in any mechanical purposes is as the quantity of matter multiplied into the square of its velocity.
> . . . I think it would not be amiss if it was called the mechanical force or mechanical momentum of bodies in motion. Elkana, 1974b, pp. 44–45

Cavendish made the intriguing observation that "in the collision of elastick bodies there is very often an increase of momentum as usually computed" (Elkana, 1974b, p. 45). Of the above cradle pendulum situation (Fig. 88) he wrote: "The body D will move after the stroke [B hitting C, C in turn hitting D] with a much greater momentum that what B acquired by the fall" (Elkana, 1974b, p. 45).

That the accomplished experimenter, Cavendish, was making such heavy weather of the conservation laws is a salutary reminder for teachers who expect students to absorb, in one or two lessons, foreign concepts that great minds wrestled with for decades.

THE PENDULUM IN NEWTONIAN PHYSICS

Newton used the pendulum to establish a value for the gravitational constant, to disprove the aether, to distinguish mass from weight, to determine relative viscosities of fluids, to perform controlled collision experiments, to illustrate the conservation laws, to serve as a precision timekeeper, and to determine the speed of sound in air.[29] In the decades after Newton, the pendulum was used widely in illustrating or "proving" the fundamentals of Newtonian science, or classical physics, as we know it. The humble pendulum became more sophisticated, conical, compound, articulated, and played a more refined role in mechanics. Much of Newtonianism could be demonstrated by suitable manipulation of the pendulum— the conical pendulum, for instance, representing idealized planetary circular motion with constant velocity and yet a constant force and acceleration towards the center.

The Cradle Pendulum

The cradle pendulum, for instance, was commonly used to illustrate Newton's puzzling third law (for every action there is an equal and opposite reaction), and to illustrate varying coefficients of restitution (relative elasticities) of bodies.

It was a commonly used pedagogical device. It consists of polished steel balls whose coefficient of restitution is almost 1.0 (hence having almost perfectly elastic collisions). When the ball at one end is lifted and let fall, the impulse is transmitted undiminished along the train, and the ball at the other end flies up almost to the height of the first ball. All combinations of ball numbers and weights can be used, but always the number of balls hitting one end of the train, are ejected at the other end.

The cradle pendulum nicely illustrates how both momentum *and* kinetic energy are conserved in elastic collisions. As is standardly demonstrated to classes, the number of balls hitting one end constrain the number leaving at the other. If just momentum were conserved one might expect two balls hitting to cause one ball to fly off at twice the speed at the other end, or one ball hitting to cause two balls to fly off the other end at half the speed. This does not happen. The latter combinations satisfy the conservation of momentum, mv, but not the conservation of kinetic energy, $\frac{1}{2}mv^2$.

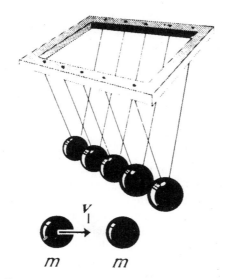

Figure 89. Cradle pendulum (Freier, 1965, p. 45).

The Ballistic Pendulum

The pendulum was also used to measure the velocity of bullets and other projectiles. Today high-speed cameras and special movie film enables bullet velocity to be directly measured, but Benjamin Robins in his 1742 *New Principles of Gunnery* described how a ballistic pendulum could, indirectly, do the same thing (Fig. 90). The speed of a pendulum at its lowest point is solely a function of the vertical height through which it has descended in describing its arc. Its speed does not depend on its mass or the length of its arc. The formula is exactly the same as that for velocity of a body freely falling from height h:

$$v = \sqrt{2gh}$$

Conversely, the height to which a pendulum rises is solely a function of its speed at its lowest point. So if a bullet of known mass is fired into, and embeds in, a lead or timber pendulum bob, then the velocity imparted to the bob can be measured by the height to which it rises. Then the conservation law states that momentum before impact is equal to momentum after impact:

$$mu + MU = (M + m)v$$

Because the ballistic pendulum's bob is stationary,

$$mu = (M + m)v = (M + m)\sqrt{2gh}$$

And so the bullet's velocity, u, before impact can be calculated. Because the height risen, h, is usually small and difficult to measure directly, the horizontal displacement of the ballistic pendulum after impact is easier to measure and h can be calculated indirectly by simple geometry. (Modern school classes can easily enough construct ballistic pendulums and measure the impact speed of arrows, sling shots, etc. (See Allchin et al., 1999, pp. 628-629).

Foucault's Pendulum

One of the most famous uses of the pendulum was that of the French physicist Léon Foucault (1819–1868) who used it in 1851 to establish dynamically the fact of the earth's rotation. That the earth rotates on its axis was first proposed in the ancient world by Aristarchus of Samos[30] Copernicus revived the idea, and Galileo defended it against the opposition of the church and Aristotelian physicists. Newton assumed it and built a physics upon the assumption. But until Foucault, no one had really proved the hypothesis against the rival hypothesis of a stationary earth and everything else in the heavens rotating accordingly. Certainly numerous

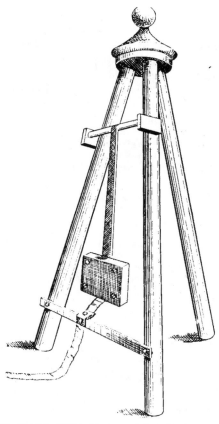

Figure 90. Robin's 1742 ballistic pendulum (Taylor, 1941, p. 205).

apparent problems with the Copernician hypothesis had been overcome—Galileo with his idea of circular inertia, and then Newton with his idea of linear inertia, had accommodated the rotating earth hypothesis to the anomalous facts of clouds and birds remaining overhead, of arrows shot vertically into the air coming down where they were shot from, of bodies dropped from towers landing at the base of the tower and not a distance to the west, and so on. All of this made the Copernican hypothesis more plausible, but there was no proof; no smoking gun.

The Copernican heliocentric hypothesis more easily accommodated astronomical and terrestrial phenomena, but with sufficient elaboration, the rival Ptolemaic geocentric hypothesis could accommodate the phenomena. The Copernican hypothesis needed a dynamical proof for the reality of the earth's rotation. Three centuries after Copernicus proposed his hypothesis, Foucault, provided this with his pendulum experiment.

On Newton's theory, a pendulum set swinging in a particular plane, should continue to swing indefinitely in that same plane. The only forces on the bob being the tension in the cord, and its weight directed vertically downward. Foucault—described as "a mediocre pupil at school, [but] a natural physicist and an incomparable experimenter" (Dugas, 1988, p. 380)—"saw" that if a pendulum were placed exactly at the north pole (Fig. 91), and suspended in such a way that the point of suspension was free to rotate (that is, not constrain the pendulum's movement by applying torque), then:

> The motion of the Earth, which forever rotates from west to east, will become appreciable in contrast with the fixity of the plane of oscillation, whose trace on the ground will seem to be actuated by a motion conforming to the apparent motion of

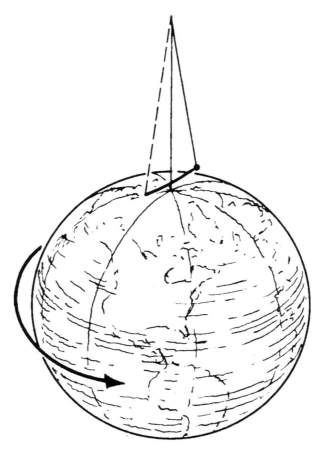

Figure 91. Foucault's intuition (PSSC, 1960, p. 340).

the celestial sphere. And if the oscillations can continue for twenty-four hours, in this time the plane will execute a whole revolution about the vertical through the point of suspension. . . . at the pole, the experiment must succeed in all its purity. (Dugas, 1988, p. 380)

Foucault realized no one was about to travel to the north pole to conduct the crucial experiment, and so he had to ascertain the expected effect in lower latitudes.

. . . when our latitudes are approached, the phenomenon becomes complicated in a way that is rather difficult to appreciate. To the extent that the Equator is approached, the plane of horizon has a more and more oblique direction with respect to the Earth. The vertical, instead of turning on itself as at the pole, describes a cone which is more and more obtuse.
From this results a slowing down in the relative motion of the plane of oscillation. This becomes zero at the Equator and changes its sense in the other hemisphere. (Dugas, 1988, p. 381)

Foucault intuited the relation between latitude and "rotation" of the plane of oscillation:

angular displacement of the plane = the product of the angular displacement
of a pendulum's oscillation   of the earth in the same time with the sine
                                   of the latitude.

Or, if T^1 is the time in which the plane of the pendulum rotates 360°, and T is the period of rotation of the earth, and b is the latitude where the experiment is being conducted, then:

$$T^1 = T/\sin b$$

From the formula, it can be seen that at the poles, $T^1 = T$ (as $\sin b = \sin 90° = 1$); whereas at the equator $T^1 = \infty$ (or infinity, as $\sin 0° = 0$), thus there is no rotation of the plane of oscillation at the equator. The following table gives some figures for how many degrees a Foucault pendulum rotates in 24 hours at different locations:

North Pole	360°
Edinburgh	298°
London	282°
New York	241°
Sydney	−201°
South Pole	−360°

Using the above formula connecting latitude to amount of rotation in one day, the latitude of a location can be ascertained once the rotation of a local Foucault pendulum is known –an exercise that school classes can usefully perform.[31]

Foucault began looking for rotation effects in Paris using a 5 kg ball swinging on a 2 m wire. He moved up to a 11 m wire, and finally in the Panthéon in Paris, he used a 28 kg ball on the end of 67 m of steel wire. After light objects were placed in a circle at the extremity of the swings, the ball was seen to progressively knock them down! Either the plane of the pendulum was rotating or the earth was rotating underneath. Newton's analysis of the forces on the pendulum, and care in the construction of the experiment so as to eliminate torque forces at the suspension point, ruled out the former, thus the earth must be rotating on its axis. Although the effect mystified the Parisian public, who flocked to see it, for scientists, Copernicus was at long last vindicated. And that by "a pupil who was mediocre at school!"

It was not only the Parisian public who turned out to see Foucault's pendulum: Replicas were set up in numerous countries and were greeted with the same enthusiasm. In the United States tens of thousands read accounts of the pendulum, and thousands attended demonstrations. Scores of newspapers editorialized on the subject. Cartoonists exercised their craft upon it. A version of "pendulum wars" raged between scientists convinced of Foucault's demonstration of the movement of the earth, and a minority of skeptical scientists and more than skeptical public (Conlin, 1999). As Michael Conlin notes, The *Boston Morning Commonwealth* identified the two questions raised by the Foucault pendulum that vexed the popular mind: How could the pendulum move independently of the earth's rotation if its point of suspension was attached to the earth? And, why did the plane of the pendulum take 34 hours to rotate and not 24? (Conlin, 1999, p. 195) One letter writer pleaded for a scientist to explain the experiment in a "manner adapted to the understandings of those who…have more zeal for science than capacity for imbibing it" (Conlin, 1999, p. 196).

Stokes' Pendulum Experiments

Foucault demonstrations were the public tip of a scientific pendular iceberg. Nearly all areas of mechanics were informed and refined by exacting pendulum experiments. The British physicist G.G. Stokes, for instance, used the pendulum to investigate the properties of fluids. In the opening paragraph of a 1851 paper he remarked:

> The great importance of the results obtained by means of the pendulum has induced philosophers to devote so much attention to the subject, and to perform the experiments with such a scrupulous regard to accuracy in every particular, that pendulum observations may justly be ranked among those most distinguished by modern exactness. (Stokes, 1851, p. 1)

It was *deviations* from the theoretical ideal that allowed Stokes to calculate, to astonishing degrees of precision, the various factors that mitigated against ideal pendular behavior. He used painstaking pendulum experiments to ascertain the index of friction of air, gases, and many liquids. These values which were subsequently used in many hydraulic applications, as well as enabling "us to calculate for certain forms of pendulum the correction due to the inertia of the air" (Stokes 1851, p. 127).

Stokes elaborated the distinction that Galileo enunciated between the real world and a world on paper, and how theory-informed experimental practice bridged these worlds:

> It is unnecessary here to enumerate the different methods which have been employed, and the several corrections which must be made, in order to deduce from the actual observations the result which would correspond to the ideal case of a simple pendulum performing indefinitely small oscillations in a vacuum. (Stokes, 1851, p. 1)

Stokes, and 19th century science, took for granted the hard-won methodological victory that Galileo had achieved over the empiricism, and anti-experimentalism, of medieval Aristotelian science, with its commitment to studying the "natural" motions of bodies. Indeed the history of classical mechanics can be seen as a long attempt by experimentalists to make the world conform to Newtonian theory, not the reverse.

CHAPTER 9

Clocks and Culture
The Clock Analogy in Philosophy and Theology

The clock not only played a key role in the foundation of modern physics, but it also had an influential role in philosophy by making intelligible the new mechanical world view, and in theology by being utilized in the design argument for God's existence. These extrascientific, ideological, and cultural influences of the clock are seldom mentioned in science programs, but a discussion of them can promote an appreciation of the interplay of science and culture.

As far as the study of timekeeping, or horology, is concerned, there are unfortunate divisions between science, cultural history, technical history of clockmaking and philosophy. Science texts deal with the properties of pendulum motion and perhaps make passing reference to pendulum clocks and timekeeping, but they usually ignore the cultural and methodological aspects of timekeeping. Cultural history books elaborate the origins and social impacts of timekeeping, but usually ignore the science and the crucial methodological transformations that enabled the "properties" of the pendulum to be discovered.[1] Philosophy texts mention Galileo, Huygens and Newton, and may comment on the methodological aspects of their "new" science and on their mechanical world view, but their involvement in the improvement of timekeeping, the longitude problem, and the cultural impacts of their work is usually ignored.

The bestseller, *Longitude*, by Dava Sobel (1995) well illustrates this point. Few of the thousands of readers relate the longitude story to their pendulum studies in school science, and Sobel completely left out the crucial methodological matters that lay at the heart of Galileo's discoveries. She merely repeats the chandelier story (Sobel, 1995, p. 37).[2]

These divisions are understandable from a disciplinary perspective, and from the need for scholars to be specialists. But they are regrettable from an educational perspective where efforts should be made to bridge disciplinary boundaries, especially for the majority of school students who are not proceeding with higher studies in science. Future sales clerks, lawyers, plumbers, and reporters ought to leave school with some sense of the big picture of science. It is hoped that science majors will also be taught in such a way that they develop an appreciation of the historical, philosophical, and cultural dimensions of science.

At a very general level, it is a commonplace that science has an impact on culture, and on ways of understanding ourselves and the world and vice versa. A study of the clock's influence on philosophy and theology is a way of elucidating this general claim and at the same time illustrating the function of metaphor in science. The history of the clock metaphor illustrates the fundamental point that people who form conceptions are themselves formed, that the "I think" is dependent upon the "we think," as Paulo Freire, following Hegel, often repeated. People's thinking is constrained by the thought patterns, concepts, preoccupations of their time. There is a historical and cultural dimension to cognition. Observation is theory dependent, and theory is historically and culturally dependent.

THE MECHANICAL WORLD VIEW

The clock, and then the watch, was the first machine, as distinct from tool, to be widely available and commonly utilized. Its personal impact can be gauged from John Comenius' (1592–1670) mid-17th century remark:

> How is it that such an instrument [the clock] can wake a man out of sleep at a given hour, and can strike a light to enable him to see? How is it that it can indicate the quarters of the moon, the positions of the planets, and the eclipses? Is it not a truly marvellous thing that a machine, a soulless thing, can move in such a life-like, continuous, and regular manner? Before clocks were invented would not the existence of such a thing have seemed as impossible as that trees could walk or stones speak? Yet every one can see that they exist now. (Comenius, 1657/1910, p. 96; in McReynolds, 1980, p. 99)[3]

It should not be surprising that the introduction of the clock had an effect on people's thinking, and that it entered into social consciousness, ideology, and science. This is what, 400 years later, the computer has done. The astronomer Johannes Kepler (1571–1630), for instance, used the clock metaphor to convey his understanding of the operation of the universe:

> My aim in this is to show that the celestial machine is to be likened not to a divine living thing but rather to a clockwork (*horologium*) . . . in so far as nearly all the manifold movements are carried out by means of a single quite simple magnetic corporeal force (*vis magnetica corporalis*) just as in a clockwork all the motions come from a simple weight. Moreover I show that this physical cause (*ratio*) can be determined by numbers and geometry. (Kepler to Hohenburg, 1605; in Crombie, 1994, p. 542)

Fifty years later, Comenius also used the clock metaphor to explain his understanding, not just of the world, but of human beings and their behavior:

> Just as the great world itself is like an immense piece of clockwork put together with many wheels and bells, and arranged with such art that throughout the whole structure one part depends on the other, and the movements are perpetuated and harmonised; thus it is with man . . . The weight, the efficient cause of motion, is the brain, which by the help of the nerves, as of ropes, attracts and repels the other wheels or limbs, while the variety of operations within and without depends on the commensurate proportion of the movements.
>
> In the movements of the soul the most important wheel is the will; while the weights are the desires and affections which incline the will this way or that. The escapement is the reason, which measures and determines what, where, and how far anything should be sought after or avoided. The other movements of the soul resemble the less important wheels which depend on the principal one. Wherefore, if too much weight be not given to the desires and affections, and if the escapement, reason, select and exclude properly, it is impossible that the harmony and agreement of virtues should not follow, and this evidently consists of a proper blending of the active and the passive elements.
>
> Man, then, is in himself nothing but a harmony . . . as in the case of a clock or of a musical instrument which a skilled artificer has constructed. (Comenius, 1657/1910, pp. 47–48; in McReynolds, 1980, p. 99)

Comenius is one of the major figures in the history of education and psychology and one of many who use the clock to explicate human behavior. This Cartesian mechanistic tradition culminated in Julien de La Mettrie's *Man a Machine*, published in 1748, in which he announced: "I am right! The human body is a watch" (Mettrie 1748/1912, p. 141). Ten years later Claude Helvétius (1715–1771), the French philosopher, in his summing up of the motives of his fellow sex, wrote "the love of women is . . . the main spring by which they are moved" (Helvétius, 1758/1810, p. 261).[4] His sensualist proto-Freudian account of what makes males "tick" was well elucidated by his use of the watch metaphor.

After 2,000 years of Aristotelianism, the major natural philosophers of the scientific revolution (Descartes, Boyle, Newton, Leibniz, Gassendi, etc.) were reactivating the mechanical and atomistic ideas of Aristotle's ancient opponents, especially Democritus and Epicurus. The world was understood in terms of particles (atoms, corpuscles) moving in a void and affecting each other only on contact. This mechanical world stood opposed to the organic, coordinated, harmonius world of Aristotle with its self-moving and self-actualizing natures, all perfecting themselves and seeking their proper ends. A battle for "hearts and minds" was being waged. John Donne's (1571–1631) lament, in his *Anatomy of the World*, is well known:

> *And new philosophy calls all indoubt:*
> *The element of fire is quite put out:*
> *The sun is lost, and th' earth, and no man's wit*

Can well direct him where to look for it.
And freely men confess that this world's spent,
When in the planets, and the firmament
They seek so many new; they see that this
Is crumbled out again to his atomies,
'Tis all in pieces, all coherence gone,
All just supply, and all relation.

What hope have we to know ourselves, when we
Know not the least things which for our use be?
(Heath-Stubbs and Salman, 1984, p. 79, 81)

The new mechanical world view had to be made intelligible, believable, and consistent with the overwhelmingly religious understanding of common people, clergy, and scientists: The clock metaphor was at hand to do this. Leibniz, for instance, said:

> . . . [the scholastics] saved the appearances by fabricating faculties or occult qualities, just for the purpose, and fancying them to be just like little demons or imps which can without ado perform whatever is wanted, as though pocket watches told the time by a certain horological faculty without needing wheels. (Leibniz, 1981, p. 68)

Thomas Hobbes (1588–1679), another philosophical champion of the mechanical world view, regarded not only the solar system but also the animal world, as explainable in mechanistic terms. To prepare the reader for this somewhat counterintuitive position, Hobbes in the opening lines of his *Leviathan* (1651) wrote:

> For seeing life is but a motion of limbs, the beginning whereof is in some principal part within; why may we not say, that all *automata* (engines that move themselves by springs and wheels as doth a watch) have an artificial life? For what is the *heart*, but a *spring*; and the *nerves*, but so many *strings*; and the *joints*, but so many *wheels*, giving motion to the whole body. (Hobbes, 1651/1962, p. 59)

When the clock is turned against Aristotle's world view, it is worth remembering that Aristotle himself was a great champion of metaphor. As he said, in the *Poetics,* when he discussed language, communication, and poetry:

> But the greatest thing by far is to be a master of metaphor. It is the one thing that cannot be learnt from others; and it is also a sign of genius, since a good metaphor implies an intuitive perception of the similarity in dissimilars. (Barnes, 1984, p. 2334)

It is also worth remembering that for all the apparent utility of the clock metaphor for early modern mechanical philosophers, the unanswered question was why did the clock weight fall in the first place? This was an Achilles heel of mechanical philosophy, and attempts (largely unsuccessful) to provide mechanical explanations of gravitation occupied philosophically inclined physicists right up to Einstein's proposals in the early 20th century.

THE DESIGN ARGUMENT

The argument from design for the existence of God has had a checked career in the history of ideas. Plato, in his *Laws*, was among the first to enunciate it. Plato ran together the first cause argument for God's existence and the design argument. He argued that movement required some first mover: It was inconceivable that things could just start to move without a first intelligence beginning the motion or first "stirring the pot," as one might say. He asked his audience to look to the type of movement around them and if they could ascertain whether the first intelligence, soul, was good or bad: "If the world moves wildly and irregularly, then the evil soul guides it" (*Laws* X, 897B). But when the heavens are examined, "There would be impiety in asserting that any but the most perfect soul or souls carries round the heavens" (*Laws* X, 898C). Aristotle did not pursue this argument. As has been indicated, his world was one of individual natures that came to fruition and expressed themselves in their natural motions. Intelligence, soul, form—Plato's principles of movement—were immanent within the Aristotelian world. There was no requirement to call upon *external* movers except, as has been pointed out, for violent motions.

Thomas Aquinas (1224–1274), who with enormous intelligence used Aristotelian philosophy to explicate the Christian world view, thought individual natures might well explain individual natural motions, but the manifest coordination of all natural motions nevertheless needed an explanation. Aquinas produced a most sophisticated version of the first cause argument, but he also proposed a version of the design argument. In the fifth of his arguments for God's existence, he ran through the above matters concerning the manifest coordination of animate and inanimate activites, and concluded: "Therefore some intelligent being exists by whom all natural things are ordered to their end; and this being we call God" (Aquinas, 1270/1952, p. 13).

Oresme and the Clockwork Universe

Nicole Oresme, who introduced the pendulum into European science (natural philosophy), was the first person to give the design argument a concrete, metaphorical, widely intelligible expression. This was through his novel clockwork metaphor. In his 1377 *Book of the Heavens and the World*, he described the world as a regular clockwork that was neither fast nor slow, never stopped, and worked in summer and winter by night as well as by day. He drew a direct comparison between the movements of the planets and a mechanical clock that balanced out all forces by means of its escapement: "This is similar to when a person has made an *horloge* and sets it in motion, and then it moves by itself" (Menut and Denomy, 1968, p. 282). Christianity, Islam, and Judaism already had the idea of God as creator and the world as God's creation. The clock and the clockmaker were powerful ways of conveying these theological beliefs.

Oresme lived in Paris, and his 1377 book was published shortly after Charles V had Henry de Vick's large clock installed in his palace and had instructed all other city clocks to fall into line with its ringing. When Oresme was looking for a way of communicating his understanding of the solar system and its divine origins, he did not have to look far: The clocks of Paris, ringing out everywhere, provided just the metaphor. The clocks were familiar, their operations known, and they, and their makers, were valued. Thus Oresme easily attained the agreement of his readers when he argued:

> . . . no one would say that the absolutely regular movement of a clock happens casually without having been brought about by some intellectual cause; for much stronger reasons must the movements of the heavens depend on some intellectual power higher and greater than human understanding. Thus the heavens show us the magnitude of God and his works: *Celi enarrant gloriam Dei, et opera manuum eius annuntiat firmamentum. Genesis* viii, 22 (in Crombie, 1994, p. 404)

The limitations of the mechanical clock (it did have to be wound each day, it did not keep time very accurately, it was subject to malfunction) meant the metaphor was imperfect. Nevertheless, it was good enough to begin the argument. And with the increasing precision, reliability, and availability of the clock, Oresme's clockwork metaphor and his inference to God's existence became more popular and convincing.[5]

The crucial part of Oresme's argument was that once the world was likened to a clock, then he could easily move to the conclusion that there must be a worldly clockmaker: God. The argument had the form:

1. All clocks are artifacts, and hence designed.
2. The world is like a clock.
3. Therefore, the world must have a designer.

The argument has some plausibility. Premise 1 is uncontroversial. Premise 2 is problematic. Clearly a lot hinges on how strongly we interpret "like." There are many things, for instance, that are patterned (rock faces), or regular in their movement (monsoon rains) that we do not feel compelled to call "designed." We may be prepared to say that they are like clocks, but not so alike that we have to invoke a designer. Once we admit that something is an *artifact*, then the argument goes through. But the issue is precisely whether something is an artifact, or sufficiently like an artifact to require a designer. This is the *strong* sense of "like" that is required for the validity of the argument. There are *weak* senses of "like" (the object is pretty, it is orderly, it is useful, etc.), that do not require the object having a designer.

Three centuries after Oresme, Johannes Kepler wrote: "My aim is to show that the heavenly machine is not a kind of divine, live being, but a kind of clockwork" (Crosby, 1997, p. 84). The proponents of "natural theology" welcomed the turning of each new page in the book of nature. They were confident that each and every page, paragraph, line, and word, would further reveal the "Majesty of the Lord." Most would say that their optimism was misplaced. The history of the design argument from the 17th century through its Darwin-influenced decline in the late 19th-century and on to its reinvocation by Paul Davies and others in the late 20th century is an informative example of the interplay of science with philosophy and culture.[6] A good science education could make this interplay more apparent to students and by doing so would contribute positively to their education and their understanding of the dynamics of their culture.

Newton's Argument with Leibniz about God's Involvement in a Clockwork Universe

Newton and Leibniz did not share the doubts of Boyle and Paley about the astronomical world being a clock-work.[7] Their arguments about just how the clock analogy applied and did not apply and what implications could be drawn about the actions and character of the Designer nicely illustrate the interplay of science and metaphysics.

Newton, as has been previously mentioned, wrote his *Principia* "with an eye upon such principles as might work with considering men for the belief of a Deity" (Thayer and Randall, 1953, p. 46). In the *General Scholium* added to the second edition of the *Principia*, and in the *Queries* appended to his *Opticks* (Newton, 1730/1979), Newton laid out his theistic understanding of the world, and the necessity of invoking a designer to explain the contrivance of its parts and their orderly working. In these respects his writing was a distinct Newtonian contribution to natural theology: the attempt to deduce knowledge about the attributes and actions of God from facts about the natural world.[8] In the *General Scholium,* Newton remarked on the disposition of the solar system:

> The six primary planets revolved about the sun in circles concentric with the sun, and with motions directed towards the same parts, and almost in the same plane. Ten moons are revolved about the earth, Jupiter, and Saturn, in circles concentric with them, with the same direction of motion, and nearly in the planes of the orbits of those planets. Newton, 1713/1934, p. 543

And then, in the spirit of Oresme, he commented:

> . . . but it is not to be conceived that mere mechanical causes could give birth to so many regular motions . . . This most beautiful system of the sun, planets, and comets, could only proceed form the counsel and dominion of an intelligent and powerful Being . . . and on account of his dominion he is wont to be called *Lord God.* (Newton, 1713/1934, p. 544)

Newton put forward a version of Aquinas' design argument. Aquinas' individual natures have been replaced by laws, but there is still need for an initial coordinator. Newton said:

> The planets and comets will constantly pursue their revolutions in orbits given in kind and position, according to the laws above explained; but though these bodies may, indeed, continue in their orbits by the mere laws of gravity, yet they could by no means have at first derived the regular position of the orbits themselves from those laws (ibid).

Newton's concern was that although comets and planets both obey the law of gravitational attraction, chance could not explain the coplanar and regular motion of the planets; the law of attraction is consistent with the earth, and indeed all the planets, having comet-like trajectories.

In *Query 28* of the *Optiks* Newton advanced a version of the first cause argument, saying that "the main Business of natural Philosophy is to argue from Phaenomena without feigning Hypotheses, and to deduce Causes from Effects, till we come to the very first Cause, which certainly is not mechanical." This first nonmechanical cause is necessary to account for a variety of features of the universe: planets moving all in the same way and in orbits that are concentric and planar, the earth not moving like a comet, the fixed stars that do not fall in upon each other, etc. He said this first cause must be "incorporeal, living, intelligent, omnipresent"; and that although philosophy did not bring us to immediate knowledge of this cause, "yet it brings us nearer to it, and on that account is to be highly valued" (Newton, 1730/1979, p. 369)[9]

In *Query 31* of his *Optiks,* Newton conjoined his account of the macro world, the solar system, with his corpuscular account of the micro world. Matter, motion, and attraction are the key explanatory elements in both the macro and micro worlds. He stated:

> And thus Nature will be very conformable to her self and very simple, performing all the great Motions of the heavenly Bodies by the Attraction of Gravity which intercedes those Bodies, and almost all the small ones of their Particles by some other attractive and repelling Powers which intercede the Particles.

Newton then reminded the reader that God is necessary for getting everything going:

> The *Vis inertiae* is a passive Principle by which Bodies persist in their Motion or Rest, receive Motion in proportion to the Force impressing it, and resist as much as they are resisted. By this Principle alone there could never have been any Motion in the World. Some other Principle was necessary for putting Bodies into Motion. (Newton, 1730/1979, p. 397)

His argument then followed through with a claim that scandalized his continental colleagues (and anticipated the 19th century law of thermodynamics):

> ... and now they are in Motion, some other Principle is necessary for conserving the Motion. For from the various Composition of two Motions, 'tis very certain that there is not always the same quantity of Motion in the World ... By reason of the Tenacity of Fluids, and Attrition of their Parts, and the Weakness of Elasticity in Solids, Motion is much more apt to be lost than got, and is always upon the Decay. .... And if it were

not for these Principles, the Bodies of the Earth, Planets, Comets, Sun, and all things
in them would grow cold and freeze, and become inactive Masses. (Newton, 1730/
1979, pp. 397, 399)

Gottfried Leibniz (1646–1716) was Newton's greatest and most authoritative opponent. He was an enthusiastic supporter of the new science of Galileo and an opponent of Aristotelianism, but he could never sanction Newton's recourse to action at a distance or attraction across empty space. He thought this was a return to occult philosophy, and that Newton's assumptions about space and time were simply false. Of his own philosophical system, he said:

> This overthrows attractions, properly so called, and other operations inexplicable by natural powers of creatures; which kinds of operations, the assertors of them must suppose to be effected by miracles or else have recourse to absurdities, that is, to the occult qualities of the schools; which some men begin to revive under the specious name of forces; but they bring us back again into the kingdom of darkness. This is *inventa fruge, glandibus vesci* [to feed on acorns when corn has been discovered].
>
> In the time of Mr. Boyle, and other excellent men, who flourished in England under Charles II, no body would have ventured to publish such chimerical notions. (Alexander, 1956, p. 92)

Further, Leibniz's religious sensibilities were offended by Newton's image of God the clockmaker who periodically returned to his creation to tinker with it and put things back on the right path. His argument with Newton and Newtonianism was played out in a two-year-correspondence with Samuel Clarke (1675–1729), a friend, theologian, and disciple of Newton.[10] This is one of history's great intellectual arguments: The scientific, methodological and philosophical matters at stake were significant and the antagonists, Newton (through Clarke) and Leibniz, were intellectual giants.[11] Ernst Cassirer spoke of the dispute:

> The controversy between Newton and Leibniz is one of the most important phenomena in the history of modern thought. . . . They not only had different views on the nature and properties of God, on the structure of the material universe, the concepts of space and time, and on the possibility of *action at a distance* . . . it [the debate] is rather a collision between two fundamental philosophical methods. (Cassirer, 1943, p. 366)

The controversy began with a clock-invoking letter by Leibniz to Princess Carolina of Wales (who had known Leibniz when her husband, George II of England, was Electoral Prince of Hanover. Leibniz complained:

> Natural religion itself, seems to decay (in England) very much. Many will have human souls to be material: others make God himself a corporeal being. . . .
>
> Sir Isaac Newton, and his followers, have also a very odd opinion concerning the work of God. According to their doctrine, God Almighty wants to wind up his watch from time to time: otherwise it would cease to move. He had not, it seems, sufficient foresight to make it a perpetual motion. Nay, the machine of God's making, is so imperfect, according to these gentlemen, that he is obliged to clean it now and then by an extraordinary concourse, and even to mend it, as a clockmaker mends his work; who must consequently be so much the more unskillful a workman, as he is oftner

> obliged to mend his work and to set it right…[Newton and his followers] must needs
> have a very mean notion of the wisdom and power of God. (Alexander, 1956, p. 12)

Clarke concurred with Leibniz's clockwork image but drew different theological conclusions from it. In his first reply to Leibniz, he observed that we think the better of artisans if their work does not require subsequent attention, but continues to work harmoniously without interference, then he added:

> But with regard to God, the case is quite different; because he not only composes or
> puts things together, but is himself the author and continual preserver of their original
> forces or moving powers: and consequently 'tis not a diminution, but the true glory of
> his workmanship, that nothing is done without his continual government and
> inspection. The notion of the world's being a great machine, and going on without the
> interposition of God, as a clock continues to go without the assistance of a clockmaker;
> is the notion of materialism and fate, and tend, (under pretence of making God a
> *supra-mundane intelligence*,) to exclude providence and God's government in reality
> out of the world. And by the same reason that a philosopher can represent all things
> going on from the beginning of creation, without any government or interposition of
> providence; a sceptic will easily argue still farther backwards, and suppose that things
> have from eternity gone on (as they now do) without any true creation or original
> author at all, but only what such arguers call all-wise and eternal nature. (Alexander,
> 1956, p. 14)

Leibniz is not so easily put off. In his second letter he stated:

> He who buys a watch, does not mind whether he got the several parts made by others,
> and did only put them together; provided the watch goes right. And if the workman
> had received from God even the gift of creating the matter of the wheels; yet the
> buyer of the watch would not be satisfied, unless the workman had also received the
> gift of putting them well together. In like manner, he who will be pleased with God's
> workmanship, cannot be so, without some other reason than that which the author has
> here alleged. . . .
>
>   I do not say, the material world is a machine, or watch, that goes without God's
> interposition; and I have sufficiently insisted, that the creation wants to be continually
> influenced by its creator. But I maintain it to be a watch, that goes without wanting to
> be mended by him: Otherwise we must say, that God bethinks himself again. No; God
> has foreseen everything; . . . there is in his works a harmony, a beauty, already pre-
> established. (Alexander, 1956, p. 18)

The exchange continued, with Leibniz contending, among other things, that the basic philosophical principle of sufficient reason[12]—"one of the most certain both in reason and experience; against which no exception or instance can be alleged" (Alexander, 1956, p. 74)—ruled out Newton's conception of space and time as absolute. If time is absolute, then there is no sufficient reason why God should have started the world at one time rather than another (Alexander, 1956, pp. 26–27). The physically crucial point was brought up by Leibniz as to whether the "imperfection" in the world was a loss of motion or a loss of force (Alexander, 1956, p. 88). Leibniz agreed with Newton that the former decayed, but he insisted that a distinction between motion (momentum) and force (energy) should be made, and that the latter was forever constant from the instant God first "energized" the world. Leibniz closed his correspondence and his case by reverting to the clock analogy with which he began:

> I maintained that the dependence of the machine of the world upon its divine author, is rather a reason why there can be no such imperfection in it; and that the work of God does not want to be set right again; that it is not liable to be disordered; and lastly, that it cannot lessen in perfection. (Alexander, 1956, p. 89)

God as Clockmaker

By the early 1700s, the design argument, specifically the clockwork version of it, was the stock in trade of all Christian apologists. Within a few years of the publication of Newton's *Principia*, the botanist and zoologist, John Ray, published his influential *The Wisdom of God Manifested in the Works of the Creation* (1691). The book's subtitle indicated it dealt with:

> The Heavenly Bodies, Elements, Meteors, Fossils, Vegetables, Animals (Beasts, Birds, Fishes, and Insects), more particularly . . . the Body of the Earth, its Figure, Motion, and Consistency, and in the admirable structure of the Bodies of Man, and other Animals; as also their Generation, etc., With Answers to some Objections. (Pedersen, 1992, p. 69)

This book set the tone and style for 200 years of natural theology.

Robert Jenkin, in his *The Reasonableness and Certainty of the Christian Religion* (1700), stated the widely felt view that Christian belief was supported by the discoveries of the new science:

> Indeed infidelity could never be more inexcusable than in the present age, when so many discoveries have been made in natural philosophy, which would have been thought incredible to former ages, as any thing perhaps that can be imagined, which is not a downright contradiction. That gravitating or attractive force, by which all bodies act one upon another, at ever so great a distance, even through a vacuum of prodigious extent, lately demonstrated by Mr. Newton; the Earth, together with the planets, and the sun and stars being placed at such distances, and disposed of in such order, and in such a manner, as to maintain a perpetual balance and poise throughout the universe, is such a discovery, as nothing less than a demonstration could have gained it any belief. And this system of nature being so lately discovered, and so wonderful, that no account can be given of it by a hypothesis in philosophy, but it must be resolved into the sole Power and good pleasure of Almighty God. (In Dillenberger, 1961, p. 148)

William Derham, rector of Upminster, fellow of the Royal Society, and author of the first English horological manual (*The Artificial Clock-Maker* (1696), published in 1715 the influential *Astro-Theology; or, A Demonstration of the Being and Attributes of God from a Survey of the Heavens.* He argued:

> If we consider that those Motions are wisely ordered and appointed, being as various,

and as regular and every way nicely accomplished, as the World and its Inhabitants have occasion for. This is a manifest sign of a wide and kind, as well as omnipotent CREATOR and ORDERER of the World's affairs, as that of a clock or other Machine is of Man. (In Macey, 1994, p. 663)

Robert Boyle (1627–1691), a trenchant anti-Aristotelian for whom self-moving natures were a slight upon the dominion of God,[13] thought of himself as equally a natural philosopher (scientist) and a theologian. He wrote extensively in both fields. He used the clockwork analogy when describing the world, but he expressed some ambivalence about its adequacy; it was perhaps more a metaphor than an analogy. Boyle was more circumspect than Derham and other contemporaries about inferring the existence of a heavenly clockmaker from the motions of the planets and stars. He wrote:

> The situation of the Celestial bodies afford not such strong arguments for the wisdom and design of God, as the bodies of animals and plants; for there seems more admirable contrivance in the muscles of a man's body, than in the celestial orbs; and the eye of a fly seems a more curious piece of Work than the body of the sun. (*Theological Works*, vol. II, p. 220; in Dillenberger, 1961, p. 115)

David Hume's (1711–1776) circumspection was so great that it amounted to an iconoclastic rebuttal of the whole design argument. This was put forward in his *Dialogues Concerning Natural Religion* (Hume, 1779/1963) where he criticized premise 2 of Oresme's argument (p. 220), saying although it made sense to compare one thing to another where *both* were examinable, it made no sense to do so where one of the pair was not independently examinable. In the second case, the analogy could only be weak, and not sustain any inferences. Further, there were lots of things in the world that did not work well–plagues, for instance—so why not have a design argument to an incompetent and malevolent supreme being?

William Paley (1743–1805), perhaps the most popular proponent of the design argument,[14] was little bothered by Hume's critique. Paley concluded his 600 page *Natural Theology; or Evidence for the Existence and Attributes of the Deity Collected from the Appearances of Nature* (Paley, 1802/1964) with this succinct argument:

> . . . after all the schemes and struggles of a reluctant philosophy the necessary resort is to Deity. The marks of design are too strong to be got over. Design must have a designer. That designer must have been a person. That person is God.

Paley's argument opened with the well known story of a person finding a watch on a lonely heath.[15] He says:

> In crossing a heath, suppose I pitched my foot against a *stone*, and were asked how the stone came to be there; I might possibly answer, that, for any thing I knew to the contrary, it had lain there for ever . . . But suppose I had found a *watch* upon the ground, and it should be enquired how the watch happended to be in that place; I should hardly think of the answer which I had before given, that for anything I knew, the watch might have always been there. Yet why should not this answer serve for the

watch as well as for the stone? . . . For this reason, and for no other, viz. That, when
we come to inspect the watch, we perceive (what we could not discover in the stone)
that its several parts are framed and put together for a purpose.

Paley also followed Boyle in being circumspect about the world as purposive:

My opinion of astronomy has always been that it is *not* the best medium through
which to prove the agency of an intelligent Creator, but that this being proved [by
other means], it shows beyond all other sciences, the magnificence of his operations.
(Paley, 1802/1964, chap. xxii)

His reasons for circumspection were these:

Now we deduce design from relation, aptitude, and correspondence of parts. Some
degree, therefore, of *complexity*, is necessary to render a subject fit for this species of
argument [the design argument]. But the heavenly bodies do not, except perhaps in
the instance of Saturn's ring, present themselves to our observation as compounded
of parts at all. (Paley, 1802/1964, chap. xxii)

Newton had believed that *both* astronomical and biological realms revealed
the handiwork of the Creator. In his early notebooks, composed when he was 22
years old, Newton wrote:

Were men and beasts made by fortuitous jumblings of atoms there would be many
parts useless in them, here a lump of flesh there a member too much. Some kinds of
beasts might have had but one eye some more than two. (Christianson 1996, p. 28)

Paley felt the design argument was on firmer ground when *only* biological
matters were attended to. Paley thought, as did Newton and Boyle a hundred
years earlier, that a close examination of the eye was a cure for atheism:

Were there no example in the world of contrivance except that of the *eye*, it would be
alone sufficient to support the conclusion which we draw from it, as to the necessity
of an intelligent Creator . . . Its coats and humours, constructed, as the lenses of a
telescope are constructed, for the refraction of rays of light to a point, which forms
the proper action of the organ; the provision in its muscular tendons for turning its
pupil to the object, similar to that which is given to the telescope by screws; . . . these
provisions compose altogether an apparatus, a system of parts, a preparation of means,
so manifest in their design, so exquisite in their contrivance, so successful in their
issue, so precious and so infinitely beneficial in their use, as, in my opinion, to bear
down all doubt that can be raised on the subject. (Paley, 1802/1964, chap. vi)

More generally, he thought the whole of nature testified to a beneficent Designer. In a lovely passage, so evocative of his outlook, he wrotes:

It is a happy world after all. The air, the earth, the water, team with delighted existence.
In a spring noon, or a summer evening, on whichever side I turn my eyes, myriads of
happy beings crowd on my view. "The insect youth are on the wing." Swarms of new-
born *flies* are trying their pinions in the air. Their sportive motions, their wanton mazes,
their gratuitous activity, their continual change of place without use or purpose, testify
their joy, and the exultation which they feel in their lately discovered faculties. A *bee*
amongst the flowers in spring, is one of the most cheerful objects that can be looked
upon. (Paley, 1802/1964, chap. xxvi)

Some 20 years after Paley, the *Bridgewater Treatises*, published during the 1830s, represented the final serious attempt to develop a natural theology from considerations of the structure and function of nature.[16] Darwin's scientific vision of "nature red in tooth and claw," and millenia of slow mindless adaptations, gradually drew the educated curtain on Paley's halcyon spring vision and his esteem of the eye. But prior to Darwin, it is easy to see how powerful was the effect of Paley's design argument.

Poetry and the Design Argument

John Donne, as we have seen, gave poetic expression to the sense of foreboding that the collapse of the ordered Aristotelian worldview provoked in many: "And new philosophy calls all indoubt . . . t'is all in pieces. . . . " But poets, novelists and essayists also gave more hopeful expression, through the new genre of physicotheological writing, to the new science, and especially to its design argument.[17] Where self-moving and self-perfecting natures brought order and coordination to the Aristotelian world, it was the sense of a divine clockwork, an inherent mechanism, that gave order to the post-Aristotelian world. In the 17th and 18th centuries there were no sharp boundaries between clergyman, theologian, scientist, philosopher, and poet; the "two cultures" of C.P Snow had not taken root. The contributors to the new genre frequently were all of these things.

Joseph Addison's *Ode* (1712) illustrates this genre:

> The spacious firmament on high,
> With all the blue ethereal sky,
> And spangled Heavens, a shining frame,
> Their great original proclaim.
> Th' unwearied Sun, from day to day,
> Does his Creator's pow'r display;
> And publishes, to every land,
> And work of an almighty hand.
>
> Soon as the evening shades prevail,
> The moon takes up the wond'rous tale;
> And nightly, to the listening Earth
> Repeats the story of her birth:
> Whilst all the stars that round her burn,
> And all the planets, in their turn,
> Confirm the tidings as they roll,
> And spread the truth from pole to pole.

> What though, in solemn silence, all
> Move round that dark terrestrial ball?
> What though no real voice nor sound
> Amidst their radiant orbs be found:
> In reason's ear they all rejoice,
> And utter forth a glorious voice;
> For ever singing as they shine:
> 'The hand that made us is divine.'
>   (Heath-Stubbs and Salman, 1984, p. 129)

Shaftesbury Thomson's *The Seasons* (1730) is in four long parts written over a decade. *Winter* concluded with the ringing lines:

> Nature! great parent! whose directing hand
> Rolls round the seasons of the changeful year,
> How mighty! how majestick are thy works!
> With what a pleasing dread they swell the soul,
> That sees, astonish'd! and, astonish'd sings!
>   (Jones, 1966, p. 109)

While the conclusion of *Autumn* distilled the way in which the new science revealed God:

> Oh Nature! All-sufficient! Over all!
> Enrich me with the knowledge of thy works!
> Snatch me to Heaven; thy rolling wonders there,
> World beyond world, in infinite extent,
> Profusely scatter'd o'er the void immense,
> Shew me; their motions, periods, and their laws
> Give me to scan; thro' the disclosing deep
> Light my blind way: the mineral strata there;
> Thrust, blooming, thence the vegetable world;
> O'er that the rising system, more complex,
> Of animals; and higher still, the mind,
> The varied scene of quick-compounded thought.

Henry Baker's *The Universe: a Poem intended to Restrain the Pride of Man* (1734) is another nice example of the genre, an example which indicates the cultural roots and effects of the design argument. Baker was a scientist, and author of several books on the microscope. In *The Universe* there are copious notes about Boyle, Leuwenhoek, Ray, Hooke, and Huygens. Baker began with the heavens, writing:

> Along the sky the sun obliquely rolls,
> Forsakes, by turns, and visits both the poles.
> Diff'rent his track, but constant his career,
> Divides the times, and measures out the year...
> Bring forth thy glasses: clear thy wond'ring eyes:
> Millions beyond the former millions rise:
> Look further:—millions more blaze from remoter skies.

Of the other end of the Great Chain of Being, he wrote:

> Alas, what's man, thus insolent and vain?
> One single link of nature's mighty chain.
> Each hated toad, each crawling worm we see,
> Is needful to the whole as well as he.
> Like some grand building in the universe,
> Where ev'ry part is useful in its place.
>   (Jones, 1966, p. 127)

In case the argument got clouded by the verse, Baker had a footnote where he said: "The Deity necessarily inferred from the contemplation of every object, but more especially visible in the animate creation, so infinitely diversified in the several species and kinds of fish, reptiles, quadrupeds, insects, and birds."

Alexander Pope's *Essay on Man* (1733) was a major work in the genre. He defended divine design, even when it is not so apparent, and warned against too hasty judgments of disorder:

> All Nature is but Art, unknown to thee;
> All chance, direction which thou canst not see;
> All discord, harmony not understood;
> All partial evil, universal good.

Even when appearances are chaotic and anarchic, nevertheless

> All are but parts of one stupendous whole,
> Whose body Nature is, and God the soul;
> That, chang'd thro' all, and yet in all the same;
> Great in the earth, as in th'ethereal frame,
> Warms in the sun, refreshes in the breeze,
> Glows in the stars, and blossoms in the trees,
> Lives thro' all life, extends thro' all extent,
> Spreads undivided, operates unspent ...
> To him no high, no low, no great, no small;
> The fill, he bounds, connects, and equals all.
>   (Jones, 1966, p. 141)

Not all the poets and writers eulogized the clock and its attendant notion of clockwork people and an ordered clockwork universe (Mayr, 1986). For the Romantic poets, the clockwork metaphor was generally pejorative. They rebeled against the notion of a mechanized world, and the reductive idea of people as machines. Jonathon Swift (1667–1745), dean of St. Patrick's Dublin and great satirist, lived and wrote through the height of the horological revolution. Swift's *Gulliver's Travels* was a scarcely concealed attack upon the supposed horological triumphs of the period. In Lilliput, Swift had one of the Lilliputians describe Gulliver's watch:

> *Out of the right Fob hung a great Silver Chain, with a wonderful kind of Engine at the Bottom. We directed him to draw out whatever was at the End of that Chain; which appeared to be a Globe, half Silver and half of some transparent Metal...He put this Engine to our Ears, which made an incessant Noise like that of a Water-Mill. And we conjecture it is either some unknown Animal, or the God he worships.* (Macey, 1980, p. 162)

Later, William Blake (1757–1827), in his *Jerusalem*, contrasted the pastoral hourglass to the new clock which he saw as symbolic of the dark satanic mills of Albion or England:

> And all the Arts of Life they changed into the Arts of Death in Albion.
> The hour-glass contemnd because its simple workmanship
> Was like the workmanship of the plowman, & the water wheel,
> That raises water into cisterns: broken & burned with fire:
> Because its workmanship was like the workmanship of the shepherd.
> And in their stead, intricate wheels invented, wheel without wheel:
> To perplex youth in their outgoings, & to bind to labours in Albion
> Of day & night the myriads of eternity that they may grind
> And polish brass & iron hour after hour.
>
> (Macey, 1980, p. 205)

METAPHOR IN SCIENCE AND IN EDUCATION

The clock metaphor so permeated the debate between Newton and Leibniz, the conceptions of the universe put forward by Oresme and Kepler, the vision of man articulated by Comenius, Boyle, Hobbes, that it is difficult to see what would be left of these debates, conceptions and visions, of the mechanical world view itself, if the metaphor were taken out. This raises the more general question of the role of metaphor in science.[18]

Newton's *Principia* exemplified the problem. He went to great length in the introductory definitions to point out his treatment of motion would be a mathematical treatment:

> The words "attraction," "impulse" or any "propensity" to a centre, however, I employ indifferently and interchangeably, considering these forces not physically but merely mathematically. The reader should hence beware lest he think that by words of this sort I anywhere define a species or mode of action, or a physical cause or reason.

Having derived the mathematical relation between bodies, he gave it a commonplace name—"attraction"—because that was the only way to make the relation intelligible, but stressed that this was to be understood metaphorically. It is well known how his followers and the originators of the corpuscularian worldview lost sight of the metaphorical quality of Newton's descriptions. Newton's correspondence with Richard Bentley, the first Boyle lecturer, is in part concerned with pulling Bentley back from his literalist, nonmathematical interpretation of the *Principia*'s definitions. Newton, as we have seen, urged caution:

> You sometimes speak of gravity as essential and inherent to matter. Pray do not ascribe that notion to me, for the cause of gravity is what I do not pretend to know and therefore would take more time to consider it. (Thayer, 1953, p. 53)

John Locke (1632–1704), the philosopher and self-styled "underlabourer" in Newton's garden, struggled over the same matters as Bentley. In the first three editions of his famous *Essay Concerning Human Understanding* (1690, 1694, 1695) Locke enunciated the basic tenets of the mechanical worldview, saying "bodies operate one upon another . . . by impulse, and nothing else." Increasingly he came to appreciate the contradiction between these tenets and the foundations of the *Principia*, which had been published in 1687. In a letter to Stillingfleet he admitted:

> It is true, I say, "that bodies operate by impulse, and nothing else." And so I thought when I writ it, and can yet conceive no other way of their operation. But I am since convinced by the judicious Mr. Newton's incomparable book, that it is too bold a presumption to limit God's power, in this point, by my narrow conceptions. The gravitation of matter towards matter, by ways inconceivable to me, is not only a demonstration that God can, if he pleases, put into bodies powers and ways of operation, above what can be derived from our idea of body, or can be explained by what we know of matter, but also an unquestionable and every where visible instance, that he has done so. And therefore in the next edition I shall take care to have that passage rectified. (Stein, 1990, p. 32)

Book Two of the fourth edition of the *Essay* (1700) contained these changes, which were the first changes within philosophical atomism, or the mechanistic world view, prompted by modern science.[19] This is another example of the interaction of physics and metaphysics.

The problem with which Newton grappled in the 17th century was felt more acutely after the scientific advances of the 19th and early 20th centuries. Arthur Eddington said, in 1928, that

> It is difficult to school ourselves to treat the physical world as purely symbolic. We are always relapsing and mixing with the symbols incongruous conceptions taken

from the world of consciousness. Taught by long experience we stretch a hand to grasp the shadow, instead of accepting its shadowy nature. Indeed, unless we confine ourselves altogether to mathematical symbolism it is hard to avoid dressing our symbols in deceitful clothing. When I think of an electron there rises to my mind a hard, red, tiny ball. (Eddington, 1928/1978, p. xvii)

One aid to untangling the conceptual issues is to distinguish the realms of the natural world and the theorized world with which science deals. Eddington confused the mathematical nature of the latter with the supposed mathematical nature of the former. This comment was made in the context of Eddington's idealist worldview, but the problem of clothing the unknown in comfortable and familiar garb—the problem of speaking metaphorically—is just as pressing for realists.

Darwinian theory with its "struggle for existence," "war of nature," "tree of life," and "survival of the fittest" is full of metaphorical expression that can aid and also deflect understanding. Darwin's reconciliation of his gradualist assumptions with the abrupt breaks found in the fossil record, by use of the analogy of the damaged book with chapters, pages, paragraphs, sentences and words missing, is an outstanding instance of argument by analogy in science.

How much science could proceed without metaphor is a moot point. Mach saw metaphor as indispensible for the growth and communication of science, but regarded it as only an aid to understanding that could be jettisoned:

What a simplification it involves if we can say, the fact A now considered comports itself, not in one, but in many or in *all* its features, like an old and well-known fact B . . . light like a wave motion or an electric vibration; a magnet, as it were laden with gravitating fluids. (Mach, 1894/1986, p. 240)

As with theories more generally, Mach denied any truth value to metaphorical description; such description had only heuristic value. Max Black took a similar view: "Perhaps every science must start with metaphor and end with algebra; and perhaps without the metaphor there would never have been any algebra" (Black, 1962, p. 242). An alternate view might be to accept there is a continuum between description, representation, and depiction. And consequently a continuum between description, model, metaphor, analogy, theory and so on. They are all ways of representing something about the world, and thus are all to some degree making truth claims about the world. This view is consistent with what Wartofsky calls a modest realism (Wartofsky, 1979a).

The use of metaphor and analogy in science teaching is well researched (Duit, 1991; Lawson, 1993; Dagher, 1994). What is often overlooked in this research is the degree to which metaphor is present in the content of scientific theory. In science teaching, we are frequently explaining one metaphor in terms of another; it is not pure description that then is explained metaphorically for pedagogical reasons. The interesting epistemological question is just how this metaphorical language functions in scientific theory. Is the clockwork universe just a way of stating the determinist principle? Is it a metaphysical commitment that lies underneath scientific principles? Does it just have pedagogical or communicative value? Is it an organizing principle that guides investigations, but itself makes no claims about the world? Parallel to this is the question of how metaphor and analogy function in scientific theorizing, or scientific thinking. That is, metaphor figures in the *products* of science (the clockwork universe, the struggle for existence, electron shells, etc.), but metaphorical and analogical thinking also figures in the *process* of science. Understanding this process is important for pedagogy.[20]

Metaphor is at the core of education and instruction. Although in scientific theorizing, we explain the known in terms of the less known, and even in terms of the unknowable—inertia as an explanation of why people fall forward when the bus stops, differential diffraction as an explanation of the rainbow, recessive genes as an explanation for premature balding, and so on—we nevertheless try to make sense of these explanations by analogies and metaphors. As Mach recognized, and as Aristotle did long before Mach, we try to clothe the unknown in the garb of the known; we relate the new to the old. This is a quite general point about education and the development of children's cognition. Lee Shulman, when elaborating his notion of pedagogical content knowledge (the knowledge teachers have of subject matter that goes beyond merely the subject matter and includes how that subject matter is made intelligible), wrote:

> Within the category of pedagogical content knowledge I include, for the most regularly taught topics in one's subject area, the most useful forms of representation of those ideas, the most powerful analogies, illustrations, examples, explanations, and demonstrations—in a word, the ways of representing and formulating the subject that make it comprehensible to others. (Shulman, 1986, p. 9)

Some metaphors are so powerful that they become part of a culture's self-understanding, or indeed the shared cognitive currency of many cultures: the "iron curtain," the "global village," the "dark ages," the "white man's burden," "class struggle," "cultural capital," "Asian tigers," "open mind," "even playing fields," and so on. Such metaphors form or deform ways of thinking about people, nature, and society. Kieran Egan, in his book *The Educated Mind*, discussed metaphor's role in the learning of language and the development of children's understanding and emphasizes that:

> Metaphor is one of our cognitive grappling tools; it enables us to see the world in multiple perspectives and to engage with the world flexibly. Metaphor is much more profoundly a feature of human sense-making than the largely ornamental and redundant poetic trope some have taken it to be. (Egan, 1997, p. 58)

The clockwork metaphor is itself subject to historical and cultural contingencies. From Oresme in the 14th century, Descartes in the 16th, Paley in the 19th, and up to Burgess' *Clockwork Orange*, it has had a long and noble life. Through to the 1950s and even 1960s teachers could confidently expect their students would be familiar with mechanical clockwork. Most students had one ticking away alongside their bed, most homes had one on the mantelpiece, domestic grandfather (pendulum) clocks were not uncommon, and students had mechanical watches on their wrists. Most people had seen clock mechanisms. In the 1990s all this has changed. Battery-operated, digital clocks, have almost completely replaced mechanical clocks. In 50 years time, the clockwork metaphor will itself be in need of metaphorical or analogical explanation to a generation that has never seen the instrument that Galileo, Huygens, and Hooke bequeathed to the world. Students will not be asking of their peculiar friends: "What makes him (or her) tick?" The tick of clockwork will have long ceased to be heard.

CHAPTER 10

Science and Philosophy
Some Lessons from the History of Pendulum Motion

The history of pendulum motion study well illustrates the intimate connection between philosophy and science. This connection is not often enough stressed in either science or philosophy courses. There has been a tendency for histories of science to be written as if science were not influenced by philosophy, and on the other hand there has been a tendency for histories of philosophy to be written as if science had no impact on philosophical ideas. Students in philosophy courses standardly deal with Bacon, Locke, Descartes, Leibniz, Berkeley, Hobbes, Hume, Kant, Hegel, and other philosophers but pay little attention to the scientific developments to which they were reacting.[1]

After indicating a little of the interactive history of science and philosophy, this chapter will show how the evolution of pendulum motion study bears on some central questions in philosophy of science: the definition of time, the relationship of theory to observation and experiment, the distinction between real and theorized objects, scientific laws, and objectivist accounts of science. The science of the pendulum, and time measurement, can connect with engaging philosophical investigations of time, such as those of van Fraassen (1970), Whitrow (1972), Fraser (1987), Hawking (1988), Raju (1994), and Price (1996).

From the ancient Greeks to the present, science has been interwoven with philosophy. Science, metaphysics, logic, and epistemology have been inseparable. Most of the great scientists (Democritus, Aristotle, Copernicus, Galileo, Descartes, Newton, Leibnitz, Boyle, Faraday, Darwin, Mach, Einstein, Planck, Heisenberg, Schrödinger) were at the same time philosophers. Scientists were called "natural philosophers" through to the end of the 19th century, the term "scientist" being introduced by William Whewell only in 1834. Einstein spoke of the theoretical physicist as "a philosopher in workingman's clothes" (Bergmann, 1949, p. v). Charles Sanders Peirce said in his "Notes on Scientific Philosophy," "Find a scientific man who proposes to get along without any metaphysics . . . and you have found one whose doctrines are thoroughly vitiated by the crude and uncriticised metaphysics with which they are packed."

Metaphysical issues naturally emerged from the subject matter of science: the Aristotle–Galileo controversy over final causation, the Galileo–Kepler controversy over the lunar theory of tides, the Newtonian–Cartesian argument over action at a distance, the Newton–Berkeley argument over the existence of absolute space and time, the Newton–Fresnel argument over the particulate theory of light, the Paley–Darwin argument over design and natural selection, the Mach–Planck argument over the realistic interpretation of atomic theory, the Einstein–Copenhagen dispute over the deterministic interpretation of quantum theory. These all brought to the fore metaphysical issues. Metaphysics is inherent in science.[2]

Science has always been conducted within the context of the philosophical ideas of the time. This is to be expected. Scientists think, write, and talk with the language and conceptual tools available to them. More generally, people who form opinions, are themselves formed in specific intellectual circumstances, and their opinions are constrained by these circumstances. Newton said he was able to see further than others because he stood upon the shoulders of giants. Without Copernicus, Kepler, and Galileo, not to mention Euclid's geometry, there would not have been a unified theory of terrestrial and celestial mechanics. It is often overlooked that a scientist's understanding of, and approach to, the world is formed by his or her education and milieu. This milieu is pervaded by the philosophies of the period. It is this truism that drove the poststructuralist mill of continental philosophy, and provided intellectual succor to those philosophers who attacked the supposed Cartesian–Kantian "myth of the subject." But the truism – that people who posit are themselves posited—is hardly the preserve of poststructuralists.

For example, from the 1950s to the 1980s learning theory was dominated by behaviorism. In many American universities, a Ph.D. in educational psychology required successful pigeon or rat training as a prerequisite skill. Educators' conceptualized learning in terms of concepts of "reinforcement," "stimulus," "stimulus generalisation," "response," "reward," "operant conditioning," "drive," and so on. At the time, this was how researchers thought, and how they looked upon classrooms and teaching situations. Because the education of scientists and social scientists, including educationalists, was so bereft of historical and philosophical input, most educators were ignorant of the philosophical and the ideological framework of behaviorism.[3] Most prided themselves on doing good science, in bringing the rigors of natural science into the human sciences. After the Bruner and Piaget inspired cognitive turn in psychology, and after Chomsky's savage review of Skinner's learning theory (Chomsky, 1959), academic behaviorism went into full retreat; although pockets did hold on, especially in Education Schools.

People who think, do so with concepts provided to them by their cultural and educational circumstance. And as Ernst Mach long ago observed, historical and philosophical knowledge assists people in recognizing their cultural, ideological, and ideational circumstances and being more critical of their influence. Mach, perhaps optimistically, spoke of historical knowledge setting the mind free. The lessons to be drawn from educators embrace of behaviorism in the 1950s and 1960s are relevant to today's educational embrace of constructivism. The same lack of historical and philosophical perspective traps educators into highly partisan, and very contestable, philosophical positions.[4] As with behaviorism, constructivism carries the imprint of its cultural circumstance. As one commentator observed:

> In sum, constructivism is largely a reflection of current American cultural beliefs and, as such, involves the development of instructional techniques that attempt to make the acquisition of complex mathematical skills an enjoyable social enterprise that will be pursued on the basis of individual interest and choice. (Geary, 1995, p. 32)

Science is a system of concepts, definitions, methodologies, problems, results, technologies, and professional organizations that predates the individual scientist. Inasmuch as the system of science embodies philosophical, ideological, and cultural suppositions, then the work of the scientist will be shaped by these suppositions. In Popperian, or objectivist, terms, this conceptualization of science gives priority to the third world of scientific theories, concepts, and conceptual schemes, over the second world experiences and beliefs of the scientist (Popper 1972); the latter are dependent on the former. This is one way in which Paulo Freire's, and earlier, Hegel's, comment, that the "we think," determines the "I think," can be interpreted. Of course it is only second world events (for instance, "long and hard thinking," as Newton said) that can extend or transform third world structures, but such second world events are ones that have been formed and conditioned by the third world itself. There is here an "essential tension," as Kuhn described it (Kuhn, 1959).

The connection of science with philosophy, broadly understood, is promoted in much present-day popular scientific literature. Books such as *God and the New Physics* (Davies, 1983), *The Mind of God* (Davies, 1992), *A Brief History of Time* (Hawking, 1988), *On Human Nature* (Wilson, 1978), *The Tao of Physics* (Capra, 1975), and *The Dancing Wu Li Masters* (Zukav, 1979) have been bestsellers and have conveyed, with varying accuracy, the fact that science affects and is affected by other disciplines (philosophy, psychology, theology, biology) and more generally the worldviews of a culture. The widespread impact of these books is comparable to that enjoyed in the interwar period by Arthur Eddington's *The Nature of the Physical World* (Eddington, 1928/1978), James Jeans' *Physics and Philosophy* (Jeans, 1943), and J. D. Bernal's *The Social Function of Science* (Bernal, 1939). Jeans, in the preface to his book, said:

> The aim of the present book is very simply stated; it is to discuss . . . that borderland territory between physics and philosophy which used to seem so dull, but suddenly became so interesting and important through recent developments of theoretical

physics. . . . The new interest extends far beyond the technical problems of physics and philosophy to questions which touch human life very closely. (Jeans, 1943/1981, p. i)

The pendulum falls somewhat short of "touching human life very closely," but its history can illustrate some of the borderland territory between physics and philosophy that Jeans was anxious to discuss. Classroom practical work with the pendulum can illustrate many key epistemological and methodological features of science: the relationship between events and their depiction, especially their mathematical depiction; problems of evidence and theory appraisal; the role of idealization in science; the dependence of experiment on theory and on technology, and so on.

The considerable literature generated by the philosophy for children movement suggests that children are both capable of and interested in pursuing elementary philosophical questions (Davson-Galle, 1990, Lipman, 1991, Lipman and Sharp, 1978). The science classroom provides opportunities to do this. The science must come first, but there is opportunity for children to encounter basic philosophical questions and to acquire some basic philosophical skills of an analytic and reasoning kind. Science is a made-to-order subject for children's initiation into philosophical—that is, reflective, critical, logical—thinking. The standard philosophical questions (What do you mean? How can we know?) can be encouraged at all levels of a science program. And increasingly in science–technology–society programs, the standard ethical question, "Should we do it?" can be asked. The pity is that teachers' training programs do not give teachers more familiarity and competence with these questions.

SCIENCE AND PHILOSOPHY IN THE STUDY OF MOTION

The pendulum was central to the new understanding of motion achieved in the 17th century. Understanding motion has been a central task of physics, and consequently of philosophy, from Aristotle through the medievals to Galileo, Descartes, and Newton, and up to Einstein and Whitehead.[5] The historian Herbert Butterfield, with justification, remarked:

> Of all the intellectual hurdles which the human mind has confronted and has overcome in the last fifteen hundred years, the one which seems to me to have been the most amazing in character and the most stupendous in the scope of its consequences is the one relating to the problem of motion. (Butterfield, 1949, p. 3)

Aristotle's account of motion is a clear example of the influence of metaphysics on the development of physics. Aristotle's root metaphysical conception is of an ordered cosmos in which natural motions are purposeful. His universe is intelligible and teleological.[6] Collingwood spoke of Aristotle's world view:

> The world of nature is thus for Aristotle a world of self-moving things . . . It is a living world: a world characterized not by inertia, like the world of 17th-century matter, but by spontaneous movement. Nature as such is process, growth, change. This process is a development, i.e. the changing takes successive forms α, β, γ, . . . in which each is the potentiality of its successor. (Collingwood, 1945, p. 82)

The Natural Versus Violent Motion Distinction

This conception led Aristotle to a second-order distinction between natural and violent motions, the former flowing from the "nature" of the thing, the latter occurring because of disruptive, external, nonnatural causes. This distinction in turn affected how an everyday phenomenon, such as projectile motion, was conceptualized. The metaphysics became a heuristic for the physics; it created the questions, problems, and puzzles that the physics then sought to answer.[7]

Of course, having this class of privileged basic motions (natural motions) for which no physical explanation is required, is not unique to Aristotle. For Galileo, circular motion (both terrestrial and celestial), and constant downward motion of heavy bodies, was natural.[8] For Newton and Descartes, non-accelerating linear motion was natural.[9] For Einstein, motion along geodesic lines could be seen as natural.[10] The same privileging occurred in social science. Consider that until very recent times, heterosexual behavior was regarded as *natural* and not in need of further explanation, while homosexual behavior was regarded as deviant and *unnatural* and hence in need of explanation of some kind, be it genetic, social, or moral. Identifying these classes of natural events, in physical and social science, and indicating their connection to basic cosmological, social, cultural, or physical assumptions, is an important philosophical and scientific task. "Problematizing the natural" might be a more post-modern way of expressing this task.

One philosopher, who after considering the assumptions behind Newton's laws of motion (where nonaccelerating motion is "natural" and in need of no explanation), concluded:

> We have, in fact, drawn certain lines and come to regard certain successions of states as natural and others as unnatural. But in the absence of any clear and objective reasons for drawing these lines as we do, we may wonder whether other lines could be drawn, or whether, indeed, it is necessary to draw any lines at all. (Ellis, 1965, p. 46)

Another philosopher, in commenting on Newton's mechanics, suggested a reason why classical mechanics drew the lines between natural and unnatural:

> The notion of "externally-acting forces" upon an "inert" matter would then seem to be limited to a certain anthropomorphism—that is, an imposition upon the world-picture of the experience of the macro-world, and in particular, the experience of human action, and the separation of the actor from the object of his action. (Wartofsky, 1979a, p. 99)

Hegel had much earlier criticized Newton's conception of "inert matter"—perhaps the philosophical cornerstone of the Mechanical World View—saying:

> It is essential to distinguish gravity from mere attraction . . . the primary essence of
> matter is that it has weight. This is not an external property which may be separated
> from it. Gravity constitutes the substantiality of matter, which itself consists of a
> *tendency* towards a centre which falls outside it. . . . the tendency towards the
> centre, . . . is immanent in matter. (Hegel, 1842/1970, p. 242)

Earlier this century Whitehead criticized the mechanistic assumptions of classical (Newtonian) science, saying: "The concept of the order of nature is bound up with the concept of nature as the locus of organisms in process of development" (Whitehead, 1925, p. 73). He rejected the lifeless and inert character of the Newtonian–Humean universe, and traced 17th century mechanist thought to theological commitments: "The Protestant Calvinism and the Catholic Jansenism exhibited man as helpless to co-operate with Irresistible Grace: the contemporary scheme of science exhibited man as helpless to co-operate with the irresistible mechanism of nature" (Whitehead, 1925, p. 75). Whitehead says then of his own work that: "I developed a parallel line of argument, which would lead to a system of thought basing nature upon the concept of organism, and not upon the concept of matter" (Whitehead, 1925, p. 75).

This is not the place to appraise Whitehead's account of the origin of philosophical mechanism, or the cogency of his teleological and organismic science.[11] He is mentioned, as was Hegel, simply to highlight the fact that *some* form of a natural–violent motion distinction is pervasive in science, and that the "cut" between natural and violent is subject to *intra*scientific and *extra*scientific considerations. Marx Wartofsky observed of the 17th century's debate about the nature of matter that:

> the classical characterization of matter as *inert* formed the philosophical basis of much
> of 17th Century physical science. At the same time, the counter-tradition viewed matter
> as self-active or in self-motion. . . . these two contradictory metaphysical views still
> influence contemporary physics and philosophy of science, though in entirely novel
> ways. (Wartofsky, 1979b, p. 93)

while Nancy Cartwright has commented:

> My own views are that the concept of *natures* plays a special role in modern physics.
> Laws like $f = ma$ typically express how it is in the nature of the relevant properties to
> behave, and that in general is how they *will* behave *in ideal conditions*, i.e., in whatever
> conditions are appropriate for the expression of their nature. (Cartwright, 1993, p.
> 272)

Perhaps the most general problem in the study of motion is that a *mathematical* treatment of the problem abstracts from ontological, and possibly teleological, aspects of motion. This was basically the Aristotelian reason for separating physics from mathematics. The latter was said to deal with abstract bodies, the former with real ones. Real bodies had properties that mathematics could not, in principle, elucidate. The objection is similar to that made by cognitive psychologists about behaviorism: The mathematical description of behavior left aside the mental goings-on (decisions, intentions, understandings) which were key to the behavior being observed. These Aristotelian arguments against the reduction of science to mathematics are being restated by contemporary feminists, among others.

EPISTEMOLOGY AND THE ANALYSIS OF PENDULUM MOTION

Most science programs aspire to having students appreciate something of the "big picture" of science, something of the nature of science. An important part of the "big picture" is the methodology of science—not the method (how to get data or results), but how to interpret this data, how to appraise its bearing upon claims to knowledge. This is the epistemological forest that needs to be seen after the individual scientific trees have been examined. The 17th century debates about the pendulum provide exemplary material for epistemological discussion. Understanding the scientific revolution is especially important for theories of knowledge or for epistemology. An epistemology that pays no attention to the scientific revolution, or is at odds with its achievements and its methodological innovations, is at best ill-nourished and at worst irrelevant. This, unfortunately has been the case with a good deal of epistemology in the analytic tradition of philosophy.

This view of the relationship between epistemology and science has a long and distinguished heritage, going back at least to Bacon, Spinoza and Locke in the 17th century, and including Kant in the 18th century, Whewell in the 19th century, and Popper and many others in the 20th century. All these philosophers thought it incumbent upon them to articulate their theories of knowledge in the light of their understanding of the new science of Galileo and Newton. Karl Popper is perhaps the best known 20th-century advocate of the position, saying that:

> The central problem of epistemology has always been and still is the problem of the growth of knowledge. *And the growth of knowledge can be studied best by studying the growth of scientific knowledge.* (Popper, 1934/1959, p. 15)

Hilary Putnam, while denying that science has a monopoly on human knowledge, maintained that: "scientific knowledge is certainly an impressive part of our knowledge, and its nature and significance have concerned all the great philosophers at all interested in epistemology" (Putnam, 1978, p. 20).

Putnam expressed the *moderate* position on the relationship between science and epistemology. This position does not claim that science has a monopoly on human knowledge, or that all areas of knowledge have to conform to scientific standards to be seriously thought of as knowledge. This latter view, the *extreme* position, was held by a number of logical positivists. Alfred Ayer gave it its sharpest, and most public expression, in his *Language, Truth and Logic* (Ayer 1936).[12] He adopted both Hume's classification of statements into empirical and logical categories, and the positivist criterion of meaning, with a result that putatively empirical statements were only meaningful if some empirical state of affairs was relevant to its truth or falsity. On this basis ethics, theology, politics, aesthetics, metaphysics, and much else, was simply meaningless. This left science as pretty much the sole occupant of the epistemological high ground.[13]

Having this nexus between epistemology and science and, subordinately between epistemology and the history of science, is not, of course, without its problems. It is easy to misunderstand the science from which one is to draw epistemological lessons; the history itself is affected by prior methodological commitments of the historian; the scientific episode examined may not be representative of science; and, finally, science itself moves on, its methods and methodologies change, and philosophers can be left drawing epistemological lessons from an obsolete science.[14]

These difficulties are pronounced when delineating the methodological achievements of the scientific revolution and trying to have epistemological theories fit those achievements. There are many "readings" or interpretations of the scientific revolution.[15] And grasping the mix of experiment, mathematics, *apriorism*, philosophy and empiricism that characterized the achievements of Galileo and Newton is notoriously difficult. Bacon, for example, did not think Galileo had much to offer the new science, and Spinoza valued Descartes' mechanics far more than Galileo's.

There has been, predictably enough, a large element of pygmalian projection onto Galileo. Many have noted that a good clue to a commentator's own epistemology is the epistemology they attribute to Galileo.[16] The historian Alistair Crombie, in accepting the 1968 Galileo Prize given by the Domus Galileiana Pisa, made this remark:

> Galileo has been described as a cultural symbol, transcending history. Rather it seems to me that his reputation illustrates the universal human habit of creating myths to justify attitudes taken to the present and future, myths intimately tied in Western culture to our conception of time and history. As a scientific thinker Galileo has been made by an astonishing variety of philosophical reformers whatever their hearts desired: an experimentalist contemptuous of speculation, a mathematical idealist indifferent to experiment; a positivist in fact hostile to ideas although he may not always have known this himself, an illustration of the role of ideas in scientific discovery; a Platonist, a Kantian, a Machian operationalist. . . . But our intellectual inheritance is also an essentially critical one, predisposing each generation to take to pieces the history written by its predecessors in their image, before re-writing it in its own. (Crombie, 1970, p. 361)

Mindful of this natural tendency to, as Francis Bacon put it, "believe most what we want to believe," it is still possible to both delineate Galileo's epistemological position and, 400 years later, draw lessons from it for a better understanding of the methodology and epistemology of science.[17]

GALILEO'S IDEALIZATIONS AND
THE BEGINNING OF MODERN SCIENCE

Galileo is an outstanding example of the scientist–philosopher. He made substantial philosophical contributions in a variety of areas. In ontology, with his distinction of primary and secondary qualities; in epistemology, both with his criticism of authority as an arbiter of knowledge claims and with his subordination of sensory evidence to mathematical reason; in methodology, with his development of the mathematical–experimental method; and in metaphysics, with his critique of the Aristotelian causal categories and rejection of teleology as an explanatory principle.[18] Despite his important contributions to philosophy, and despite his acknowledged influence on just about all philosophers of the 17th century and on such subsequent philosophers as Kant and Husserl, Galileo has made, at best, a cameo appearance in most histories of philosophy. And of course his philosophical achievements are ignored in science texts.

Galileo's marvelous mathematical proofs of the pendulum's properties did not receive universal acclaim: On the contrary (See Chapter 7, learned scholars were quick to point out substantial empirical and philosophical problems with them. Del Monte and others repeatedly pointed out that actual pendula did not behave as Galileo maintained. Galileo never tired of saying that *ideal* pendula would obey the mathematically derived rules. Del Monte retorted that physics was to be about this world, not an imaginary mathematical world.[19]

The empirical problems were examples where the world did not "correspond punctually" to the events demonstrated mathematically by Galileo. In his more candid moments, Galileo acknowledged that events do not always correspond to his theory, that the material world and his so-called "world on paper," the theoretical world, did not correspond. Immediately after mathematically establishing his famous law of parabolic motion of projectiles, he remarked:

> I grant that these conclusions proved in the abstract will be different when applied in the concrete and will be fallacious to this extent, that neither will the horizontal motion be uniform nor the natural acceleration be in the ratio assumed, nor the path of the projectile a parabola. (Galileo, 1638/1954, p. 251)

One can imagine the reaction of del Monte and other hardworking Aristotelian natural philosophers and mechanicians when presented with such a qualification. When baldly stated, it confounded the basic Aristotelian and empiricist objective of science, namely to tell us about the world in which we live. Consider, for instance, the surprise of Giovanni Renieri, a gunner who attempted to apply Galileo's theory to his craft, who when he complained in 1647 to Torricelli that his guns did not behave according to Galileo's predictions, was told by Torricelli that "his teacher spoke the language of geometry and was not bound by any empirical result" (Segre, 1991, p. 43).

The law of parabolic motion was supposedly true, but not of the world we experience. This was indeed as difficult to understand for del Monte as it is for present-day students. Furthermore it confounded the Aristotelian methodological principle that the evidence of the senses is, with some qualifications, paramount in ascertaining facts about the world. That is, with a healthy observer, in a normal situation, then what the eyes see is what is the case.

Wolf, the PSSC authors and most science textbook writers who have mentioned Galileo's laws of pendulum motion have suggested that any observer, old or young, who cared to would see the isochronic motion Galileo supposedly saw. Not so. Del Monte and others did not see it. The historical debate was frequently the *reverse* of what modern science textbooks made it out to be: Concerning pendulum motion (and also of course the basic Copernican theory of a rotating earth), it was the Aristotelian scientists and philosophers who appealed to experience, while the "new" scientists (Galileo and his followers) appealed beyond experience to mathematics and reason.

There are dangers with abstraction. One might label something an "accident," which is really an "essential," and thus misunderstand what is being studied. This is precisely one of the charges that Hegel will later bring against Newton's mechanics when Hegel discussed Newton's inertial principle and Oresme's pendulum. He said:

> In the postulated flight of this leaden ball into infinity, the resistance of air and friction is also turned into an abstraction. When a perpetuum mobile, no matter how correctly calculated and demonstrated in theory, necessarily passes over into rest in a certain period of time, an abstraction is made of gravity, and the phenomenon is attributed solely to friction. The gradual decrease in the motion of the pendulum, and its final cessation, is also attributed to the retardation of friction, as it is also said of this motion that it would continue indefinitely if friction could be removed. . . . Friction is an impediment, but it is not the essential obstacle to external contingent motion. Finite motion must be inseparably bound up with gravity therefore, for in its purely accidental form, it passes over into and becomes subject to the direction of gravity, which is the substantial determination of matter. (Hegel, 1842/1970, p. 250)

On this point it is pertinent to note what has previously be mentioned: Namely that Newton early in his career thought that swinging a pendulum in Boyle's vacuum jar, and comparing it to the number of swings taken for a like one to come to rest in air, would be a test of the mechanical philosophers' aether hypothesis. To his surprise *both* pendula came to rest in the same time. This experiment eliminated air pressure as one of the impediments contributing significantly to the pendulum's coming to rest. Although I do not believe he mentioned this experiment, Hegel could well say: "I told you so."

Idealization and Counter-evidence

There is a vexing *methodological* problem presented here. Galileo needed to introduce idealizations to move beyond Aristotle's science, and to have a physics that could be represented mathematically: inclined planes represented as straight lines, weights on balances represented as parallel lines, projectile flight represented as a parabola. Galileo was right to discount perturbations and accidental factors, and to keep his eye on what he regarded as the salient features, or essential properties, of the situation. This is so clearly evident in his landmark claim that projectiles follow a parabolic path.

Galileo is not deterred by the "perturbations," "accidents," and "impediments" that interfered with the behavior of the freefalling, rolling, and projected bodies with which his new science was dealing.[20] His procedure was explicitly stated immediately after the above disclaimer about the behavior of real projectiles in contrast to his ideal ones. Galileo said:

> Of these properties [*accidenti*] . . . infinite in number, it is not possible to give any exact description; hence, in order to handle this matter in a scientific way, it is necessary to cut loose from these difficulties; and having discovered and demonstrated the theorems, in the case of no resistance, to use them and apply them with such limitations as experience will teach. (Galileo, 1638/1954, p. 252)

There is, however, a problem of idealization hiding fundamental mechanisms in the world. Keeping one's eye on the essential property is scientifically commendable, provided that it is the essential property, and that concentration on it does not blind one to other significant influences or properties. This is the case when Galileo maintained the isochrony of circular motion, dismissing experimental deviations as "accidents"—due to air resistance, friction, compounding effect of the weight of the string, etc. Some of the deviation was accidental, but not all of it. The core deviation of experiment from theory was because the theory was wrong: it was the cycloid, not the circle that was isochronous. How one tests idealized theories against experience, or experiment, is a complex methodological matter that will be discussed in the following sections. For theories that explicitly deal with idealized cases or situations, there is always the danger of maintaining "my theory is right, don't bother me with the facts."

Undoubtedly there was an element of metaphysics in Galileo's adherence to the circle as the tautochrone. This same conviction perhaps led him to discuss and defend Copernicus' theory of *circular* planetary orbits, despite Kepler's *elliptical* refinement of Copernicus' views being published in 1619, 14 years before Galileo's great *Dialogue,* and Galileo having a copy of the work in his library. The same conviction perhaps led Galileo to the doctrine of *circular* inertia. Alexandre Koyré (Koyré, 1943b) and Edwin Burtt (Burtt, 1932, pp. 61–95), regarded this metaphysical conviction as evidence of Galileo's Platonism.[21] Koyré expressed the core of Platonic science this way:

> The dividing line between the Aristotelian and the Platonist is perfectly clear. If you claim for mathematics a superior status, if more than that you attribute to it a real value and a commanding position in physics, you are a Platonist. If on the contrary you see in mathematics an abstract science, which is therefore of a lesser value that

those—physics and metaphysics—which deal with real being; if in particular you pretend that physics needs no other basis than experience and must be directly built on perception . . . you are an Aristotelian. (Koyré, 1943b, p. 36)

The empirical discrepancies between circular and isochronic motion were not great. For small amplitudes (less than 2°) a circular path was almost identical to a cycloid, and thus small amplitude circular pendulums were nearly isochronic (other distorting factors being excluded). A circular pendulum with an amplitude of 5° gained 41 seconds per day, one of 6° gained 59 seconds, and one of 12° gained 236 seconds.[22]

Whatever its source, it would have been easy and acceptable to attribute the nonisochrony of Galileo's pendulums to "accidental" factors, thus maintaining the circular path of the free moving pendulum as the basis of the pendulum clock. This was *nearly* correct, but no matter how much it was refined it would not yield an accurate timekeeper.[23] The theory was wrong. But of course this was not always the case. Sometimes the results were wrong. In 1919 Albert Einstein was handed a cable relaying Eddington's finding that light rays bent when passing the sun. This was a brilliant confirmation of Einstein's relativity theory. When Einstein was asked what he would have done if the cable had *not* confirmed his theoretical predictions, he famously replied: "Then I would have to pity the dear Lord. The theory is correct anyway" (Pais, 1994, p. 113).

HUYGENS AND FALSIFICATIONIST METHODOLOGY OF SCIENCE

The intimate relationship of methodology to scientific practice is clear in the work of Huygens, particularly in his responses to the anomalous discoveries of Jean Richer in his 1672 voyage to Cayenne concerning the need to shorten the seconds pendulum in equatorial regions. The theory of constant gravitation, which was pivotal for Huygens proposal of a new international standard of length based on the seconds pendulum, had a great deal of a priori and empirical support. What one did with this theory in the light of Richer's Cayenne evidence depended upon one's methodology of science. The constant gravitation theory could have been saved. Excuses could have been offered. The first excuse could have been to "shoot the messenger." To say that Richer was incompetent or that his instruments were not properly looked after, and so on. This was Huygens' initial response. These explanations were ruled out—apart from anything else, too many messengers, including Sir Edmund Halley from St. Helena Island, kept arriving.[24] Then Huygens attributed the pendulum's slowing to climatic differences between Paris and the tropics: Humid air offered more resistance, and so on. Again this hypothesis preserved the theory of unchanging gravity, but eventually the climatic explanations had to be abandoned. The slowing of the pendulum in equatorial regions was clearly established and had to be accommodated in any scientific theory of pendulum motion and timekeeping.

The logic of this accommodation of theory to evidence constitutes the *methodology* of science, something which is useful to distinguish from the *method* of science involves how one conducts experiments, gathers data, seeks information, and selects appropriate tools, instruments and means of analysis for the conduct of an investigation. Methodology involves what one does with the data and how one relates it to hypotheses and theories: How many white swans does one have to see in order to conclude that "all swans are white"? How does finite evidence bear on the truth of scientific hypotheses that are universal in their scope? And so on. Since Aristotle, methodology has been the core subject matter of philosophy of science.

Huygens himself provided perhaps the first statement of the hypotheticodeductive methodology of science. In the preface to his *Treatise on Light* (Huygens, 1690/1945) he wrote:

> One finds in this subject a kind of demonstration which does not carry with it so high a degree of certainty as that employed in geometry; and which differs distinctly from the method employed by geometers in that they prove their propositions by well established and incontrovertible principles, while here principles are tested by the inferences which are derivable from them. The nature of the subject permits of no other treatment. It is possible, however, in this way to establish a probability which is little short of certainty. This is the case when the consequences of the assumed principles are in perfect accord with the observed phenomena, and especially when these verifications are numerous; but above all when one employs the hypothesis to predict new phenomena and finds his expectations realized.

Huygens is saying that natural philosophy (science) is characterized by having principles (laws, theories, hypotheses, conjectures) from which inferences (predictions) are made about states of affairs, and then these predicted states of affairs are checked against reality or experimental results. In other words, from a theory T we infer certain observations O and then investigate to ascertain whether O is the case.

$$T \rightarrow O$$
$$O \, ?$$

In Huygens' case, the theory of constant gravitational attraction (T) implied the observation (O) that the length of a seconds pendulum in Cayenne would be the same as one in Paris. In this hypotheticodeductive schema, there can be two outcomes: O is the case (T is confirmed), or O is not the case (T is falsified, as Karl Popper would later say). Huygens concentrated on the positive outcome, the confirmation of T. He realized, as Aristotle and the medievals had before him, that a positive outcome did not entail the truth of the theory T. To believe so is to commit the Fallacy of Affirming the Consequent. There may be any number of other theories apart from T from which the same observations can be inferred.

The hypothesis that it rained last night implies that the road will be wet; but equally the hypotheses that the sanitation truck went past, that a watermain broke, or that a lawn hose was turned on, also imply the same thing. Thus the mere observation that the road is wet does not prove any particular hypothesis. This is why Huygens spoke of confirmation, not verification; and why he spoke of probability, not certainty.

Just how much probability positive confirmations provide for a theory has been bitterly argued over.[25] For centuries the hypothesis that "all swans are white" had countless confirmations, but as soon as a black swan was seen in western Australia, the hypothesis collapsed. All the probability in the world was of no consequence to its truth value. What value was it then to have given any credence to the "all swans are white" hypothesis? Should we not retreat, as many deductivists have (Musgrave, 1999), to Plato's distinction between knowledge (which is certain) and opinion (which is all on a par)? On the other hand, inductivists have tried to differentiate on probabilistic grounds the epistemic value of opinions or hypotheses.

For these reasons, others have preferred to focus not on the positive, but the negative outcome in Huygens characterization of scientific theory appraisal. This is the situation when the prediction fails, the inferred observation is not forthcoming. Huygens did not give explicit attention to this, but in his scientific practice he was clearly a falsificationist. This is the methodology, that in the 20th century, has been given greatest prominence by Karl Popper (Popper, 1963).

Care is needed in discussing Popper and falsificationism. Popper was not the simple falsificationist many make him out to be. He said on one occasion that "Criticism of my alleged views was widespread and highly successful. [but] I have yet to meet a criticism of my views" (Popper, 1963, p. 41). In Popper's first major publication, *The Logic of Scientific Discovery* (Popper, 1934/1959), he sanctioned the creation of auxiliary hypotheses to rescue theories which were apparently refuted by the evidence, provided these hypotheses were not ad hoc. In 1948 he acknowledged: "Testing proceeds by taking the theory to be tested and combining it with all possible sorts of initial conditions as well as with other theories, and then comparing the resulting predictions with reality" (Popper, 1972, p. 360). After recognizing that falsification was more complicated than many might hope, Popper reasserted the core of his position as one in which: "The proper reaction to falsification is to search for new theories which seem likely to offer us a better grasp of the facts" (Popper, 1972, p. 360).

The type of situation faced by Huygens—the revision of a theory in the light of contrary evidence—recurs constantly in the development of science. This *revision* or *augmenting* of theory is at odds with simple interpretations of falsificationism as the methodology of science. According to falsificationists, when the facts or observations (*O*) are contrary to a theory (*T*), then the theory is falsified and should be abandoned. But the death sentence is rare in the court of theory appraisal.

According to falsificationist methodology:

If T implies O	$T \rightarrow O$
And if O is not the case	$\sim O$
Then, T is false	$\therefore \sim T$

But as Huygens recognized, the simple statement "T implies O" hides a host of auxiliary assumptions: assumptions about initial conditions, the state and accuracy of experimental apparatus, the competence of the observer, and so on. Behind the bald "T implies O," there are always a host of *ceteris paribus,* or other things being equal, clauses (C).

Thus the above schema needs to be elaborated as follows:

If T and C imply O	$T.C \rightarrow O$
And if O is not the case	$\sim O$
Then, T is false or C is false	$\therefore \sim T \vee \sim C$

Initially Huygens thought, when confronted with Richer's evidence, that one or other of the C clauses was at fault: the apparatus was faulty, climate affected the pendulum swing, etc. All of this saved T (that gravitational acceleration was constant across the earth's surface), but in turn each of the postulated problematic C clauses turned out not to be problematic. Thus finally T had to be abandoned.

Indeed the testing situation is even further complicated. Metaphysics, as we will see in Chapter 10, plays a role in theory appraisal. So the situation is as follows:

If background metaphysics (M) and theory (T) and conditions (C) imply O	$M.T.C \rightarrow O$
And if O is not the case,	$\sim O$
Then, M is false, or T is false, or O is false	$\sim M \vee \sim T \vee \sim C$

Willard Quine, and before him Pierre Duhem, elaborated this methodological point about theory appraisal. Quine memorably wrote that:

> The totality of our so-called knowledge or beliefs, from the most casual matters of geography and history to the profoundest laws of atomic physics or even of pure mathematics and logic, is a man-made fabric which impinges on experience only along the edges. . . . But the total field is so underdetermined by its boundary conditions, experience, that there is much latitude of choice as to what statements to reëvaluate in the light of any single contrary experience. (Quine, 1953, p. 42)

OBSERVATION, THEORY, AND EXPERIMENT

The crucial methodological point at issue in the Galileo–del Monte debates over pendulum motion, a point which is characteristic of the whole Galilean–Newtonian methodology, is captured in Richard Westfall's observation that:

Beyond the ranks of historians of science, in my opinion, the scientific revolution is
frequently misunderstood. A vulgarized conception of the scientific method, which
one finds in elementary textbooks, a conception which places overwhelming emphasis
on the collection of empirical information from which theories presumably emerge
spontaneously, has contributed to the misunderstanding, and so has a mistaken notion
of the Middle Ages as a period so absorbed in the pursuit of salvation as to have been
unable to observe nature. In fact medieval philosophy asserted that observation is the
foundation of all knowledge, and medieval science (which certainly did exist) was a
sophisticated systematization of common sense and of the basic observations of the
senses. Modern science was born in the sixteenth and seventeenth centuries in the
denial of both. (Westfall, 1988, p. 5)

Children have the same difficulty seeing the properties of pendulum motion
that the 16th century Aristotelians had. Even with highly refined school laboratory pendulums, they do not see isochrony of large and small amplitude swings,
and their cork pendulums soon cease swinging, while the brass ones continue
much longer. All of this experiential evidence is hard to reconcile with the "laws"
of pendulum motion. Children can either think they are stupid and need to take
everything on authority, or they can conclude, as one German student did in a
survey that "physics is not about the world" (Schecker, 1992, p. 75). This is a case
of children being in the position of the early pioneers of a science. No amount of
looking will reveal isochronic motion. Looking is important, but something else
is required, namely a better appreciation of what science is and what it is aiming
to do, an epistemology of science. The historian E.J. Dijksterhuis recognized this:

To this day every student of elementary physics has to struggle with the same errors
and misconceptions which then [in the seventeenth century] had to be overcome . . . in
the teaching of this branch of knowledge in schools, history repeats itself every year.
(Dijksterhuis, 1986, p. 30)

Dijksterhuis went on to make a fundamental point. Classical mechanics is
not only not verified in experience, but its direct verification is fundamentally
impossible: "one cannot indeed introduce a material point all by itself into an infinite
void and then cause a force that is constant in direction and magnitude to act on it; it
is not even possible to attach any rational meaning to this formulation." (ibid)[26]

Laura Fermi and Gilberto Bernardini drew attention to the same methodological point regarding the centrality, for Galileo's science, of abstracting from
everyday experience. They put the matter this way:

In formulating the "Law of Inertia" the abstraction consisted of imagining the motion
of a body on which no force was acting and which, in particular, would be free of any
sort of friction. This abstraction was not easy, because it was friction itself that for
thousands of years had kept hidden the simplicity and validity of the laws of motion.
In other words friction is an essential element in all human experience: our intuition
is dominated by friction. (Fermi and Bernardini, 1961, p. 116)

In a similar vein, Alexandre Koyré, speaking of inertial motion, said:

> As a matter of fact, even today, the conception we are describing is by no means easy
> to grasp, as anyone who ever attempted to teach physics to students who did not learn
> it at school will certainly testify. Common sense, indeed is—as it always was—medieval
> and Aristotelian. (Koyré, 1968, p. 5)

Galileo did not develop a system of rational mechanics in the way medieval scientists constructed mathematical models of physical systems, and then proceeded no further. In contrast, Galileo's theoretical objects were the means for engaging with and working on the natural world. For him the theoretical object provided a plan for interfering with the material world, and where need be, for making the real world in the image of the theoretical. When del Monte told Galileo that he had done an experiment with balls in an iron hoop and the balls did not behave as Galileo asserted, Galileo replied that the hoop must not have been smooth enough, that the balls had not been spherical enough and so on. These suggestions for improving the experiment were driven by the theoretical object that Galileo had already constructed. Without the theoretical object he would not have known whether to correct for the color of the ball, the material of the hoop, the diameter of the hoop, the mass of the ball, the time of day, or any of a hundred other factors. It is this aspect of Galileo's work which moved Immanuel Kant to say:

> When Galileo caused balls, the weights of which he had himself previously determined,
> to roll down an inclined plane; when Torricelli made the air carry a weight which he
> had calculated beforehand to be equal to that of a definite volume of water . . . a light
> broke upon all students of nature. They learned that reason has insight only into that
> which it produces after a plan of its own, and that it must not allow itself to be kept, as
> it were, in nature's leading-strings, but must itself show the way with principles of
> judgment based upon fixed laws, constraining nature to give answer to questions of
> reason's own determining. Accidental observations, made in obedience to no previously
> thought-out plan, can never be made to yield a necessary law, which alone reason is
> concerned to discover. . . . It is thus that the study of nature has entered on the secure
> path of a science, after having for so many centuries been nothing but a process of
> merely random groping. (Kant, 1787/1933, p. 20)

Koyré drew out this Kantian moral concerning experiment, and its role in the development of science,[27] when he wrote:

> . . .observation and experience—in the meaning of brute, common-sense observation
> and experience—had a very small part in the edification of modern science; one could
> even say that they constituted the chief obstacles that it encountered on its way. . . . the
> empiricism of modern science is not *experiential*; it is *experimental*. (Koyré, 1968, p. 90)

Koyré's fellow countryman, Gaston Bachelard, stressed these matters in his influential 1930s work. For Bachelard:

> empirical notions derived from ordinary experience have to be revised and modified
> repeatedly before they can be of any use to microphysics, which defines reality by
> *inference* rather than *discovery*. (Bachelard, 1934/1984, p. 160)

In contrasting modern, post-Galilean, science with its Aristotelian forebears, he noted:

> Scientific observation is always polemical; it either confirms or denies a prior thesis, a preexisting model, an observational protocol. It shows as it demonstrates; it establishes a hierarchy of appearances; it transcends the immediate; it reconstructs first its own models and then reality. And once the step is taken from observation to experimentation, the polemical character of knowledge stands out even more sharply. (Bachelard, 1934/1984, p. 13)

The reason why experimentation had this impact on knowledge development was that:

> Now phenomena must be selected, filtered, purified, shaped by instruments; indeed, it may well be the instruments that produce the phenomenon in the first place. And instruments are nothing but theories materialized. The phenomena they produce bear the stamp of theory throughout. (Bachelard, 1934/1984, p. 13)

And Bachelard further argues that:

> When one has fully comprehended . . . that experimentation is always dependent on some prior intellectual construct, then it is obvious why one should look to the abstract for proof of the coherence of the concrete. (Bachelard, 1934/1984, p. 41)

These epistemological considerations are not irrelevant to education. Some accounts of teaching for students' conceptual change innocently maintain it is discrepant *events* or *phenomena* that students need to be confronted with in order to change their understandings. The foregoing account of experimentation, and its role in the development of science, suggest that this pedagogical strategy is almost precisely the wrong one. School experiments rarely worked, and when they do, the phenomena exhibited are most likely to *confirm* wrong ideas, and *falsify* the correct scientific formulation.

REAL VERSUS THEORETICAL OBJECTS

One way to conceptualize the methodological revolution of the 17th century is to recognize the difference between real objects and these objects when described from the standpoint of some scientific theory. Newton's theory of mechanics, for example, provided definitions of key concepts: momentum, acceleration, average speed, instantaneous speed, weight, impetus, force, point mass and so on. These concepts were hard-won theoretical constructs and were utilized in his account of pendulum motion. At the beginning of the scientific revolution, these concepts were only seen through a glass darkly. Acceleration, for instance, was initially defined by Galileo and all his predecessors, as rate of increase of speed with respect to *distance* traversed—a natural enough definition given that accelerating bodies increase speed over both time and distance, and that the passage of distance was both more measurable and more easily experienced by sight and feel. It was only in Galileo's middle age that he changed the definition to the modern one of rate of change of speed with respect to *time* elapsed. This was in Day Three of his *New Sciences*, where he wrote: "A motion is said to be uniformly accelerated, when starting from rest, it acquires, during equal time-intervals, equal increments of speed" (Galileo, 1638/1954, p. 162). Without such a change of definition, the important laws of freefall would not have been discovered. A good deal of a child's school life would pass while waiting for him or her to construct this modern definition of acceleration, and hence most teachers routinely commence mechanics classes by giving students the modern definition. The alternative is not even considered.

Although less clearly defined than with Newton, interlocking concepts formed the *conceptual structure* of Galileo's physics, and provided the meaning of key terms. This is Galileo's "world on paper," as he referred to it, or his prototheoretical system. This is the indispensable scaffolding of science that, as I have maintained against constructivists, children need to be given; in some sense, the scaffolding needs to be transmitted to them. There is also a real world of material and other objects which exists apart from Galileo theorizing. We can see in Galileo's practice a most important intervening layer emerging between theory and the real world—the realm of theorized objects. These are natural objects *as conceived and described by* the relevant theoretical concepts. Planets and falling apples have color, texture, irregular surfaces, heat, solidity and any number of other properties and relations. But when they become the subject matter of mechanics they are merely point masses with specified accelerations. When thus conceptualized, they are no longer natural objects, but theoretical objects. In a similar way, when apples are considered by economists they become theoretical objects of a different sort: commodities with specific exchange values. When botanists consider apples they create yet other theoretical objects. For Galileo, a sphere of lead on the end of a length of rope swinging in air became, in his mechanical theory, a pendulum conceived as a point mass at the end of a weightless chord suspended from a frictionless fulcrum. Galileo did not have the Newtonian concept of void. Galileo's heavy bodies fell because they were attracted to the earth. For all his idealization, he did not entertain the idea of bodies moving by themselves in an infinite void. Galileo's universe was populated and bounded.

Historically, grasping the significance of this abstractive or theoretical domain between conceptual schemes and the natural world has not been easy. This, in part, is what the debate between Galileo and del Monte was about. Pierre Duhem (1861–1916), the French physicist, historian and philosopher, made a similar distinction to that being made here, but did so in terms of concrete and abstract objects. Speaking of mathematical physics, he said:

> Thus at both its starting and terminal points, the mathematical development of a physical theory cannot be welded to observable facts except by a translation. In order to introduce the circumstances of an experiment into the calculations, we must make a version which replaces the language of concrete observations by the language of numbers; in order to verify the result that a theory predicts for that experiment, a translation exercise must transform a numerical value into a reading formulated in experimental language.

> But translation is treacherous . . . There is never a complete equivalence between two texts when one is a translated version of the other. Between the concrete facts, as the physicist observes them, and the numerical symbols by which these facts are represented in the calculations of the theorist, there is an extremely great difference. (Duhem, 1906/1954, p. 133)

At one point he wrote of the scientist examining Boyle's law, a law which relates the volume and pressure of a given mass of gas at a constant temperature, saying:

> Let us put ourselves in front of a real, concrete gas to which we wish to apply Mariotte's [Boyle's] law; we shall not be dealing with a certain concrete temperature embodying the general idea of temperature, but with some more or less warm gas; we shall not be facing a certain particular pressure embodying the general idea of pressure, but a certain pump on which a weight is brought to bear in a certain manner. No doubt, a certain temperature corresponds to this effort exerted on the pump, but this correspondence is that of a sign to the thing signified and replaced by it, or of a reality to the symbol representing it. This correspondence is by no means immediately given; it is established with the aid of instruments and measurements, and this is often a very long and very complicated process. (Duhem, 1906/1954, p. 166)

He went on to say:

> It is impossible to leave outside the laboratory door the theory that we wish to test, for without theory it is impossible to regulate a single instrument or to interpret a single reading; we have seen that in the mind of the physicist there are constantly two sorts of apparatus; one is the concrete apparatus in glass and metal manipulated by him, the other is the schematic and abstract apparatus which theory substitutes for the concrete and on which the physicist does his reasoning. (Duhem, 1906/1954, p. 182)

Duhem was an instrumentalist about theoretical abstractions. For him, they did not designate real properties. But his distinction, and that of theoretical versus material objects introduced above, is neutral on this ontological point. That "force" is an abstraction does not mean, of itself, that there are no forces.[28] But Duhem's observation, that without theory we cannot interpret a single reading, is important for pedagogy. Instruments and measuring devices should not be dropped into a classroom without due attention to the theory that they embody and presuppose. School experiments routinely fail. Teachers and students need, as Duhem recognized, some understanding of how theory relates to the world so they can correctly interpret these failed experiments.

OBJECTIVISM

The 17th century's analysis of pendulum motion support objectivist views of scientific theory and of epistemology.[29] These objectivist views are in opposition to all empiricist epistemologies and theories of scientific methodology. Empiricists, since Bacon, Locke and Berkeley in the 17th and 18th centuries, through to Alfred Ayer in the 20th century,[30] maintain that:

1. There is a distinction between basic, observational or intuited knowledge and theoretical knowledge.
2. Basic, observational or intuited knowledge does not involve theory.
3. Basic, observational or intuited knowledge is available or given to individual observers or thinkers (or knowers, or subjects).
4. Theoretical knowledge is derived from, or ultimately justified by reference to, basic, untheoretical knowledge.

Rationalism is the other side of the empiricist coin. Both empiricist and rationalist epistemologies are foundationalist: they seek foundations for knowledge in the experience of the individual, but whereas traditional empiricism defines experience as the 'outer' experience of the senses, rationalism extends the definition of experience to include the 'inner' experience of the individual's mind. In both epistemologies, it is the experience of the individual knower that is primary. Once this empiricist problematic is accepted, in either its empiricist or rationalist guise, then there is only a short, and much taken, step to skepticism and relativism. Clearly both 'outer' and 'inner' experience is affected by an individual's circumstances, language, culture, ideology and theory. The 'theory dependence of observation' has been much written upon, and when coupled with empiricist assumptions about knowledge, it standardly results in relativist or skeptical epistemologies. If experience is the foundation of knowledge, then the foundation is on shaky ground; and doubts about the soundness of the building are easy to induce. One modern, and energetic, variant of this empiricist problematic is constructivism which embraces the epistemological trifecta of individualism, experience and, inevitably, skepticism.[31]

Karl Popper signaled an objectivist break with the empiricist problematic in epistemology when, in his 1934 *The Logic of Scientific Discovery*, he wrote:

And finally, as to *psychologism*: I admit, again, that the decision to accept a basic statement, and to be satisfied with it, is causally connected with our experiences—especially with our *perceptual experiences*. But we do not attempt to *justify* basic statements by these experiences. Experiences can *motivate a decision*, and hence an acceptance or a rejection of a statement, but a basic statement cannot be *justified* by them – no more than by thumping the table. (Popper, 1934/1959, p. 105)

Much later he would say:

Traditional epistemology has studied knowledge or thought in a subjective sense – in the sense of the ordinary usage of the words 'I know' or 'I am thinking.' This, I assert, has led students of epistemology into irrelevances: while intending to study scientific knowledge, they studied in fact something which is of no relevance to scientific knowledge. For *scientific knowledge* simply is not knowledge in the sense of the ordinary usage of the words 'I know.' While knowledge in the sense of 'I know' belongs to what I call the 'second world,' the world of *subjects*, scientific knowledge belongs to the third world, to the world of objective theories, objective problems, and objective arguments.' (Popper, 1972, p. 108)

Most objectivists in theory of knowledge share Popper's convictions that first, the growth of scientific knowledge is the core subject matter of epistemology; and second, that knowledge is something other than beliefs or psychological states, it transcends individual consciousness. Popper delineated 'three worlds': the *first* world of objects, processes and events (material and otherwise); the *second* world of subjective, individual, mental operations (the life of the mind or private consciousness); and the *third* world of scientific, and other, theories, constructions and problem situations which are a part of culture and which, although created by people, nevertheless exist independently of first and second world events (Musgrave 1974). At the outset of his lecture on 'Epistemology Without a Knowing Subject' he stated the matter as follows:

> The main topic of this lecture will be what I often call, for want of a better name, '*the third world.*' To explain this expression I will point out that, without taking the words 'world' or 'universe' too seriously, we may distinguish the following three worlds or universes: first, the world of physical objects or of physical states; secondly, the world of states of consciousness, or of mental states, or perhaps of behavioural dispositions to act; and thirdly, the world of *objective contents of thought*, especially of scientific and poetic thoughts and works of art. (Popper, 1972, 105)

One need not give an *ontological* interpretation of Popper's three-worlds doctrine; the worlds need not be taken as separately existing realms. To say that the propositional content of beliefs is distinct from psychological states of belief, is not to say that they exist independently, only that they are independent, have different kinds of properties, and are subject to different forms of analysis. The third world can be interpreted as *logically*, not physically, distinct from the first and second worlds (Irzik 1995, p. 84). The claim that 'the cycloid is an isochronous curve' is different from the physical letters on paper that constitute the claim (first world things), and it is also different from the belief states of Huygens, or whoever else might entertain it (second world things). Questions of truth or falsity, and evidence or lack of it, can be asked of the isochronic claim; these questions cannot be asked of first and second world entities. Thus Popper, and objectivists more generally, say that propositional claims, and scientific theories, occupy a distinct logical niche with their own appropriate questions and puzzles. In a similar way, institutions exist apart from, but not alongside, their embodiment. A university exists apart from it buildings, faculty and students, but it does not exist alongside them; a marriage exists apart from a husband and wife, but it does not exist alongside them.

This objectivist tradition of Popper, Althusser, Suchting, Baltas, Sneed and Mittelstrass, emphasizes, first, the separation of cognitive or theoretical discourse from the real world. The world is neither created by the discourse (as in idealism), nor does it somehow create the discourse (as in various reflection, or imprinting, theories from Locke to Lenin), nor does it anchor or provide foundations for the discourse (as in empiricism and positivism). Theoretical discourse and the world are each autonomous. In this sense, theory exists independently of individuals. Thus scientific knowledge is, contrary to the claims of many constructivists, external to individuals. Second, objectivist views distinguish within theory between (a) the conceptual foundations of the discourse containing the definitions of theoretical and observational terms (there is no decisive distinction made between these kinds of terms); (b) the conceptual structure of the theory which is the elaboration and manipulation of the basic concepts by techniques (mathematical and logical) that produce the structure of the theory (Galileo's theorems and propositions); and (c) the theoretical objects of the theory, which are objects in the world as they are conceived and described by the theory – the balance treated as a uniform line with parallel weights suspended from it, the pendulum treated as a point mass on a weightless string, billiard balls treated as perfectly hard bodies, and so forth.

It is no great distinction to be critical of empiricist epistemology. Hegel in the opening chapters of his *Phenomenology of Mind* (1810), Wittgenstein in his *Philosophical Investigations* (1958), Quine in his 'Two Dogmas of Empiricism' (Quine 1953), and a host of post-Kuhnian philosophers, recognised that sensory input does not carry its meaning on its sleeve, (or on its neurones, as one might say). As Hanson memorably stated it: 'There is more to seeing than meets the eye ball' (Hanson 1958, p. 7).

In the Popperian tradition, the concepts and techniques of a theoretical system enable natural events to become scientific events, and as such they can then be considered and analysed by the scientist in accord with the canons of the appropriate theory. It is as a *scientific* object that objects or events are stripped of their everyday guises and become data or evidence in theoretical debate. The properties ascribed to these theoretical objects are derived from the conceptual foundations of the theory: instantaneous velocity, momentum, acceleration and so on. Likewise some properties cannot be ascribed because well-developed conceptual foundations exclude them; the particular theoretical discourse does not contain the concepts. Thus Galilean discourse excluded *levitas* [lightness] as a property of bodies, whereas in Aristotelianism it is one of the fundamental properties of bodies.

In objectivist accounts of science, scientific concepts have an essentially *twofold* meaning: empirical and theoretical. Concepts such as 'mass', 'force', 'inertia', 'momentum', 'electron', 'intelligence', 'wave', 'energy', 'drive', and so on, relate on the one hand to the world (the *empirical* or vertical component) and on the other hand to other concepts and definitions within an appropriate theoretical system (the *theoretical* or horizontal component). The vertical relation is not one of reflection: apples do not reflect or imprint themselves as the concept of a point mass on the mind of the scientist; the exertion of effort in pushing a barrow is not reflected or imprinted as the concept of Newtonian force; the sensation of speeding up on a horse does not generate the classical idea of acceleration; the arrary of Galapagos finches did not, mechanically, produce the idea of adapation in Darwin's mind. The theoretical terms are related to the phenomena, but the relation is more *ascription* rather than *description*. Objects, events and processes become scientific objects when they are identified or named as point masses, forces, acclerations, adaptations and so on.

The conceptual foundations of a theoretical discourse or research program, to use Lakatos's terminology, are slowly built up and refined in a process of theoretical production. Conceptual foundations generally contain elements, assumptions and aspects of pre-existing conceptual schemes,[32] and also elements of commonsense understandings. Eclecticism is a common problem with under-developed conceptual foundations, as is the carryover of everyday and ideological concepts into the conceptual foundations of science—the use in science of everyday notions of force, acceleration, species, atoms and so on. Despite Galileo's break with so much of the Aristotelian tradition, his conceptual scheme retained Aristotelian notions of natural movement (accelerated movement without an external force), and also circular inertia. It is only very mature theoretical discourses that manage to identify all extraneous elements within them, and then either redefine or jettison these elements. And of course the same concept often has an everyday meaning alongside its theoretical meaning, which can be a cause of confusion. As Baltas points out:

> Each concept has, as it were, two faces resembling in that respect a coin. For example 'force' is both the cause of 'acceleration' and what we exercise when we try to push some body; and 'acceleration' means simultaneously both the 'derivative of instantaneous velocity over time' and what we sense when our car speeds up. (Baltas, 1988, p. 214)

One should not strive to *replace* the commonsense concept with the scientific (a mistake of much constructivist and 'conceptual change' teaching), but rather to have children recognise when one concept is appropriate and when the other is appropriate. A child's idea of an animal is something that is large, furry and usually having four legs. The scientific idea includes ants and flies. One is not more correct than the other, it is only more, or less, correct in its appropriate domain. To say 'there is an animal in the tent', in virtue of there being a fly in the tent, is not to be scientific; it is to be mistaken about when scientific, and when everyday conceptualisations, are required. Arthur Eddington's famous question about what is the real table: the solid, coloured, one he is sitting at and writing on, or the largely empty, and colourless one that science tells him about (Eddington 1928/1978, pp. xi-xiii)—exploits just such a confusion about the legitimate occasions for the use of particular languages.

There is, however, a second sense of 'objectivism,' especially common in science education literature. This is objectivism in the sense of 'objective truth,' or 'certainty of belief.' This is *epistemological* objectivism. This is completely orthogonal to the Popperian sense (indeed Popper denied the possibility of absolute truth). Sandra Harding, for instance, writes that: 'The absolutists, such as objectivists, say that there is one and only one defensible standard for sorting belief . . . ' (Harding 1994, p. 89), and 'objectivism, with its ideal of results of research that were socially neutral' (Harding 1994, p. 99), and 'The notion [of truth] is inextricably linked to objectivism and its absolutist standards' (Harding 1994, p. 100). In science education, objectivism, as well as having these absolutist overtones, is frequently interpreted as a version of positivism.[33]

There is nothing wrong with this. But it is important to distinguish Popperian, ontological objectivism, from this second absolutist and epistemological objectivism: Aristotelian mechanics was objective, but completely mistaken; Creationist science is objective in the Popperian sense, but also completely mistaken. There is no link between the objectivity and the truth of a theory. The epistemological import of Popperian objectivism is that it is theories, and theoretical claims, that are the subject matter of epistemology, not individual beliefs or states of mind. Epistemology is concerned with, for instance, the intellectual status of Newton's third law, the soundness of Newtonian methodology, and the adequacy of the Newtonian conceptual scheme. Its subject matter is not the adequacy of Newton's beliefs or states of mind.

SCIENTIFIC LAWS

The regularity account of scientific law has been popular since David Hume's 1739 *Treatise on Human Nature*, where he attempted to refute the necessitarian view of law. For Hume, and those following him, scientific laws stated constant relations between observables, they stated what is uniformly seen to be the case. The pendulum laws presented an overwhelming problem for the Humean account: The regularities did not occur. Under very refined experimental conditions (small oscillations, heavy weights, minimum air and fulcrum resistance) they almost occurred, but "almost occurring" is not what regularity accounts of law are about. Moreover it was a commitment to the truth of the apparently disproved laws that enabled the approximate conditions for the law's applicability to be devised. That is, the truth of the law is a presupposition for identifying approximations to it, and for identifying when some behavior is to be seen as "almost law-like." Norwood Hanson expressed the matter well:

> The great unifications of Newton, Clerk Maxwell, Einstein, Bohr and Schrödinger, were pre-eminently discoveries of terse formulae from which descriptions and explanations of diverse phenomena could be generated. They were not discoveries of undetected regularities. (Hanson, 1959, p. 300)

Michael Scriven once wrote that: 'The most interesting thing about laws of nature is that they are virtually all known to be in error' (Scriven, 1961, p. 91). Nancy Cartwright, in her *How the Laws of Physics Lie* (Cartwright, 1983) made a similar point: If the laws of physics are interpreted as empirical, or phenomenal, generalizations, then the laws lie. As Cartwright stated the matter: "My basic view is that fundamental equations do not govern objects in reality; they govern only objects in models" (Cartwright, 1983, p. 129). The world does not behave as the fundamental equations dictate. This claim is not so scandalous: The gentle and random fall of an autumn leaf obeys the law of gravitational attraction, but its actual path is hardly as described by the equation $v = u + at$. This equation refers to idealized situations. A true description, a phenomenological statement, of the falling autumn leaf would be complex beyond measure. The law of fall states an *idealization*, but one that can be experimentally approached. These laws are usually stated with a host of explicit *ceteris paribus*, or "other things being equal," conditions.[34] For the laws of pendulum motion, the *ceteris paribus* conditions would be:

> the string is weightless (so no dampening occurs);
> the bob does not experience air resistance;
> there is no friction at the fulcrum;
> all the bob's mass is concentrated at a point;
> the pendulum moves in a plane and does not experience any elliptical motion;
> that gravity and tension are the only forces operating on the bob.

But these conditions can only be approached, never realized—something that Ronald Giere has been at pains to point out (Giere, 1988, pps. 76–78; 1994; 1999, chaps. 5). Giere believes that not only are scientific laws false, they are also neither universal or necessary (Giere, 1999, p. 90). His account of the laws of pendulum motion is close to what has been advanced above in discussion of the real and theoretical objects of science. He says of the pendulum laws:

> On my alternative interpretation, the relationship between the equations and the world is *indirect*. . . . the equations can then be used to construct a vast array of abstract mechanical systems . . . I call such an abstract system a *model*. By stipulation, the equations of motion describe the behavior of the model with perfect accuracy. We can say that the equations are exemplified by the model or, if we wish, that the equations are *true*, even *necessarily* true, for the model. (Giere, 1999, p. 92)

DATA, PHENOMENA, AND THEORY

Nearly all of the foregoing considerations concerning the real and theorized object distinction, idealizations, and non-Humean accounts of law point toward the importance of distinguishing data from phenomena, and both of these from theory. The data come from observations, phenomena are at one remove from observations. Scientific laws and theories are about the phenomena, not about the data.[35] If this is understood, a number of things about science, and especially studies of pendulum motion, become clearer.

Real objects (processes, events, occurrences, states) are observed either in natural (Aristotle's preference) or experimental (Galileo's preference) settings. The observation can be immediate (with eyes, microscopes, etc.) or inferred (meter readings, instrument displays, etc.). All of this occurs in the realm of the real, not the realm of discourse. The observations are then verbalized, described, written, or tabulated. This has to be done in a language (including mathematics), and according to some theoretical standpoint. This is all done in the realm of discourse. These descriptions are characteristically sifted, sorted, and selected. Lots of readings and descriptions are simply thrown away, or ignored. The result is scientific data. These then are the raw representations of real objects (processes, events, occurrences, states). This step is clearly theory dependent. A range of falling red apples, or swinging weights on a string are, in physics, represented as points on a graph, as printouts on a tickertape, as lines on a screen. These representations are not meant to mirror or copy the real. They are precisely meant to represent the real. And, as mentioned earlier, economists, artists, and farmers have different ways of representing the same real events. Adequacy of representation simply does not mean correspondence of representation, in the sense of the representation mirroring the object.

Scientific representations can change. Leonardo represented the pendulum diagrammatically, Galileo and Huygens represented it in geometric form, Newton represented it algebraically. The variously theorized pendulums are not meant to correspond to real objects: What does it mean for a sentence to correspond to a real object? For a point to correspond to an apple? Likewise the idea of a group's average age may not correspond to anything in the sense that no one may be the average age. Yet the notion of a group's average age, weight, intelligence, longevity etc. is perfectly respectable and usable, and is the "thing" that social scientific theories have to explain, and are judged against. Representations are in the domain of discourse and are separate from the domain of the real. Thus they cannot, in any serious sense, mirror or correspond to real states of affairs. Their adequacy and theoretical utility does not depend upon correspondence.

For pendulums, even highly refined experimental apparatus will give a scatter of data points. The laws of pendulum motion are not meant to, and cannot, explain these data points. They are too erratic. However in science, from data comes phenomena, and it is the phenomena which is the subject of scientific laws and theories. Often a line of best fit is put through the data points, and the line is then taken to represent the phenomena being investigated. Thereafter it is the phenomena which are discussed and debated, not the data. Any number of individual telescopic observations, when corrected and selected, constitute astronomical data. From this we infer, construct, invent planetary phenomena: circular orbits, elliptical orbits, heliocentric or geocentric orbits. The latter are not seen. They are not observational. But this is no scientific impediment. Once we settle on the phenomena, it becomes the subject matter of our scientific theories. New-

ton, in Book II of his *Principia*, after laying out his rules of reasoning in philosophy (our science), had a section on phenomena. Among six phenomena that he
believed his system of the world has to account for, were:

> That the fixed stars being at rest, the periodic times of the five primary planets, and
> (whether of the sun about the earth, or) of the earth about the sun, are as the 3/2th
> power of their mean distances from the sun.
>   That the moon, by a radius drawn to the earth's centre, describes an area proportional
> to the time of description. (Newton, 1729/1934, pps. 404–405)

These were not observational statements, and they were not data in the terms
we are using. They were statements of the phenomena to be explained. As Newton acknowledged, these phenomena came from the work of the giants on whose
shoulders he stood: Galileo, Kepler, and Brahe.[36]

Kepler's "elliptical planetary paths" were, in turn, phenomena separate from,
and not necessarily implied by his astronomical data and measurements. As William Whewell noted in the 19th century in his critique of Mill's inductivist account of science, the concept of an elliptical path was supplied by Kepler's mind,
not by his data. There is usually no univocal inference from data to phenomena.
Phenomena is underdetermined by data, just as theory is underdetermined by evidence. In the above case, the data is probably consistent with periodic times
of 5/4 power of mean distance.

In social science it is notoriously hard to establish the phenomena, even when
data is uncontroversial. Data from IQ testing is consistent with phenomena of low
intelligence, low motivation, low reading ability, and so on. In this case, just what is
the phenomena that theory has to explain, is up for grabs, or up for ideological contest. Are we to explain a group's low intelligence, their low motivation or their
low reading ability? It all depends on what phenomenon we take the data as revealing. Is the phenomenon to be explained in the Gulf War, defense of democracy or
defense of petroleum interests? Does a child's behavior reveal attention deficit disorder or spoiled brat syndrome? Is a person acting morally or serving their self-inter-
est? Are the police maintaining law and order or furthering the interests of the
ruling class? Is a woman exercising her right to choose or is she killing an infant? And
so on. The theoretical explanations will differ depending on how the phenomenon
is described and conceptualized. The road from data to phenomena is rocky and
strewn with methodological, theoretical, ideological, and cultural obstacles.

Data is idiosyncratic. Different scientists, using different equipment, test
procedures, statistical analyses, will generate different data. But this does not
necessarily imply different phenomena. Pooling idiosyncratic data, triangulating,
is meant to establish more firmly the relevant phenomena. One of Galileo's major
achievements was to "reduce" the observationally different motions of freefall,
levers and inclined planes to that of the balance. The balance became a "model of
intelligibility" of all these motions (Machamer and Woody, 1994). The motions
were "seen" as examples of the balance, and of Archimedean balance principles.

Aristotle was the prophet of observation in science; he elevated observation to a position of epistemological primacy in natural philosophy from which it has seldom been moved. The British empiricists, with their commitment to tracing back all meaningful statements to sense impressions, continued, under a different name, this Aristotelian orientation.[37] The 20th century positivists and logical empiricists were also Aristotelian as regards the primacy they accorded observation, and observation statements in their logic of science. Ernest Nagel, in his influential *The Structure of Science*, wrote:

> Scientific thought takes its ultimate point of departure from problems suggested by observing things and events encounted in common experience; it aims to understand these observable things by discovering some systematic order in them; and its final test for the laws that serve as instruments of explanation and prediction is their concordance with such observations. (Nagel, 1961, p. 79)

Postpositivist critics (Kuhn, Feyerabend, Rorty, etc.) took aim at this empiricist account of observation, but only to say that, contra positivist hopes, observation was infected with theory. Most postpositivists did not dispute the Aristotelian-derived primacy accorded to observation by the positivists and empiricists that they were railing against. Sandra Harding, for instance, wrote:

> what the sciences actually observe is not bare nature but always only nature-as-an-object-of-knowledge—which is always already fully encultured. (Harding, 1994, p. 89)

Despite the often strong words, argument about the theory dependence, or otherwise, of observation was basically an inhouse dispute. The pendulum story suggests that it is the fundamental commitment, by both sides, to observation sentences as the foundation of science, that is problematic. Recall Huygens comment on how he discovered the isochronism of the cycloid curve: "This we did by following in the footsteps of the geometers." And Galileo's constant recourse to "accidents" and "impediments" to maintain his pendulum laws against del Monte's empirical evidence. Wallis Suchting, in a paper on "The Nature of Scientific Thought," commented:

> Thus the *key* inadequacy of empiricism has really nothing to do with the centrality it accords to sense-experience; in particular, the controversy over whether the "basic language" of science should be "phenomenonalistic" or "physicalistic" is irrelevant to the main question, a mere internal family dispute, as it were. The central deficiency of empiricism is one that it shares with a wide variety of other positions, namely, all those that see objects themselves, *however they are conceived*, as having epistemic significance *in themselves*, as inherently determining the "form," as it were, of their own representation, rather than as determining the degree of applicability of representations of a given "form," and hence, conversely, that the nature of what is represented can be more or less *directly* "read off its representation." (Suchting, 1995, p. 13)

Research on Scientific Representation and Learning

The question of scientific representation of phenomena, and issues raised for science learning by the choice of representation, has recently attracted the attention of science education researchers. For instance, Wolf-Michael Roth, Kenneth Tobin, and Kenneth Shaw, as part of a larger research project, observed a routine physics lecture on the dynamics of a ball rolling on an inclined plane. The lecture employed demonstrations, verbal descriptions, graphs, algebra and calculus. The researchers were concerned to isolate the reasons why this topic was so poorly learned and understood. They related that:

> Experienced physics teachers know that newcomers to the field have considerable difficulties understanding the presumed isomorphism between a natural phenomenon such as a rolling ball and the inscriptions used to re-present it. (Roth et al., 1997, p. 1088)

They went on to say that:

> As part of our overall research program, we wanted to understand why lectures which attempt to illustrate these isomorphisms are difficult to understand. (Roth et al., 1997, p. 1088)

Using the terminology introduced above, we can see that these researchers have data (video recordings, their notes, interview transcripts etc), and their characterization of the phenomenon that they are investigating is the lecturer's attempt "to illustrate isomorphisms between scientific representations and phenomena." But is this really the phenomenon? In terms of the philosophical position developed in this book, there is not meant to be any isomorphism between representation and objects. There is a relationship of adequacy or inadequacy between representations and objects, but this is not the same as isomorphism or lack of isomorphism. The idea of isomorphism between representation and object is, as has been shown above, entrenched in empiricist understanding of science. And it remains there even with avant garde empiricists who held up the isomorphic standard in order to criticize science for failing to meet it. The point is that the standard is irrelevant to science, and consequently to judgments about science. Maybe the lecturer thought he or she was providing isomorphic representations to students, but maybe the researchers were just projecting their empiricist understanding of science on to the teacher. They were creating the phenomenon that their research program was meant to explain, but the data did not unequivocally point to this particular phenomenon.

This interpretation is strengthened when we further read that:

> Our analysis was based on the assumption that the isomorphism between phenomena and inscriptions does not exist *a priori* but is constructed through scientific practice. (Roth et al., 1997, p. 1088)

That is, crude empiricism—the British empiricists who thought that objects when sensed impressed their "faint images" on the understanding—thought that isomorphism was direct (a priori is the term used by the researchers). But Roth and colleagues saw that the representations had a more complex origin, they were "constructed through scientific practice." This of course is entirely true. The problem seemed to be that the researchers remained committed to the crude empiricist idea that the representations were nevertheless meant to be isomorphic with the real objects and events. This mistaken commitment, coupled with their correct recognition that forms of scientific representation were produced in scientific practice, enabled them to argue for constructivist and relativist interpretations of science. But these conclusions simply did not follow from recognition of the complexity and creativity behind scientific representations. The latter story is fully compatible with realist and fallibilist accounts of science.

REDUCTIONIST ACCOUNTS OF TIME

Philosophers in ancient Greece concerned themselves with the definition of time, and of the relationship between time and its putative measures—what we might call the "validity of the measuring instrument." Plato briefly discussed time in his *Timaeus* dialogue. He suggested that time only came into existence when the world was ordered by the *Demiurge*. In the preexisting unordered world, time (unlike space) did not exist. Time thus has a beginning, and possibly an end:

> For there were no days and nights, months and years, before the Heaven came into being; but he planned that they should now come to be at the same time that the Heaven was framed. . . . as they were brought into being together, so they may be dissolved together, if ever their dissolution should come to pass. (*Timaeus* 37–38, Allen, 1966, p. 274)

Time exists because planets have motion:

> In virtue, then, of this plan and intent of the god for the birth of Time, in order that Time might be brought into being, Sun and Moon and five other stars—"wanderers," as they are called—were made to define and preserve the numbers of Time. (*Timaeus* 38, Allen, 1966, p. 275)

Aristotle's Reductionist Account of Time

Aristotle was a reductionist about time. Time did not exist apart from change in, or movement of, bodies. If there were no bodies that moved, there would be no time. His analysis was more detailed than Plato's, whose views he regarded as "too naïve" (*Physics* 218^{b}5, Barnes 1984, p. 370). In the *Physics* he began his discussion of the topic by noting Plato's views, and then observing:

> But neither does time exist without change; for when the state of our minds does not change at all, or we have not noticed its changing, we do not think that time has elapsed, any more than those who are fabled to sleep among the heroes in Sardinia do when they are awakened; for they connect the earlier "now" with the later and make them one, cutting out the interval because of their failure to notice it. . . . [because] when we perceive and distinguish we say time has elapsed, evidently time is not independent of movement and change.

> We must take this as our starting-point and try to discover—since we wish to know what time is—what exactly it has to do with movement. (*Physics* 218^b 21–30, Barnes, 1984, p. 371)

And further:

> But we apprehend time only when we have marked motion, marking it by before and after; and it is only when we have perceived before and after in motion that we say that time has elapsed. For time is just this – number of motion in respect of "before" and "after." (*Physics* 219^a 25, 30, Barnes, 1984, p. 371)

He listed some "attributes" of time, namely its countability and its continuity:

> . . . time is not movement, but only movement in so far as it admits of enumeration. (*Physics* 219^b 1, Barnes, 1984, p. 372)

And,

> Just as motion is a perpetual succession, so also is time. (*Physics* 219^b 10, Barnes, 1984, p. 372)

And some things that could not be properties of time:

> It is clear, too, that time is not described as fast or slow, but as many or few and as long or short. For as continuous it is long or short and as number many or few; but it is not fast or slow – any more than number with which we count is fast or slow. (*Physics* 220^b 1–5, Barnes, 1984, p. 373)

He then stated :

> Not only do we measure the movement by the time, but also the time by the movement, because they define each other. (*Physics* 220^b 15, Barnes, 1984, p. 373)

The crucial question then arose of what movement will measure time?

> One might also raise the question of what sort of movement time is the number of. Must we not say "of *any* kind"? For things both come into being in time and pass away, and grow, and are altered, and are moved locally; thus it is of each movement *qua* movement that time is the number. (*Physics* 223^a 30, Barnes, 1984, p. 377)

But he ruled out all movements, bar one, as the correct measure of time:

> Now there is such a thing as locomotion [local movement], and in locomotion there is included circular movement, and everything is counted by some one thing homogeneous with it, units by a unit, horses by a horse, and similarly times by some definite time, and, as we said, time is measured by motion as well as motion by time (this being so because by a motion definite in time the quantity both of the motion and of the time is measured): if, then, what is first is the measure of everything homogeneous with it, regular circular motion is above all else the measure, because the number of this is the best known. Now neither alteration nor increase nor coming into being can be regular, but locomotion can be. This also is why time is thought to be the movement of the sphere, viz. Because the other movements are measured by this, and time by this movement. (*Physics* 223^b 11–24, Barnes, 1984, p. 378)

Movement, or local motion, could be either rectilinear or circular or some combination of both. Elsewhere in the *Physics*, rectilinear motion is ruled out as a measure of time. First, because:

> For the line traversed in rectilinear motion cannot be infinite; for there is no such thing as an infinite straight line; and even if there were, it would not be traversed by anything in motion; for the impossible does not happen and it is impossible to traverse an infinite distance. On the other hand rectilinear motion on a finite line is composite if it turns back, i.e., two motions, while if it does not turn back it is incomplete and perishable; and in the order of nature, of definition, and of time alike the complete is prior to the incomplete and the imperishable to the perishable. Again, a motion that admits of being eternal is prior to one that does not. Now rotatory motion can be eternal; but no other motion, whether locomotion or motion of any other kind, can be so, since in all of them rest must occur, and with the occurrence of rest the motion has perished. (*Physics* 265ᵃ 15–25, Barnes, 1984, p. 442)

And second, because:

> Again, rotatory motion is also the only motion that admits of being regular. In rectilinear locomotion the motion of things in leaving the beginning is not uniform with their motion in approaching the end, since the velocity of a thing always increases proportionately as it removes itself farther from its position of rest; on the other hand rotatory motion alone has by nature no beginning or end in itself but only outside. (*Physics* 265ᵇ 11–15, Barnes, 1984, p. 443)

Having established what the inprinciple requirements were for a measure of time, and what kinds of motion could satisfy them, Aristotle, given his overall cosmology, could look only one way to find a circular movement that satisfied them: the heavens. In his cosmology, everything in the sublunar realm was subject to imperfect motion. Things came into being and passed away; and through the time of their existence, they did not manifest perfect, regular, continuous motion. Only the heavenly realm, the supralunar realm, exhibited perfect motion. Thus the rotation of the spheres, carrying the planets, stars and sun, instantiated and measured time. The true clock was the heavenly clock. Terrestrial clocks (hour glasses, clypesedra, etc.) were mere practical approximations to a clock. For Aristotle, the movement of the heavenly bodies not only measured time, but this movement constituted time. Time was, for Aristotle, identical to its measure. He was a temporal reductionist.[38]

Saint Augustine's Antireductionism

Saint Augustine, 800 years after Aristotle and at the beginning of the Middle Ages, put most strongly the nonreductionist case. This was done in Book XI of his *Confessions*, a book, which along with his *City of God*, had enormous impact on the development of Western thought, religion, and culture. Saint Augustine, at the beginning of his classic discussion of time in the *Confessions*, asked: "What, then, is time? I know well enough what it is, provided that nobody asks me; but if I am asked what it is and try to explain, I am baffled" (Augustine, 400/1961, p. 264). On this, Augustine spoke for many. Nevertheless Augustine attempted to answer his question, and did so in a way that laid out the main concerns about time that philosophers have entertained in the subsequent 1500 years. As one contemporary philosopher of time said, "In a modern translation much of Saint Augustine's work on time would pass for 20-century philosophy" (Price, 1996, p. 12).

After Augustine's introductory self-deprecatory apology for knowing what time is only so long as no one asked him, he said that:

> I once heard a learned man say that time is nothing but the movement of the sun and the moon and the stars, but I did not agree. Would it not be more likely that time was the movement, not only of heavenly bodies, but of all the other bodies as well? If all the lights of the sky ceased to move but the potter's wheel continued to turn, would there not still be time by which we could measure its rotation? (Augustine, 400/1961, p. 271)

And then, befitting the intensely religious and spiritual nature of this writing, Augustine went on to say:

> O God, grant that men should recognise in some small thing like this potter's wheel the principles which are common to all things, great and small alike. (Augustine, 400/ 1961, p. 271)

He entertained the *theoretical* possibility that the sun might vary in the duration of its orbit, a possibility that was, at the time, contrary to science and common sense. Augustine said that it was perfectly sensible to recognize that whatever the putative standard of time, "The same body might move at different speeds, and sometimes it is at rest, and we measure not only its motion but also its rest by means of time. . . . Time, therefore is not the movement of body" (Augustine, 400/1961, p. 273). As Augustine proceeded to grapple with his understanding of time, he said: "I do indeed measure it [time], but I do not know what I measure" (Augustine, 400/1961, p. 274).[39] After pointing out that judgments of a long and short poem are made not according to the length of pages, but according to the duration of recitation, and that this depended on subjective choices, Augustine articulated a view of time that has cast a long shadow:

> It seems to me, then, that time is merely an extension, though of what it is an extension I do not know. I begin to wonder whether it is an extension of the mind itself. (Augustine, 400/1961, p. 274)

Augustine granted that we measure time. Yet he recognized that we could not measure the past, because it is gone, nor the future, as it is not yet, nor even the present, as this has no duration (Augustine, 400/1961, p. 275). Just as the past does not exist, so also the beginning of events whose duration we are attempting to measure, has ceased to be while we measure. Thus the temporal dimension of objective, external events is not measurable.[40] What is measurable, is the impression that these external things leave in our mind, in our memory. After discussing the temporal measurement of recited poetry, Augustine concluded:

> It is in my own mind, then, that I measure time. I must not allow my mind to insist that time is something objective. I must not let it thwart me because of all the different notions and impressions that are lodged in it. I say that I measure time in my mind. For everything that happens which happens leaves an impression on it, and this impression remains after the thing itself has ceased to be. It is the impression that I measure, since it is still present, not the thing itself, which makes the impression as it passes and then moves into the past. When I measure time it is this impression that I measure. Either, then, this is what time is, or else I do not measure time at all. (Augustine, 400/1961, p. 276)

The Scientific Revolution's Undermining of Reductionism

Reductionism was attractive, or at least viable, so long as some process was demonstrably regular. As has previously been mentioned, the favored celestial processes for the reduction of time (the daily rotation of the earth, or the annual orbit of the earth) proved not to be regular. Tycho Brahe's sighting in 1572 of a new star, and his subsequent proof that comets were not sublunar phenomena, put paid to the perfect heavens of Aristotelian cosmology. Novas, or new stars, contradicted the idea of unalterable heavens. Comets punctured, so to speak, the ancient idea of heavenly spheres. Brahe said:

> . . . all the comets observed by me moved in the ethereal regions of the world and never in the air below the moon as Aristotle and his followers have tried without reason to make us believe for so many centuries; . . . the results pertaining to comets, the true ethereal nature of which I prove conclusively, show that the entire sky is transparent and clear, and cannot contain any solid and real spheres. For the comets as a rule follow orbits of a kind that no celestial sphere whatever would permit . . . we have found that there is no such thing as penetration of spheres and limits of distance, as the solid spheres do not really exist. (in Ariotti, 1973, p. 40)

When Kepler found that the planets moved not in circles but in ellipses, with their speed varying depending upon their distance from the sun, then the Aristotelian conviction that the heavens could provide us with a valid clock, with a reliable and regular measure of time, had to be abandoned.

Galileo's isochronic pendulum seemed to be an ideal substitute, not only was it regular, but it was tangible and immediate. However, Mersenne and Descartes cast doubt over the empirical adequacy of this option. Huygens' mathematical analysis killed it off. Huygens, however, thought his cycloid pendulum was an adequate substitute, that time could be reduced to its motions. He thought the cycloid pendulum could both measure and instantiate time—for "all times and all nations."

As we have seen, Robert Hooke disputed with Huygens, or at least with

Huygens' supporters in the Royal Society, over a number of Huygens' pendula claims. Crucially he opened up an in-principle problem with all pendula measures of time:

> . . . for if the gravity of the earth be altered, either by time or place, all endeavours of making a standard this way will be in vain; for whensoever and wheresoever the gravity of the earth is stronger, there must the length of a seconds pendulum be much longer; and when and where it is weaker, there must it be shorter. (in Ariotti, 1972, p. 386)

Hooke's inprinciple problems became Richer's and Halley's real problems. Together they buried the cycloid pendulum as the perfect measure and instantiation of time. But this also amounted to a burial of Aristotle's original claim that time and its measure were identical. The assumption was that if the measures (originally celestial, and then terrestrial) were flawed, then time had to be rescued from them.

Newton and Mach on Absolute Time

Newton, with his idea of absolute time, cut time loose from its measures, from attempts to reduce it to movement of any kind. In the famous Scholium appended to the 1717 second edition of his *Principia*, he wrote:

> Absolute, true, and mathematical time, of itself, and from its own nature flows equably without relation to anything external, and by another name is called duration: relative, apparent, and common time, is some sensible and external (whether accurate or unequable) measure of duration by the means of motion, which is commonly used instead of true time; such as an hour, a day, a month, a year. (Newton, 1729/1934, p. 6)

Some paragraphs later Newton elaborated the distinction:

> Absolute time, in astronomy, is distinguished from relative, by the equation or correction of the apparent time. For the natural days are truly unequal, though they are commonly considered as equal, and used for a measure of time; astronomers correct this inequality that they may measure the celestial motions by a more accurate time. It may be, that there is no such thing as an equable motion, whereby time may be accurately measured. (Newton, 1729/1934, p. 7)

Great though Newton's authority was, his idea of absolute, suprasensible, space and time did not go unchallenged. Leibniz and Bishop Berkeley were the foremost contemporary critics. Subsequently, Ernst Mach (1838–1916) was perhaps the most vocal and prominent critic of Newton's account of absolute space and time. Writing in his monumental *Science of Mechanics*, Mach said:

> It would appear as though Newton in the remarks here cited still stood under the influence of the mediaeval philosophy, as though he had grown unfaithful to his resolves to investigate only actual facts. When we say a thing A changes with the time, we mean simply that the conditions that determine a thing A depend on the conditions

that determine another thing *B*. The vibrations of a pendulum take place *in time* when its excursion *depends* on the position of the earth. Since, however, in the observation of the pendulum, we are not under the necessity of taking into account its dependence on the position of the earth, but may compare it with any other thing . . . the illusory notion easily arises that *all* the things with which we compare it are unessential. . . . Time, accordingly, appears to be some particular and independent thing, on the progress of which the position of the pendulum depends . . . It is utterly beyond our power to *measure* the changes of things by *time*. Quite the contrary, time is an abstraction, at which we arrive by means of the changes in things. (Mach, 1883/1960, p. 273)

Mach, from his positivist perspective, then declared the notion of absolute time to be metaphysical and lacking in scientific interest:

A motion may, with respect to another motion, be uniform. But the question whether a motion is *in itself* uniform, is senseless. With just as little justice, also, may we speak of an "absolute time"—*of a time independent of* change. This absolute time can be measured by comparison with no motion; it has therefore neither a practical nor a scientific value; and one is justified in saying that he knows aught about it. It is an idle metaphysical conception. (Mach, 1883/1960, p. 273)

Mach thought Aristotle's reductionism was the only way to deal with abstract concepts such as time (and also space and length). They had to be reduced to, not merely indicated by, something concrete. The fact that, contrary to Aristotle's beliefs, there was no constant motion in the world, or heavens, to reduce time to, was a problem with which science had to learn to live.

In his 1882 lecture on *The Economical Nature of Physics*, Mach elaborated these reductionist ideas, and criticized Newton and Kant for not breaking with the essentially primitive inclination to induce abstractions from phenomena, and then to reify them. He wrote:

For the natural inquirer, determinations of time are merely abbreviated statements of the dependence of one event upon another, and nothing more. When we say that acceleration of a freely falling body is 9.810 meters per second, we mean the velocity of the body with respect to the centre of the earth is 9.810 metres greater when the earth has performed an additional 86400th part of its rotation—a fact which itself can be determined only by the earth's relation to other heavenly bodies. Again, in velocity is contained simply a relation of the position of a body to the position of the earth. Instead of referring events to the earth we may refer them to a clock, or even to our internal sensation of time. Now, because all are connected and each may be made the measure of the rest, the illusion easily arises that time has significance independently of all. (Mach, 1898/1986, pp. 204–205)

Thus for Mach, time was always local. It was the relationship of one event to another, with duration between the events being measured by some arbitrary standard which itself was just another regularly recurring event. Time was thus an *intra*-universe phenomenon. It made no sense to speak of time in relation to the universe as a whole. Questions, such as how long ago did time begin? or even, how old is the universe? were thought to be meaningless. There was no cosmic time, only local time. These criticisms by Mach of the conceptual foundation of

Newton's mechanics, despite their positivist-inspired shortcomings,[41] neverthe-less undermined the dogmatism and closed mindedness of 19th century physics and prepared the way for Einstein's overturning of the very idea of time as an independent quality. Mach's *relational* and *reductionist* views of time became the ruling orthodoxy in positivist philosophy;[42] and, usually without acknowledgment, in operationalist-inspired science textbooks (Assis and Zylbersztajn, 2001). Richard Feynman, for instance, in his *Lectures in Physics*, wrote that putative time standards could not be labeled more or less accurate, only more or less reliable. By this he meant, more or less regular and convenient. This is a restatement of Mach's claim against Newton, a claim that has become so orthodox that its provenance need not be given.

CHAPTER 11

Teaching and Learning about Time and Pendulum Motion
Some Theoretical Considerations

Investigation of the concept of time brings together historical, cultural, anthropological, psychological, educational, and philosophical considerations. Time is a rich and multifaceted concept. It has been noted earlier that time measurement—which is different from timekeeping—could not occur until people thought that time was something measurable. That is, time measurement depended upon an understanding of time as an independent variable, and as something with metric (measurable) properties. This understanding was, so to speak, a long time coming. Indeed it did not crystalize until the 16th century and then only in Europe.

SOCIAL AND PERSONAL EMERGENCE OF THE TIME CONCEPT

The medievals regarded the study of motion as the heart of science, or natural philosophy.[1] "Ignorance of motion means ignorance of nature," was an oft repeated motif. Yet, as Marshall Clagett and other historians of medieval science have noted, what bedeviled the efforts of the medievals was their inability to identify time as an independent variable against which other changes could be measured. It was not just the *technical* problem of their not being able to measure time accurately, but it was a *conceptual* problem of not having the idea of time as something independent of motion that could be measured. In Gaston Bachelard's terms, this lack of the concept of independent and measurable time was an *epistemological obstacle* to the progress of medieval mechanics (Bachelard, 1934/1984). Charles Gillispie pointed to this when he identified Galileo's primary achievement in physics:

... what was revolutionary in Galileo's law of falling bodies [was] that he treated time as an abstract parameter of a purely physical event. This enabled him to do what no Greek had done, to quantify motion with numbers. Galileo spent twenty years wrestling with the problem before he got free of man's natural biological instinct for time as that in which he lives and grows old. Time eluded science until Galileo. (Gillispie, 1960, p. 42)

Gillispie added: "Again and again, Galileo himself tried to find a general expression for the velocity of a body in relation to the distance it had fallen. And always he failed" (ibid.).

Music and the Concept of Metrical Time

Galileo was a genius, but even geniuses are dependent upon their circumstances. Galileo did not just sit in a room and by great powers of concentration think up the new science. Nor did he begin *de novo* by playing around with objects. He brought great intelligence, patience, courage, and insight to his circumstances, but he was nevertheless still dependent upon them. There has been enormous scholarship devoted to identifying the constellation of factors that precipitated the scientific revolution in western Europe in the late 16th and early 17th centuries.[2] The birth of modern science, and of the scientific way of thinking, was historically a very contingent event. It did not happen earlier, and it did not happen outside of a small part of western Europe.

Music was one important factor in the rise of modern science.[3] Galileo's father was a musician and the author of innovative treatises on musical theory (Brown, 1992). Kepler, Galileo, Mersenne, and Huygens were all musical theorists and conducted experiments on aspects of the physics of music. The influence of music on the birth of modern science can be seen in the titles of Mersenne's major work, *Harmonie Universelle* (1636) and Kepler's monumental *Harmonice Mundi* (1619). Music is an especially important consideration for the subject matter of this book. Stillman Drake, for instance, is of the opinion that:

without Galileo's having been present at this father's musical experiments in 1588, he probably would not have gone on to his own study of pendulum motion. Without musical training, Galileo would hardly have been able to make his very first timings nearly exact. Music played not only a unique, but an essential role in leading Galileo to his new physics, a science of precise measurements, for music is an art demanding precise measurement and exact divisions. (Drake, 1992, p. 15)

More specifically, with respect to the emergence of the idea of time as an independent and measurable quality, Geza Szamosi nominates the "theory and praxis of polyphonic music and its measured notations" as the crucial factor. He said of polyphonic music that it:

... combines two essentially different but simultaneously sounding melodic lines. ... It was in the very nature of this music that it could not develop without a notational system which allowed for the symbolic representation of measured time

> intervals . . . the composer had first to . . . regulate the temporal profile of each melody by the same time unit, a time standard. . . . Time, in other words, has to be measured—in the proper sense of this word—exactly. (Szamosi, 1990, p. 184)

In polyphonic music and singing, voices are not in unison but in different and independent parts. This unique kind of singing arose in Europe in the 11th century. The parts required written notation and timings to release them from the bounds of mere memory and to enable them to be repeated by different groups in different places. With this notation and timing, written composition could replace improvisation in the creation of musical works (Grout and Palisca, 1960, Crosby, 1997, Chap. 8).

Szamosi made a particular point about the place of symbolism, language and culture in the measurement of time. This is not so much the theory dependence of observation (hearing), but the theory dependence of *articulating* observations. It is the latter which is fundamental to science.

> Let me emphasize this point again: The necessity for using exactly measured time in composition arose in *polyphonic music only*. Time units are often needed in monophonic music as well. A singing or dancing or marching group cannot stay together without using the same time unit. They have to keep time, in other words. But . . . keeping time is not time measurement. *Horses can easily be trained to keep time. They cannot be trained to measure time.* Measurement is a symbolic operation involving the conscious use of units or standards, the comparison of simultaneously passing time intervals. This was an operation which evolved in polyphony and in polyphony only. (Szamosi, 1990, p. 187)

When discussing the development of concepts it is important to distinguish historical or cultural development from personal development. The types of questions that can be asked of both, are different. In the first case we are working with the origin of novel ideas—the emergence of the concepts of metric time, linear inertia, mass-velocity dependence, animal evolution, the unconscious, women's liberation, class struggle, and so on. There is an element of *discovery,* or *innovation,* or *creativity* involved here. In the second case (an individual's acquisition of the concepts of metric time, linear inertia, animal evolution, the unconscious, women's liberation, class struggle), what we are overwhelmingly dealing with is *transmission,* or *learning,* or *initiation:* in brief, *education.* There are elements of creativity and discovery, but phylogenetic emergence of ideas, and ontogenetic emergence are different processes. Newton's idea of linear inertia was genuinely novel; no one had previously thought of it. When students first learn about linear inertia in a mechanics course, it is a new idea for them, but it is not a novel idea for the species. There is a parallel consideration when considering the metric time concept.

Aristotelian Empiricism and the Origin of Concepts

For Aristotle, a person looked at something and the mind identified, through a process of abstraction, an essence which typified or characterized the phenomenon. It is more and more prone to error as the judging mind, *nous*, ascends from the immediate sensory properties appropriate to the sensory organ. Aristotle distinguished a foundational and vertical level of perception by individual sense organs of states of affairs appropriate to them (the ear hearing, the eye seeing, the finger touching, etc.) from a higher level, "commonsense" perception, where the individual perceptions are aggregated and elements of judgment enter.[4] This is an important consideration because the empiricist tradition wished to identify a "bedrock" level of perception that is free of inference and free of judgment. With Aristotle such "bedrock" perception is below the level of common sense, because there concepts, language, and judgment are involved. As he wrote:

> Perception of the special objects of sense is never in error or admits the least possible amount of falsehood. Next comes perception that what is incidental to the objects of perception *is* incidental to them: in this case certainly we may be deceived; for while the perception that there is white before us cannot be false, the perception that what is white is this or that may be false [e.g. is a man or not, 430b, 24]. Third comes the perception of the common attributes which accompany the incidental objects to which the special sensibles attach (I mean e.g. of movement and magnitude); it is in respect of these that the greatest amount of sense-illusion is possible. (*De Anima* iii, 428b, 20–25; Barnes, 1984, p. 681)

Aristotle's theory of perception and the origin of ideas is *individualist*: The observer stood before an object or event and as the sense organs were stimulated, the mind progressively tried to ascertain the *essence* of the object or event. His theory is also *passivist;* in as much as the observer interfered with the object being observed, then distortion entered into judgment; the true essence of the object could not be manifest because the *natural* behavior of the object was being disturbed by the observer's interference.[5]

The British empiricists, Locke and Hume, merely modified Aristotle's account when they discussed the origin of ideas or concepts: first we have sensations, and then faint images of these are etched in the mind. "Nothing [sensible, intelligent, truthful] in the mind that was not first in the senses," was the claim. John Locke in his *Essay Concerning Human Understanding* gave a classical empiricist statement of the origin of ideas:

> Let us then suppose the mind to be, as we say, white paper, void of all characters, without any ideas; how comes it to be furnished? Whence comes it by that vast store, which the busy and boundless fancy of man has painted on it with an almost endless variety? Whence has it all the materials of reason and knowledge? To this I answer, in one word, from EXPERIENCE; in that all our knowledge is founded, and from that it ultimately derives itself. Our observation, employed either about external sensible objects, or about the internal operations of our minds, perceived and reflected on by ourselves, is that which supplies our understandings with all the materials of thinking. These two are the fountains of knowledge, from whence all the ideas we have, or can naturally have, do spring. (Locke, 1690/1924, pp. 42–43)

Hume continued this analysis, becoming more of a sensationalist than Locke and more committed to the epistemological role of impressions. Sense impressions are the chief component of Hume's famed "philosophical microscope." He divided the content of the mind into two classes; thoughts (ideas) and impressions (Hume, 1777/1902, p. 18). By the term "impression" Hume meant "all our more lively perceptions, when we hear, or see, or feel, or love, or hate, or desire or will" (ibid.). Hume continued, "to express myself in philosophical language, all our ideas or more feeble perceptions are copies of our impressions or more lively ones" (Hume, 1777/1902, p. 19). This quintessential empiricist way of posing the problem of perception has cast a long shadow. One can detect its influence in the writings of 20th-century sense data theorists such as Alfred Ayer, H.H. Price, and Gilbert Ryle.[6]

Although Locke and Hume championed the new science, Galileo's work on the pendulum hardly fits their empiricist scheme for the origin of ideas or even the vindication of ideas. As one commentator observed:

> . . . the concept, say, of a "material point with a determinate mass" does not constitute the common essence of apples, planets and projectiles. It is rather a concept of the conceptual scheme of physics which is *produced* together with the other concepts of this system and which is, precisely, *attributed* to such real objects so that their movement may be accounted for by this system as a whole. (Baltas, 1988, p. 216)

Looking at the bob of a pendulum could never, as Hume and Locke suggested, produce a faint image of a material point. Nor does watching an accelerating body produce the concept of acceleration as rate of change of velocity with respect to time elapsed. This consideration has pedagogical consequences both for the acquisition of concepts and for conceptual change, some of which are discussed in Schecker (1992). This is related to Popper's recognition:

> In science it is *observation* rather than perception which plays the decisive part. But observation is a process in which we play an intensely *active* part...An observation is always preceded by a particular interest, a question, or a problem—in short, by something theoretical. (Popper, 1972, p. 342)

Transmission or Construction of Concepts? A Problem for Constructivism

The result of observation is a *theorized object* or *event,* to use the terms introduced above. In economics, the falling apple is described as a wasted *commodity*; in physics, as an *accelerating point mass*; in agriculture, as *cattle feed*; in poetry, as *nature's gift*, and so on. One implication is that the theoretical structure that precedes observation is something that students need to receive from teachers. The quality of teaching is a function of the grasp that teachers have of the relevant theoretical systems and their ability to impart it in clear and understandable terms. This is the point Richard Dearden made in his criticism of Discovery Learning programs of the 1960s:

> Far from being like trotting around a garden, learning science and what is characteristic of scientific inquiry involves initiation into a social tradition of inquiry and is therefore something which, one way or another, has to be taught. (Dearden, 1967, p. 143)

Note that Dearden said "in one way or another" science has to be taught. There is no automàtic connection between the transmission model of education, and authoritarian, nonengaging, pedagogy. Just as there are many ways to skin a cat, so, too, there are many ways to teach something. But at the heart of science are concepts, and these need to be understood first. Scientific concepts are social and historical constructions; they are *defined.* And definitions are not discovered or constructed. Students do not discover, much less construct, what *momentum, power, acceleration, valency, force, mass, weight, oxidation,* and so on, mean: They *learn* what these terms mean. They may learn more or less badly depending on how prepared they are and how well the concepts are presented, and they have to put effort into their learning, but all of this is a long way from students constructing their own definitions of scientific concepts.

This point was expressed by the Canadian Jesuit priest and philosopher, Bernard Lonergan who in an address on "The Teaching of Physics" said:

> Teaching physics without students knowing the relevant mathematics is not teaching physics. What does a scientist meant by acceleration? He means d^2s/dt^2. If you know what is meant by those symbols from the differential calculus, you know exactly what is meant by acceleration and velocity. . . . It is possible to give students who have not done the mathematics some approximate notion of it, but it will take them a great deal of time to understand that approximate notion, and when they get it, they will be able to do very little with it, because it is not accurate, and its implications do not stand out. . . . the teaching of physics without a proper account of the fundamental notions...gives an illusion of knowledge, a false idea of what science is. And it clutters the mind. (Lonergan, 1993, p. 145)

Carl Sagan has written of going to the University of Chicago after attending a less-than-satisfactory New York City high school where "there was no soaring sense of wonder, no hint of an evolutionary perspective, and nothing about mistaken ideas that everyone had once believed" (Sagan, 1997, p. 4). Of Chicago, he said:

> College was the fulfillment of my dreams: I found teachers who not only understood science, but who were actually able to explain it. . . . I was a physics student in a department orbiting around Enrico Fermi . . . I discovered what true mathematical elegance is from Subrahmanyan Chandrasekhar . . . teachers were valued for their teaching, their ability to inform and inspire the next generation. . . . I witnessed at first hand the joy felt by those whose privilege it is to uncover a little about how the Universe works. (Sagan, 1997, pp. 4–5)

That learning depends on transmission, and that the quality of learning is dependent upon the quality, clarity and intelligibility of what is transmitted is hotly debated among educators. Catherine Fosnot, in the preface to a book advocating constructivist pedagogy, wrote: "Teachers who base their practice on constructivism reject the notion that meaning can be passed on to learners via symbols or transmission" (Fosnot, 1996, p. ix). Likewise, other science educators have explained:

> Teaching based on this traditional view of science attempts to transmit to learners concepts which are *precise* and *unambiguous*, using language capable of *transferring*

ideas from expert to novice . . . The new paradigm [constructivism] regards science as a human and social construct, and views learning as the personal construction of new knowledge. (Carr et al., 1994, pp. 147, 149)

Ann Howe and Harriet Stubbs, in a recent award-winning article, rejected transmission views and embraced the constructivist account of knowledge development. They asked what is the source of knowledge, and answered:

> Theory and practice in science education have emphasized experience with phenomena as they occur in nature or in the laboratory followed by reflection and discussion as the source. Having experienced the event or made the observation, the learner works through the cognitive dissonance that results and, in the process, constructs new knowledge. (Howe and Stubbs, 1997, p. 170)

Of course by "new knowledge" what is meant is "new beliefs." Given general constructivist principles, the distinction between knowledge and belief usually collapses.

There are, to put not too fine a point on it, major problems with the foregoing antitransmission positions. Children can look at pendulums, as children and adults did for thousands of years, without seeing any of the scientific properties of the pendulum, without experiencing any dissonance. They will begin to have knowledge of the pendulum not when they look and discuss, but when they are effectively taught by teachers who have mastered the science and mathematics of pendulum motion and can convey this knowledge in a meaningful and engaging manner. This could mean giving lectures, setting pages of a text to read, organizing practical work and setting investigative questions, arranging small group discussions and so on. There is some crucial "soft-focus" wording in the Howe and Stubbs thesis concerning the type of discussion that is allowed. If discussion is peer discussion, then the thesis is manifestly false. But if discussion allows for the participation of discussants who know something, in effect teachers, but including textbook writers, then the thesis may be true. But not necessarily true: Discussion, or indeed teaching, does not necessarily lead to new knowledge. This is something every frustrated teacher and every learner knows.

There are well-identified tensions within constructivism on just how "personal" this personal construction of knowledge is supposed to be. For instance, Rosalind Driver and colleagues wrote:

> The objects of science are not the phenomena of nature but constructs that are advanced by the scientific community to interpret nature. (Scott et al., 1994, p. 6)

They then referred to Russell Hanson's discussion of Galileo's novel concept of acceleration to illustrate the importance of proper theoretical concepts for scientific advance and understanding:

> Hanson (1958) gives an eloquent illustration of the difference between the concepts of science and the phenomena of the world in his account of Galileo's intellectual struggle to explain free-fall motion. For several years Galileo collected measurements

of falling objects representing acceleration in terms of an object's change of velocity over a given distance, a formulation that led to complex and inelegant relationships. Once he began to think about acceleration in terms of change of velocity in a given time interval, then the constant acceleration of falling objects became apparent. (Scott et al., 1994, p. 6)

In addition,

The notion of acceleration did not emerge in a non-problematic way from observations but was imposed upon them. . . . The point is that, even in relatively simple domains of science, the concepts used to describe and model the domain are not revealed in an obvious way by reading the "book of nature." Rather, they are constructs that have been invented and imposed on phenomena in attempts to interpret and explain them, often as results of considerable intellectual struggles. (Scott et al., 1994, p. 6)

Having recognized these points about scientific concepts, constructivists are then left with a difficulty. How do children acquire the concepts if they, according to constructivist claims, have to construct them themselves? Undoubtedly the constructs are the result of earlier intellectual struggles (Huygens' replacement of Galileo's circle with the cycloid, for instance). Most would say that it is the teacher's job to teach these concepts in ways that are engaging and intelligible.

Initially the more individual-focused Piagetian constructivists thought children would construct a scientific understanding of the world for themselves. When this was seen as too ambitious a goal, but transmission of meaning was still seen as unacceptable, then constructivists embraced a more social constructivist position: not the individual alone, but the individual as part of a group, constructed scientific knowledge.[7]

One educational psychologist sympathetic to constructivism has addressed some of the foregoing points in explicitly Popperian terms. Carl Bereiter wrote:

The scholarly disciplines are distinguished from most other occupations by their concentration on World 3. They are concerned with producing and improving World 3 objects, such as theories, explanations, historical accounts, problem formulations and solutions, proofs and disproofs. (Bereiter, 1994, p. 22)

He concluded his discussion with an observation that echoed the distinction made earlier in this chapter between the social and personal origin of concepts:

To be of maximum help to students in this kind of endeavour, teachers need an epistemology that helps them distinguish between efforts directed toward the construction of knowledge and efforts directed towards changes in students' minds. Constructivist and sociocultural approaches are both of value, but neither one quite provides the tool to do that job. (Bereiter, 1994, p. 23)

Natural versus Social Origin of Concepts

The empiricist tradition, beginning with Aristotle and including the 17th century British empiricists and 20th century sense data theorists, ignored the contribution of society to the development of an individual's store of knowledge and to their conceptual world. Their epistemological problematic was that of a cognitive subject who observed the world and came to an understanding (or misunderstanding) of it. Jean Piaget (1896–1980), with one very important qualification, belongs to this tradition. His work warrants attention because he explicitly investigated children's ideas of time (Piaget, 1970) and their understanding of pendulum motion (Inhelder and Piaget, 1958, Chap. 4). Also because debate between Piagetian radical constructivists, such as Ernst von Glasersfeld, and more social constructivists, such as Rosalind Driver, in part hinge on the adequacy of Piaget's theory and methodology.

From the 1920s to the 1970s Jean Piaget advanced his philosophical program of *Genetic Epistemology*[8] and in this context he, and collaborators, investigated extensively the development of children's thinking. He took pains to *exclude* from his investigations, those aspects of a child's thinking that arose from cultural sources. In his 1929 *The Child's Conception of the World* he recognized the two sources of ideas, or "convictions" as he called them, saying that there are:

> . . . two very different types of conviction among children which need to be distinguished. Some are . . . influenced but not dictated by the adult. Others, on the contrary, are simply swallowed whole, either at school, or from the family, or from adult conversations which the child overhears, etc. These naturally have not the slightest interest. (Piaget, 1929, p. 28; in Russell, 1993, p. 73)

If in the investigation of the growth of ideas, the influence of school, family, and adult conversation—in other words, culture—is ignored, then Piaget does belong to the empiricist tradition. He investigated the mental processes that occur when a subject confronts the world and develops ideas about it. Piaget did not investigate as a psychologist who was interested in the origin of *ideas*, he investigated as an epistemologist who was interested in the origin and justification of *knowledge*.[9] He was interested in what we might call "unmediated knowledge": knowledge of nature unmediated by culture. Piaget did qualify the empiricist problematic, and wrote of "The Myth of the Sensory Origin of Scientific Knowledge" (Piaget, 1972, Chap. 4). There, after a brief survey of the empiricist position, he wrote: "Thus we may conclude that knowledge never derives from sensation alone, but from what action adds to this datum" (Piaget, 1972, p. 46). Piaget went on to say that, "The basic defect in an empiricist interpretation is that of neglecting the activity of the individual . . . [logicomathematical] activity of the individual remains indispensable in order to assure an objective reading of factual data" (Piaget, 1972, p. 62). That is, Piaget abandoned the *passivist* part of the Aristotelian–empiricist tradition, and recognized the importance of human activity, action and, speaking loosely, experimentation for the growth of ideas and knowledge. However Piaget retained the *individualism* of the Aristotelian program.[10]

Kieran Egan, in his *The Educated Mind*,[11] acknowledged the role of language in cognition, but still longed for the immediate sensory knowledge that is the hallmark of empiricism. As Egan stated the matter:

> Ours is, for much of the time, a peculiar *languaged* understanding of the world. . . . While we are, willy nilly, committed to languaged understanding, it seems to involve some loss of the instinctive, vivid, intimately participatory involvement with the natural world that characterizes our fellow mammal's understanding. The educational task is to make languaged understanding as rich as possible while losing as little as possible of the "oneness with nature" that is our birthright as animals. (Egan, 1997, p. 67)

Piaget offered only a limited criticism of empiricism. He allowed for agent's activity in the world, and even for the agent's reflection on their mental operations. However it was the individual who had ideas and constructed knowledge. The alternative, social constructivist tradition, traces its roots to the Soviet psychologist Lev Vygotsky who emphasized the social formation of mind and recognized the importance of language and participation in culture for the development of an individual's concepts and cognitive operations (Vygotsky, 1934/1962, 1978, Wertsch, 1985). Even so, this social constructivist research program is not epistemology. The study of mental life, even the social–historical–cultural study of mental life or consciousness, leaves aside considerations of how adequate or truthful are the conceptions, ideas, and understandings; and how rational is our assent to them. These latter investigations constitute epistemology, the former constitute psychology or social psychology.

Figure 92 illustrates the three sources of an individual's ideas or concepts— experience, education, culture. All three sources contribute to ideas about pendulum motion, but clearly *scientific* understanding of pendulum motion, and acceleration as Bernard Lonergan observed, will be dependent upon formal instruction, either from teachers, books or some other external source.

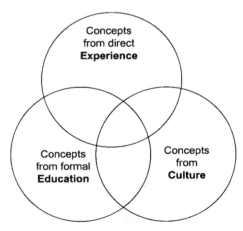

Figure 92. Direct experiential, cultural, and educational sources of ideas (Based on Russell, 1993, p. 71).

WHAT IS A MISCONCEPTION?

Teaching about pendulum motion focuses issues raised by the numerous studies of children's alternative frameworks (Hewson, 1985), preconceptions (Osborne and Freyberg, 1985) and misconceptions (Novak, 1983). For instance, did del Monte have misconceptions about pendulum motion? In one sense he did not. The pendulum behaved, approximately, as he and other Aristotelians said it did and not as Galileo claimed. Nevertheless del Monte had a major misconception in that his thinking was tied to the Aristotelian explanatory framework. His conceptions were, by the standards of the new science, mistaken.

Some educators, particularly constructivists, are reluctant to tell a student that they have a misconception, or a misunderstanding or, in brief, that they are wrong. One recent study of student experiences in a constructivist classroom, reported a student as saying:

> I never felt there was a sense of right or wrong. Occasionally you would put forward some idea and Mr. V. would say "Right that is fine," write it up on the board and collate it with all the other ideas. Occasionally you did see that perhaps your idea didn't quite fit into the chain of what was going on. But you were never actually told, no that is wrong, that is not right. It was always put up there with the other ideas and it might have sorted itself out. (Hand et al., 1997, p. 569)

This is not just a *tactical* or *pedagogical* reluctance to tell someone their ideas are wrong on the grounds that it might discourage a learner, that it might diminish their self-confidence, and so on. The reluctance derives from the constructivist theoretical notion that knowledge is an individual construction and that, provided ideas make sense to a student, they cannot be wrong. Some postmodernists express this as the idea that there are multiple realities, and hence multiple, and equally valid, ways of making sense of reality.

However to expect students to construct anything Newtonian, for instance, by playing around with objects and putting their ideas on the board or brainstorming their ideas, is to underestimate the intellectual revolution inaugurated by Galileo and Newton. It also underestimates the pedagogical problems in getting children to comprehend the classical scientific world-view. There is an important educational role for "messing about," as Hawkins has described it, for being acquainted with the phenomena, as Mach demanded, or for tinkering around, as Feynman has suggested, but this role is not sufficient for producing contemporary scientific concepts and understanding. Richard Dearden commented about "mucking around" that:

> With very young children especially there is an important place for this kind of learning, but such limited and undirected curiosity does not amount to science. All of this could and did and does go on where science had never been heard of. Such finding out does not even begin to resemble science until problems start to present themselves which cannot be solved without putting forward, and then testing experimentally, suggested solutions of a non-obvious kind. (Dearden, 1967, p. 143)

Galileo's conceptual scheme did not emerge from his playing with objects, it emerged by great mental effort using borrowed concepts and learned logical and mathematical techniques. Newton attributed, at one point, his success in discovering the law of universal gravitation to "industry and patient thought" (Westfall, 1980, p. 105), and at another point, to "standing on the shoulders of giants." Maybe both are lessons that can be relearned in science education: Learning takes effort and patient thought and is best done when there are shoulders to stand on.

It is notorious how poorly science students and teachers understand pendulum motion. Most teachers, and some students, can repeat the four laws of pendulum motion (period is independent of amplitude, independent of mass, varies as square root of length, and oscillations are isochronic for a given length), but their conceptualization of the movement is weak. Frederick Reif conducted revealing studies on this conceptualization. Students and teachers were given the drawing shown in Figure 93 for the pendulum's movement and asked to draw in the acceleration of the bob at each of the points A, B, C, D, E. Reif discovered 80% of physics students got the answer wrong. The correct answer to the question is given in Figure 94.

One can appreciate the intellectual problems involved in getting the correct answer. At the bob's lowest point, when its tangential speed is greatest, its acceleration is vertically upwards. When the bob is at its highest point and apparently stationary, it is nevertheless accelerating along the tangent of its arc. Figure 93 represents, phenomenologically and atheoretically, the real pendulum. Figure 94 represents the pendulum as described by the theory of classical mechanics, the theorized pendulum. Being able to represent the second correctly depends upon a sophisticated grasp of mechanics. Such a grasp is not generated by observing the pendulum. It develops only by good teaching done by knowledgeable teachers or from other sources, such as well-written textbooks.

The classroom problems encountered when the theoretical object of science is discordant with everyday experience of material objects is illustrated in the

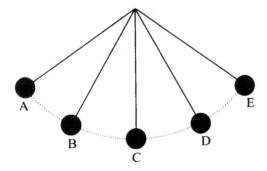

Figure 93. Moving pendulum (Matthews, 1994a, p. 133).

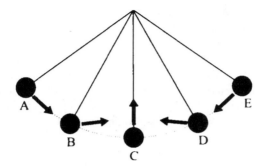

Figure 94. Acceleration of bob (Matthews, 1994a, p. 133).

following classroom exchange. Here a teacher, who previously introduced the concept of acceleration talks to three students who are struggling to apply the concept to the motion of a body thrown upwards that seems to come to a halt and then falls back down. This is precisely the pendulum situation.

Teacher: Suppose I toss a ball straight up into the air like this. (*demonstrates*) What is the ball's acceleration at the top of the trajectory?
Student 1: Zero.
Student 2: Yeah, zero.
Teacher: Why is it zero?
Student 1: Well, at the top the ball stops moving, so it must be zero.
Teacher: OK. If I place the ball on the table so that it does not move, is it accelerating?
Student 2: No. It is not moving.
Teacher: What if I roll the ball across the table so that it moves at a constant velocity (*demonstrates*). Is the ball accelerating in this case?
Students 1 and *2*: Yeah.
Student 3: No way! If the ball is rolling at a constant speed it does not have any acceleration because its speed does not change.
Teacher: So it appears that an object can have a zero acceleration if it is standing still or if it is moving at a constant velocity. Let's reconsider the case where the ball is at the top of its trajectory. (*demonstrates again*) What is the ball's acceleration when it is at the top?
Student 3: It would be zero because the ball is standing still at the top. It's not moving—it has to turn around.
Student 2: I think it might be accelerating because it gets going faster and faster.
Student 1: Yeah, but that doesn't happen until it gets going again. When it is standing still it is not accelerating.

(Mestre, 1991, p. 58)

This episode is difficult to reconcile with the previously-mentioned claim of Ann Howe and Harriett Stubbs that new knowledge results from observation plus discussion: "Theory and practice in science education have emphasized experience with phenomena as they occur in nature or in the laboratory followed by reflection and discussion as the source [of knowledge" (Howe and Stubbs, 1997, p. 170). The above discussion is not leading to knowledge because, among other things, the students have not grasped what acceleration means in scientific terms. Without this information they are just groping in the dark.

Idealization in Science and the Learning of Science

Idealization in science has been recognized as one of the major stumbling blocks to meaningful learning of science (Nersessian, 1992). This is in part because intuitive beliefs are so strongly influenced by everyday concrete experience. Lewis Wolpert, in his book, *The Unnatural Nature of Science* (1992), correctly remarked:

> Scientific ideas are, with rare exceptions, counter-intuitive: They cannot be acquired by simple inspection of phenomena and are often outside everyday experience . . . doing science requires a conscious awareness of the pitfalls of "natural" thinking. (Wolpert, 1992, p. xi)

He essentially repeats what Pierre Duhem said at the turn of the century, when he warned against grounding science instruction in common sense:

> Now is it clear merely in the light of common sense that a body in the absence of any force acting on it moves perpetually in a straight line with constant speed? Or that a body subject to a constant weight constantly accelerates the velocity of its fall? On the contrary such opinions are remarkably far from common-sense knowledge; in order to give birth to them, it has taken the accumulated efforts of all the geniuses who for two thousand years have dealt with dynamics. (Duhem 1906/1954, p. 263)

The history and philosophy of science (HPS) can contribute a good deal to student teachers' understanding of this fundamental issue. Schecker (1992) addressed some of these questions in an interesting way. He asked 254 high school students to comment upon the following statement:

> In physics lessons there are often assumptions or experiments of thought, which obviously cannot be realized in actual experiments, like completely excluding air resistance and other frictional effects or assuming an infinitely lasting linear motion.

The students were asked to tell whether the method was useful or not useful. There were 11% who said it was useless. "Why should I consider something that does not exist?" A large group, up to 50% said it was useful, but only for physics because physics did not deal with reality. "I don't need to refer everything to reality. I am simply interested in physics." Only about 25% had any comprehension of the method of idealization in science. Champagne and others expressed the matter this way: "The arduousness of learning mechanics is expressed in the effort required as students shift their thinking from one paradigm to another. Paradigm shifts are not accomplished easily, neither in the scientific enterprise nor in the minds of students" (Champagne et al., 1980, p. 1077).

History and philosophy can make the idealizations of science more understandable and can explain them as scientific tools of trade, or instruments, whereby the complex concrete world can be investigated. Just as material tools (spanners, beaters, microscopes) can be more or less useful, or more or less elegant, so also can the particular idealizations in science. To represent a planet as a point mass and its orbit as an ellipse, was the precondition for Newton's masterful mathematical analysis of planetary motion. In the light of refined measurements and observations, the initial assumptions can be altered and the picture of planetary motion made more realistic or concrete. Newton dealt first with the two-body problem in analyzing planetary motion, and then struggled a long time with the three-body problem. Eventually, the idealizations allowed a person to step out of a rocket and walk on the moon's surface. Some understanding of this is important for students being introduced to the "world of science."

A failure to appreciate what idealization is and is not has been at the basis of a lot of antiscience criticism. It was, of course, Newtonian idealization that the Romantic reaction was directed against. For them (Keats, Goethe, and so on) the rich world of lived experience was not captured by the colorless point masses of Newton. In the 20th century, Sartre, Marcuse, Husserl, Tillich, and others repeated versions of this charge (Dahlin, 2001). Sartre at one point said that evil is "the systematic substitution of the abstract for the concrete" (Passmore, 1978, p. 70). This substitution is precisely the charge of del Monte against Galileo. However this very substitution is the *raison d'être* of the scientific revolution. Although it might be fashionable to agree with Sartre, the price is the rejection of the unequalled methodology of the 17th century's new science. There is obviously a problem here. Aldous Huxley, at the end of World War II made an intelligent comment on the matter:

> The scientific picture of the world is inadequate, for the simple reason that science does not even profess to deal with experience as a whole, but only with certain aspects of it in certain contexts. All of this is quite clearly understood by the more philosophically minded men of science. . . . [Unfortunately] our times contains a large element of what may be called 'nothing but' thinking. (Huxley, 1947, p. 28)

The prominent German science educator, Martin Wagenschein, wrote on this dichotomy at the heart of an education in science:

> My deepest motive force is the cleavage between an original feeling-at-home within nature and a strong fascination by physics and mathematics, and the ensuing irritation by the growing alienation between man and nature effected by science, starting early at school. My pedagogical aim is to overcome—still better to avoid—the cleavage by an educationally centered humanistic physics teaching. (Wagenschein, 1962, p. 9, W. Jung trans.)

A historically and philosophically literate science teacher can assist students to grasp just how science captures, and does not capture, the real subjective lived world (Dahlin, 2001). An HPS-illiterate teacher leaves students with the unhappy choice between disowning their own world as a fantasy, or rejecting the world of science as a fantasy.

PIAGET ON CHILDREN'S UNDERSTANDING OF TIME

The development of children's concepts of time is an exemplary field for ascertaining the contributions of nature, culture, and education to the development of cognition. It is also an important field for education. Constructivists emphasize how children's prior understanding affects the internalization of new concepts. What teachers mean to convey and what they do convey are not necessarily the same thing.

Einstein initiated, and Piaget carried out, the first investigations of children's ideas of time. As Piaget recounted in the foreword to his classic *The Child's Conception of Time* (Piaget, 1946/1970):

> This work was prompted by a number of questions kindly suggested by Albert Einstein more than fifteen years ago [1930] when he presided over the first international course of lectures on philosophy and psychology at Davos. Is our intuitive grasp of time primitive or derived? Is it identical with our intuitive grasp of velocity? What, if any, bearing do these questions have on the genesis and development of the child's conception of time? Every year since then we have made a point of looking into these questions. (Piaget, 1946/1970, p. ix)

Piaget proceeded to comment on the problem of getting, so to speak, "unadulterated" data. He said that his team looked:

> . . . at first with little hope of success because, as we quickly discovered, the time relationships constructed by young children are so largely based on what they hear from adults and not their own experiences. (Ibid.)

But by applying a grouping technique, used in his investigation of number and quantity, Piaget thought he could tease out the child's natural, intuitive, or spontaneous temporal concepts. He certainly thought there was something to find, as he later stated in his *Psychology and Epistemology*:

> The fundamental concepts of physical space, of time, of velocity, of causality and so on, are indeed, borrowed from a kind of common sense well in advance of their scientific organization. And since the intellectual pre-history of human societies will probably remain for ever unknown, it is essential that we should study the formation of these concepts in the child, thus resorting to a kind of embryology of the mind. (Piaget, 1972, p. 52)

Piaget recognized this study was of importance to education, because "teachers and educational psychologists constantly come up against problems raised by the failure of school children to grasp the idea of time" (Piaget, 1946/1970, p. x). And he was enough of an historian of science to recognize that the Newtonian, absolutist idea of time is not intuitive. At the beginning of his landmark 1946 study he wrote:

> The aim of this section is to set the development of the concept of time in the kinetic context outside which it can have no meaning. We are far too readily tempted to speak of intuitive ideas of time, as if time, or for that matter space, could be perceived and conceived apart from the entities or the events that fill it . . . It is only once it has already been constructed that time can be conceived as an independent system. (Piaget, 1946/1970, p. 1)

What prompted Einstein's initial question to Piaget was the fact that in Einsteinian physics, time was no longer an independent fundamental variable. Time depended on and was derived from velocity. Velocity was fundamental, not time. Einstein was interested in ascertaining whether or not this was a basic intuition, whether or not it represented a return to some primitive cognition that had been overlaid by the enculturation of Newtonian understanding. In brief, Piaget regarded the Einsteinian conception of time as a return to the primitive. He concluded his initial study with the observation:

> Relativistic time is therefore simply an extension, to the case of very great velocities and quite particularly to the velocity of light, of a principle that applies at the humblest level in the construction of physical and psychological time, a principle that, as we saw, lies at the very root of the time conceptions of very young children. (Piaget, 1946/1970, p. 279)

This root principle was:

> . . . time is the coordination of motions at different velocities—motions of external objects in the case of physical time, and of the subject in the case of psychological time. [Before] the conception of time [has been] grasped operationally, i.e., as the ratio of the distance covered to the velocity, the temporal order is confused with the spatial order, and duration with the path traversed. . . . The construction of time proper therefore begins with the correlation of velocities, be it in the case of human activity or of external motions. (Piaget, 1946/1970, p. 255)

Of course the Einsteinian notion is not "primitive" in any literal sense: It is a highly technical conception and comes about only as a consequence of full immersion in Newtonian science. In the same way, the Newtonian ideas came about only as a result of rigorous testing of Aristotelian reductionist views. The relativistic temporal ideas may have a primitive "form," but their "substance" is not primitive. There is a recycling, but at a vastly more sophisticated level than the first appearance.

Piaget, in *Psychology and Epistemology*, reflected on 40 years of investigating the concept of time and summarized his findings:

> A further example of this embryology of the mind is provided by the development of the concept of time. Indeed Einstein once suggested that we should try to determine

whether, in the development of intelligence, the intuition of time precedes that of speed, or the reverse. To solve a problem of this kind, we can present children with movements which are synchronous, in whole or in part (running figures, flowing liquids and so on), at equal or unequal speeds, and ask them to determine the orders of temporal succession, or ask them to compare durations. When the paths are parallel and the moving objects leave together from neighboring points at equal speeds, it seems at first sight that the concept of time presents no difficulty, because all the temporal judgements are then really spatial judgements in disguise: the order of events is confused with the order of points on the trajectory, duration confused with space traversed, and so on.

On the other hand, we only have to make the speeds unequal, and all the temporal intuitions are disrupted. Small children do not admit, for example, that moving objects have stopped simultaneously if one of them has overtaken the other while they were moving: there is no concept that time is the same in the case of the two different speeds. Alternatively, while admitting the simultaneous departures and arrivals of two movements AB and AB', they will deny the equality of the synchronous durations if the path AB' is greater than the path AB. The order of events is transposed in order to be reconciled with the order of spatial succession, and so on. Above all, the subjects will not establish any relationship between the order of temporal successions and the inclusion of temporal sequences: knowing that Paul is older than he is, Peter will refuse to deduce from this that Paul was born first. At about eight to nine, however, we observe a grouping together of temporal relationships: a sequence of events $ABCD$ is ordered in time independently of speeds and spatial positions, and the duration AB is then thought of as shorter than the duration AC within which it is included, the latter (AC) as shorter than AD, and so on. At this stage, and only at this stage, the setting up of a temporal metric becomes possible, whereas previously the movements of the clock or the hourglass could not be synchronized with the others for lack of common speeds.

We can see immediately the relevance of such findings for physical epistemology. The relationship $v = d/t$ implies that v is a relationship and that both d and t are straightforward intuitions. The truth, however, is that some intuitions of speed, such as those of overtaking, actually precede those of time. . . . In the macroscopic universe, however, the subordination of time with respect to velocity remains fundamental since at high velocities relativistic time comes up against the same difficulties as does the young child's idea of time, and also presupposes a subordination of temporal relationships with respect to certain velocities. (Piaget, 1972, p. 78)

In 50 years of research, during which the tasks and questions have been refined so as to minimize the compounding effects of memory load, verbal misunderstanding, distracting cues, and the like, the core of Piaget's thesis concerning an individual's acquisition of the time concept seems not to have been challenged.[12]

CHAPTER 12

Teaching and Learning about Time and Pendulum Motion
Some Pedagogical Considerations

The pendulum has many qualities that make it an admirable device for science teaching. It is simple, tangible, cheap, and it also embodies so much rich science, history of science, cultural history, and scientific methodology (epistemology). It is easily manipulated, and thus dependencies of one variable (mass, amplitude, period, length, shape) upon another can be investigated along with skills such as data collection, timing, graphing, interpreting and so on. The pendulum can be refined, and data interpretation made more sophisticated, to take into account air resistance, friction at the support, and compounding effects of the wire. Thus some appreciation of experimental design, method and sources of error can be attained. With suitable elaboration, some basic methodological issues in science can be understood. The pendulum manifests many of the foundational laws of modern mechanics, particularly the conservation laws, and so provides a way of making these laws intelligible. By refining concepts, and refining technology (compound pendulums, conical pendulums, etc.), the pendulum can take students a considerable way into the absorbing world of classical mechanics.

The pendulum's historical connection with clockmaking, time measurement, the longitude problem, and navigation make it a particularly fertile subject for crossdisciplinary, contextual study. The analysis of pendulum motion is connected to fundamental philosophical and methodological considerations. It is a window on the scientific revolution. Analysis of the pendulum enables students to gain a sense of the experimental and idealized nature of modern science and thus to contrast it with ancient and medieval science, and with common sense. The pendulum can be used to illustrate the historical interrelatedness of science and philosophy in the Western tradition. Finally, but not least, if the pendulum is taught in a historical–investigative manner, then the idea of participation in a tradition of thought can be engendered.

THEMES IN CONTEMPORARY SCIENCE EDUCATION REFORMS

The treatment of pendulum motion that I propose (historical, philosophical, cultural and technological) fits well with a number of proposals for the improvement of science education:

- *Science–Technology–Society (STS) Emphasis*. The pendulum is a very clear case, albeit hardly ever examined, of the interactions of science, technology and society (Bybee, 1985, 1990; Aikenhead and Solomon 1994).
- *History–Philosophy–Sociology of Science Emphasis*. The value of history and philosophy in elucidating the science of pendulum motion and timekeeping is clear. For this reason it is a wonderful case study for HPS proponents (Rutherford and Ahlgren, 1990; Matthews, 1994a).
- *Integrated Curricula*. Just as school science is being urged to be more contextual, school social science, technology and history programs are urged to be more attentive to scientific matters . The approach to the pendulum advocated here fulfills this concern. The connections between history, geography, music, religion, commerce, mathematics, technology, and science are manifestly useful to explore (Bybee, 1990; AAAS, 1989, 1990).
- *Story Line*. One strength of old historically based science texts, such as those of Lloyd Taylor (Taylor, 1941), was that they had a storyline that linked scientific discoveries, personalities, social circumstances, and conceptual development. Since the 1950s storylines have been progressively removed from science textbooks. However the storyline idea in pedagogy has had a renaissance since the publication of Kieran Egan's *Teaching as Story Telling* (Egan, 1986).

Some science educators have argued for the utility of story telling in science instruction (Arons, 1989, 1990; Kenealy, 1989; Stinner and Williams, 1993; Wandersee, 1990; Wandersee and Roach, 1998). Study of the pendulum provides a nice opportunity for story telling. The characters—Leonardo, Galileo, del Monte, Huygens, Newton, Descartes, Hooke, Harrison—are exceptional. The plot is engaging: philosophers versus experimentalists, mathematicians versus observers, astronomical versus mechanical approaches to timekeeping, the expansion of the pendulum model into more and more areas of mechanics, craft refinement of clock mechanisms. The conclusion is marvelous: Newton's *Principia* with its unification of terrestrial and celestial mechanics, and Harrison's long-sought-after solution to the problem of longitude.

- *Historical-Investigative approach.* It is recognized that a danger for historical approaches to science instruction is that they can be overly "bookish" and over emphasize the student as spectator rather than participant. There is frequently too much reading and listening and not enough doing and participating in scientific procedures. However it is also acknowledged that there are dangers in approaches that just occupy students with practical work and laboratory exercises (Hodson, 1992, and some contributions to Woolnough, 1991). One obvious solution to the weaknesses of both approaches is to combine historical study with practical work: have students do the experiments that were central to the development of ideas in science. This is a relatively old idea. It can be found in Mach's late 19th-century proposals for the betterment of German science instruction (Matthews, 1990) and was revived in the late 1950s in Conant's *Case Studies in Experimental Science* (Conant, 1957).
- *Generative Learning.* For both Galileo and Newton the pendulum generated a number of concepts and understandings in what was to become the core of classical mechanics: freefall motion, inclined plane motion, circular motion and conservation of momentum and energy. For Galileo and Newton, the pendulum was a model, the understanding of which was generalized to a variety of domains. The pendulum provides fruitful links into technology and design topics (replication of experiments, simple clock mechanisms), links into historical topics (exploration, origins of commerce, colonialism) and links into theological, philosophical, and methodological topics (the design argument, the mechanical world view, idealization and experiment). Peter Machamer and Andrea Woody have provided one extensive example of how such generative learning can occur using the balance, another simple Galilean model (Machamer and Woody, 1994).
- *Less is Best.* The introduction to *Science for All Americans* says that "A fundamental premise of Project 2061 is that the schools do not need to be asked to teach more and more content, but rather to focus on what is essential . . . and to teach it more effectively...The present curricula in science and mathematics are overstuffed and undernourished" (AAAS, 1989, p. 14). Most programs, and texts, reduce what is essential about the pendulum to the equation $T = 2\pi\sqrt{l/\mathbf{g}}$. This is essential but it is notorious how quickly the equation is forgotten. If other things are also regarded as essential, then long after the equation has been forgotten something of value can be retained. Teaching the pendulum in a contextual manner allows for more thorough learning *of* science and *about* science, but it will require other topics to be eliminated. This raises issues of curriculum theory and politics: What is the purpose of the curriculum? What will be in it? Is it a local, provincial, or national responsibility? What is most worth learning?

The more contextual treatment of fewer but heuristically rich topics is important given the well documented flight from science. Increasingly school science is the last contact that most students have with formal science instruction. In most Western countries, the number of students studying science beyond the years in which the subject is compulsory is diminishing.

These trends seem unlikely to be reversed in the immediate future. Given the overwhelming cultural, social, and intellectual significance of science, this is disturbing. This is all the more so when the trends are coupled with a comparable decline in the study of school history.[1] Clearly something needs to be done at school and university to give citizens a better appreciation of science and the scientific tradition.

The issue of an appropriate science program, especially for nonscience majors, has been long discussed. James Conant's Harvard Report, *General Education in a Free Society* (Conant, 1945), is a landmark document.[2] The issue was prominent during the 1960s curriculum upheavals, with an editorial in *Science* (December 28, 1962) saying:

> Science courses are not taught with a broadening function in mind. They are designed to train the science major in specialized theory and technique from the start . . . Under these circumstances the non-science major finds his encounter with science a torment of meaningless detail, providing little that he may use for a wider purpose than satisfying an academic requirement. (in Ronneberg, 1967, p. 152)

The case for a contextual and historical, but nevertheless investigative and experimental, approach to science for the nonscience major is fairly easy to make. And, given that so many students cease contact with science after school, a similar case can be and has been argued for mainstream school science. However, in addition there is no reason to accept the standard overstuffed, purely technical curriculum that characterizes so many professional or science degree, programs. In the late 1940s as Alfred North Whitehead said :

> The antithesis between a technical and a liberal education is fallacious. There can be no adequate technical education which is not liberal, and no liberal education which is not technical: that is, no education which does not impart both technique and intellectual vision. (Whitehead, 1947, p. 73)

Perhaps the most through and extensively used example of a contextual approach to high school physics—although, unfortunately, it pays very little attention to the pendulum—was the Harvard Project Physics course developed in the early 1960s by Gerald Holton, James Rutherford, and Fletcher Watson (Rutherford, Holton, and Watson, 1970).[3]

The American Association for the Advancement of Science, in its *Science for All Americans,* maintained that:

> . . . the scientifically literate person is one who is aware that science, mathematics, and technology are interdependent human enterprises with strengths and limitations;

understands key concepts and principles of science; is familiar with the natural world and recognizes both its diversity and unity; and uses scientific knowledge and scientific ways of thinking for individual and social purposes. AAAS, 1989, p. 4

The pendulum fits beautifully into this scheme of contextual and liberal education; it is a wonderful vehicle for realizing the integrative aims expressed by the AAAS (Rutherford, 2001).

TIME STANDARDS

One of the important things to teach about time, is the problem of determining a unit for its measurement. This problem has been central to the history of time-keeping and time measurement. Identifying a standard for the measurement of duration has been recognized as a problem since ancient Babylonians devised calendars in the fourth millennium BC.[4] The first difficulty to be confronted was that none of the common natural astronomical units (year, month, or day) fit with each other. The solar day is a natural unit because it is the duration from one "high noon" to the next; that is, the duration from a stick's shadow being shortest to when it is again shortest. The month is a natural unit, being the time taken for one orbit of the moon around the earth—that is, from one new moon to the next new moon.[5] The year is a natural unit, being the period for the earth to complete one orbit around the sun; that is, from one spring (or autumn) equinox to the next. These are all natural units, as distinct from conventional units such as the week and the hour.

There can be as many or as few, weeks in a year as we wish, and as many or as few, hours in a day as we wish. The Babylonians, for instance, had 24 "hours" in their day, the Chinese 12, and the Hindus 60. The lack of fit of the natural units meant that if the day is taken as the standard, then the month and year consist of fractional days. If the month is taken as the standard, then there are fractional months in a year. Where people insisted on making the units fit, then the seasons quickly got out of synchrony with the calendar. The Sumerians, for instance, were committed to a sexagesimal number system and thought that the year *ought* to have 360 days, being twelve months of 30 days. This simplified drawing calendars and calculating durations, but these constructions soon ceased to mirror natural occurrences—winters that used to occur in January began occurring in July. Babylonians prized the month and made their year consist of twelve months (or 354 days), 11 days short of the solar year on which seasons are based. The Dead Sea Scroll sect gave primacy to the Jewish week of seven days and chose 52 weeks (364 days) to be the year. When the seasons began to move from their customary calendar months—the rains coming later and later—the sect blamed this on the sinfulness of their neighbors (O'Neil, 1975, p. 46).

Against the fixed stars, or the zodiac, the sun moves easterly through a cycle

each 365 days. But its path is not in a fixed plane: The plane moves up and down
throughout the cycle. Thus at noon, the sun's daily highest spot, is sometimes
higher or sometimes lower in the sky. On March 21, the vernal equinox, the sun is
directly overhead at noon for places on the earth's equator. Then it moves north
each day, until about June 21 (northern hemisphere's summer solstice, or longest
day) when it is directly overhead for places at latitude 23½° north (Tropic of Cancer). The sun then moves south so that by about September 23 (autumnal equinox), it is directly overhead at noon on the equator again. It continues south so
that by about December 21 (northern hemisphere's winter solstice, or shortest
day), it is overhead at noon for places 23½° south of the equator (Tropic of Capricorn). At the solstices the sun reverses its direction, (the term "Solstice" comes
from the Latin for "sun standing still"). Figure 95 illustrates the following:

Vernal Equinox: where the sun's path (ecliptic) crosses the celestial equator
on its way north, around March 21. Days and nights are of equal duration.

Autumnal Equinox: where the sun's path (ecliptic) crosses the celestial equator on its way south, around September 23. Days and nights are of equal duration.

Summer Solstice (northern hemisphere): the most northerly point of the ecliptic, around June 22. The northern hemisphere's longest day.

Winter Solstice (northern hemisphere): the most southerly point of the ecliptic, around December 22nd. The northern hemisphere's shortest day.

These macro problems of devising a calendar that was both convenient and
natural (convenient proposals were not natural, natural proposals were not convenient) were only the beginnings of the problem of setting a time standard. Not

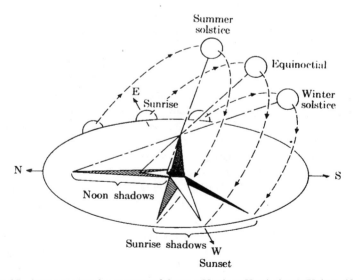

Figure 95. Apparent annual movement of the sun (Northern Hemisphere) (Holton, 1985, p. 5).

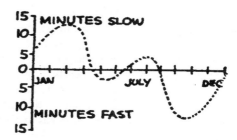

Figure 96. Variation in length of solar day
(Jespersen and Fitz-Randolph, 1982, p. 61).

only did the natural units not fit with each other, but they were, also irregular in duration.

The *solar* day is the interval between successive crossings of the meridian by the sun and was an obvious standard. Early on it was seen that the solar day varied in length through the year (Fig. 96).

Sundials that in May had their shortest shadow showing noon, had in February their shortest shadow showing 15 minutes before noon, and in September 15 minutes after noon. On only four days of the year (about April 16, June 14, September 2 and December 24) does the shortest shadow occur at noon by the clock.[6] Thus no particular day could be a standard; instead the *mean solar* day, or the average duration of the solar days was proposed. This is taken to be 24 hours. The difference between the solar day and the mean solar day is called the *equation of time* and is shown in Figure 96.

This variation in length of solar days results from (1), the earth orbiting the sun in an ellipse and not a circle and thus speeding up at the perihelion and slowing at the aphelion; and (2), from the earth's axis of rotation being inclined to the plane of its orbit. These variations were sidestepped by defining the day as the

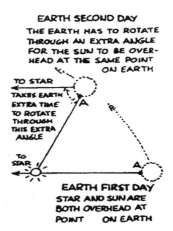

Figure 97. Difference between sidereal and solar days
(Jespersen and Fitz-Randolph, 1982, p. 61).

duration between successive passages of a *star* over the meridian, the *sidereal* day, which is constant in duration (Figure 97).

However the duration of a sidereal day is different from that of a solar day, being about four minutes shorter. Currently a sidereal day is about 23 hours, 56 minutes, and 4 seconds as measured by a clock which has been calibrated to the mean solar day. The difference between solar and sidereal days results from the fact that the earth orbits the sun, at the rate of a little under 1° per 24 hours as it rotates, so any point on the earth has to rotate an extra 1° after the star is directly overhead in order for the sun to be directly overhead. It takes approximately 4 minutes to rotate this extra degree (24 × 60 ÷ 360). Because of its uniformity, the sidereal day is used by astronomers and space scientists for their calculations.

The time that it takes for the earth to do a complete circuit of the sun, as judged by the vernal equinox reference point in space, is the *tropical year*, and its duration is about 365.2564 mean solar days.

This itself is not fixed as the vernal equinox moves very slowly in space against the background of the fixed stars. This is known as the *precession of the equinoxes*, and results from the earth's axis very slowly "wobbling" as it rotates (Figure 98). Because of this wobble, the earth's axis of rotation traces out a circle, that subtends an angle of 47°, each 25,800 years. This precession effect was discovered by the Greek astronomer, Hipparchus of Nicaea, in about 130 BC.

The ancients knew of this precession, and Hipparchus estimated it to be 40 seconds of arc per annum, a figure remarkably close to the modern estimate of 50.3 seconds of arc per annum.

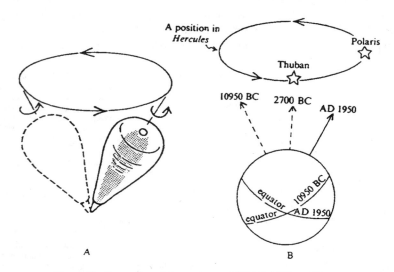

Figure 98. Precession of the equinoxes (O'Neil, 1975, p. 21).

A fundamental problem with any astronomical standard for time measurement is that the earth's rate of rotation is slowing. A year in the Devonian period (350–410 million years ago) had 400 days. That is, in the Devonian period the earth rotated 400 times in each of its orbits of the sun. Because of a slower rotation rate, it now makes about 365 complete turns in one orbit.

The progressive uncovering of irregularities in astronomical time standards (the solar day, sidereal day, tropical year, etc.) was accompanied, as we have seen, by progressive improvement and refinement of nonastronomical time measures. In the 1890s Siegmund Riefler (1847–1912) of Munich was making pendulum clocks accurate to one-hundredth of a second per day. By the middle of the 20th century, the best pendulum clocks used in observatories to time astronomical processes were accurate to within 1/10 of a second per day. William H. Shortt, in the early 1920s, constructed a number of free-pendulum clocks, where the inherent frictional problems associated with pendulum motion in clockwork were overcome by simply dissociating the pendulum from the clock! The pendulum swung freely in a vacuum, and its movement was conveyed *electrically* to a separate clock. Such a clock was installed in Greenwich Observatory in 1924 and was accurate to a few thousandths of a second per day, or about one second per year (Wilson 1980, p. 123; Burgess 1998, p. 278). When pendulum clocks received their impulse not from a falling weight but from an electrical impulse mediated by a magnetic field, their accuracy increased to about 3/100 of a second per day over periods of years (Cooper, 1948, pp. 124–126). But even this accuracy, the greatest obtained by the heirs of Huygens' pendulum clock, was a hundred-fold less than the then newly created quartz clocks.

In 1929 the first quartz crystal clock was made. Its timekeeping was dependent upon the ultra regular vibrations occasioned by applying an electric voltage to a quartz crystal cut in specific directions. Originally these clocks were 3 meters high, 2½ meters wide, and 1 meter deep. They have now been reduced to wrist watches. When properly kept, they are accurate to one or two milliseconds per day (Cooper, 1948, pp. 126–128). In 1949 the United States National Bureau of Standards announced the world's first time source linked to the natural frequency of atomic particles, specifically the ammonia molecule. It was recognized that these atomic vibrations were more regular than the heavenly motions that they were being calibrated against. In recognition of the variability of astronomical periods (the day and the year), in 1956 an international agreement was reached to define the second as a proportion of one given year, the year 1900. The second was then defined as 1/31,556,925.9747 of the tropical solar year 1900.

Finally, in 1967, international time was finally cut loose from astronomical standards. In that year, the second was defined as the time elapsed during 9,192,631,770 oscillations of the undisturbed cesium atom. Electronic devices can count these oscillations in the same way that the escapement motion in grandfather clocks counted the oscillations of the pendulum. Cesium oscillations were the basis for Coordinated Universal Time (UTC), which in 1972 replaced Greenwich Mean Time as the universal time standard. The latest cesium clocks are accurate to one second in three million years (Barnett, 1998, p. 160). Because such a time unit is independent of the motion of the earth, atomic time and earth time can get out of synchrony. The world body that oversees UTC decreed that UTC and astronomical time would not be allowed to diverge by more than 0.9 seconds. When the difference gets to this amount a "leap second" is added to UTC. There have been 21 leap seconds added to UTC over the past 25 years, the last in 1998. These are added to the final hour of the December 31. The adding of leap seconds is necessitated by the earth's slowing.[7] This is all a very long technological distance from Huygens original "seconds pendulum." But for all the technological sophistication, the increased "accuracy" is in the order of seconds, or perhaps a minute, per day over the simple pendulum. It is this level of precision that is demanded by modern science—to say nothing of cultured reasons such as keeping records for the Olympic Games 100 meter sprint.

One basic methodological issue raised by this progressive refinement of the time standard is whether or not time measurement is getting more accurate or just more convenient? On the Newtonian, and realist, assumption that time flows equably and evenly and is separate from those periodic events that can be taken for its measure, we might answer that our measures are getting more accurate. On the Machian and positivist view that there is no independent time flowing equably or inequably, our measures are not more accurate, only more convenient. There is no absolute standard against which to measure time and judge accuracy; there are only periodic recurrences of events (earth's rotation, sun's rotation, pendulum swings, atomic vibrations) that we choose as a standard against which to compare other motion. We just assume that the periodicity chosen as a standard is regular and equable until its use suggests that it is not. Then we look for another standard.

THE PENDULUM AND SIMPLE HARMONIC MOTION

The standard Newtonian analysis of the forces on a *simple pendulum* (one where the bob is heavy and the supporting wire is very light) is as shown in Figure 99. If the bob, of mass m, is pulled aside from the vertical to an angle q, then its weight, mg, pulls vertically downwards. This force is resolvable into a restoring force F that acts tangentially to the arc, and another force N that lies along the extension of the wire. This force N, which is perpendicular to the direction of movement, does not contribute to the bob regaining its equilibrium position, it only sets up an equal tension in the wire, expressed as $-N$. The value of the forces are:

$$F = -mg \sin \theta$$
$$N = mg \cos \theta$$

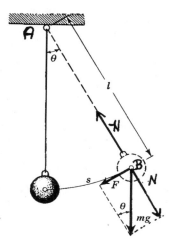

Figure 99. A simple pendulum (Weber, White and Manning, 1959, p. 137).

F has a negative value because its direction is opposite to that of the bob's displacement, which is positive along the arc. Its value is given by $s = l\theta$. Hence the restoring force F is proportional to sin q, while the displacement is proportional to q. Thus the restoring force is not exactly proportional to the bob's displacement, but if q is small, $<3°$, then $\sin \theta \cong \theta$ (in radians). For instance, if $\theta = 3°$, then it has the value 0.05236 radians, while $\sin \theta = 0.05234$. For smaller θ the agreement is even better.

Simple Harmonic Motion (SHM) was been introduced in Chapter 7 in the discussion of Hooke's work. It was defined as motion where the restoring force, and hence acceleration, is proportional to the body's displacement but opposite in direction. Thus for small displacements the vibration of a pendulum is simple harmonic motion. And, as θ (in radians) $= s/l$ (from the definition of a radian):

$$F = ma = -mg \sin \theta \cong mg\theta = mgs/l$$

So,

$$-s/a = l/\mathbf{g}$$

And it can be shown that for SHM:

$$T = 2\pi \sqrt{-s/a}$$

Hence for the pendulum, with small oscillations:

$$T = 2\pi \sqrt{l/g}$$

Which is the equation Huygens first formulated in 1673. Alternatively, one can see from the foregoing equation that:

$$\mathbf{g} = 4\pi^2 l/T^2$$

This formula then provides a way of measuring *g* at any point, and as such was widely used in geodesic surveying and in determining the shape of the earth. Using a standard seconds pendulum, $T = 2$ and $l \approx 1$ *m*, while $\pi^2 \approx 9.86$. Thus $g = 4 \times 9.86 / 4 = 9.86$ *m/s²*. Deviations from this indicate changes in the earth's shape or other influences.

Another way of approaching this topic is to consider the simple pendulum of mass *m* suspended from *O* by a length *l* and pulled out to point *M* from its equilibrium point *A* (Fig. 100). When the bob is at *M*, it is higher than at *A* by the height *AP*, where $AP = AM^2/AB$, where *AM* is the chord of the arc *AM* and $AB = 2l$.

If *m* is the mass of the bob and *g* the acceleration due to gravity, then the bob's weight will be *mg*, and the work done is lifting the bob from *A* to *M* will be *mgAP*, which is also the potential energy of the bob (using *A* as the zero datum). From similar triangles, we can write the bob's potential energy as:

$$PE = mg/2l \times AM^2$$

From which it can be seen that the potential energy of a bob when displaced through any arc, varies as the square of the chord of that arc. If it had varied as the arc, then its motion would have been strictly isochronous. But for very small vibrations, the difference between the arc and the chord can be neglected. If the arc is *s*, then the potential energy of the bob is given by:

$$PE = mg/2l \times s^2$$

But for simple harmonic vibrations, the potential energy is given by:

$$PE = 2\pi^2 ms^2/T^2$$

From these two formula for potential energy, we see that:

$$g = 4\pi^2 l/T^2$$

Yet another way of understanding the dynamics of the pendulum is to consider that the bob at all points of its oscillation acts like a bob on an inclined plane that is tangential to the bob's path at all points. This situation is represented in Figure 101.

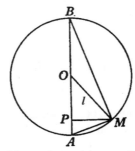

Figure 100. Pendulum and potential energy (Maxwell, 1877, p. 97).

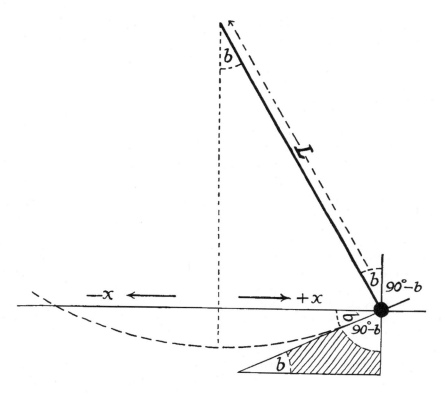

Figure 101. Pendulum and inclined plane (Hogben, 1940, p. 286).

When the pendulum's amplitude is b, then the tangential plane is also inclined at b, and the linear displacement from the vertical is x. The acceleration along the plane is $g \sin b$, If the pendulum's length is l, then $\sin b$ equals x/l, and the bob's acceleration down the tangential plane equals $-(g/l)x$ (minus sign because its direction is opposite the positive direction). This actual acceleration to the left, at right angles to the bob, can be resolved into a vertical component and a horizontal component. The horizontal acceleration is $a \cos b$, so for small displacements (small b) the horizontal acceleration is approximately equal to the tangential acceleration (e.g., $\cos 5° = 0.9962$).

Remember that at the lowest point there is *no* horizontal component of the vertical weight force, and so there is no horizontal acceleration, only vertical acceleration toward the point of suspension—another counterintuitive result of applying a scientific conceptual scheme to the behavior of real bodies. This is one of the conceptual things that so befuddled the students in Reif's study mentioned in Chapter 11. When $b = 90°$, the bob is horizontal and the situation is equivalent to freefall. Then the vertically downward acceleration is g sin b, that is g sin $90°$, or g (as sin $90° = 0$).

A HISTORICAL–INVESTIGATIVE APPROACH
TO TEACHING PENDULUM MOTION

A problem with "hands-on" science is that while the hands are on, the mind is frequently off. *Benchmarks for Science Literacy* observed that "Hands-on experience is important but does not guarantee meaningfulness. It is possible to have rooms full of students doing interesting and enjoyable hands-on work that leads nowhere conceptually" (AAAS, 1993, p. 319). Too often in science museums, or science centers, this degenerates to "gee-whiz" or "fun" science, where the gee-whiz experience dissolves like the cotton candy purchased from the center's snack bar and leaves about as much nourishment. Good laboratory classes rise above this level, but too frequently laboratory work reduces to a cookbook exercise without students knowing a great deal about either the ingredients, utensils, or methods.[8]

Another approach to practical work, as already mentioned, is to try to wed laboratory classes to historical stories—that is, to follow along the path of experimental science, to follow in the footsteps of the masters, as one might say. While doing this, it is possible to reproduce something of the intellectual puzzles or scientific debates that originally prompted the experiments. Participation in this sort of journey can give students a much richer appreciation of the achievements, techniques, and intellectual structure of science, while developing their own scientific knowledge and competence. Ernst Mach (1838–1916), perhaps the first science educator as distinct from teacher,[9] recognized this at the end of the 19th century:

> . . . every young student could come into living contact with and pursue to their ultimate logical consequences merely a *few* mathematical or scientific discoveries. Such selections would be mainly and naturally associated with selections from the great scientific classics. A few powerful and lucid ideas could thus be made to take root in the mind and receive thorough elaboration. (Mach, 1886/1986, p. 368)

Mach's ideas were in large part ignored: 20th-century science curricula and textbooks have been filled by an avalanche of facts, laws, formulas, theories, and exercises.[10] They are akin to forcing students to march through unknown country without giving them time to look sideways. Frequently students only aspire to hang on in the march until they are able to fall over the finish line and then put it all behind them. Over 100 years ago, Mach decried this propensity to over-stuff curricula, saying:

> I know nothing more terrible than the poor creatures who have learned too much. What they have acquired is a spider's web of thoughts too weak to furnish sure supports, but complicated enough to produce confusion. (Mach, 1886/1986, p. 367)

The great English science educator, Henry Armstrong, shared Mach's view of this matter:

> I have no hesitation in saying that at the present day the so-called science taught in most schools, especially that which is demanded by examiners, is not only worthless, but positively detrimental. (Armstrong, 1903, p. 170)

Some people of course saw the wisdom of Mach's advice. In the 1960s and 1970s some proposals for laboratory-based, historically-informed science program were advanced (Shamos 1959; Ronneberg, 1967; Devons and Hartmann, 1970). At the present time, in Germany, Falk Riess and colleagues have successfully used reconstructed historical apparatus and experiments in the teaching of physics (Riess, 1995, 2000). They promote these historical reconstructions as a way of teaching both science and the history of science. Nahum Kipnis has promoted this historical–experimental approach, with less emphasis on authentic historical reconstruction and more on following through the process of historical development of branches of science (Kipnis, 1996). Kipnis has, for example, based a course on optics around retracing the classic, and usually very simple, experiments and demonstrations in the history of the subject (Kipnis, 1992).

The topic of timekeeping, and more specifically the pendulum, is ideally suited to this historical–investigative approach to teaching. The technology is simple and cheap: shadow sticks, primitive sun dials, simple and even compound pendulums. The classic experiments are clear and easy to replicate, and they allow a lot of individual improvisation, elaboration, and to checking sources of error.

Reproducing Galileo's 1602 Law of Chords Experiment

Galileo in his 1602 letter to del Monte, (see pp. 102–104) had recourse to geometry to prove his law of chords—that the times of free descent along all chords of a circle terminating at the nadir are the same, and are equal to the time of free fall from the highest point of the circle, that is along the diameter. Galileo's physical intuition, when depicted geometrically, was rich in implications for the investigation of freefall, inclined-plane motion and pendulum motion. It can also be put to experimental test. The chords experiment, like his more famous inclined-plane experiment, readily lends itself to a historical–investigative reproduction, something that has been done in a British Open University course (Pentz, 1970).

A modern proof of Galileo's chord claims can, in one form or another (assignments, projects, lectures), be conveyed. Consider Figure 102, where a ball of mass m moves down a plane CB under just the influence of gravity and reaches a velocity of v at B. The exercise is to calculate the time of descent along CB and to compare it to the time of fall down AB.

We know, as Galileo did but in different terms, that the kinetic energy of the ball at B is equal to its potential energy at C.[11] That is the kinetic energy at B is equal to the energy used in lifting it to C through a height of h. Thus:

$$\tfrac{1}{2}mv^2 = mgh$$

or

$$v^2 = 2gh$$
$$v = \sqrt{2gh}$$

As the ball is subject to a constant force, and hence constant acceleration, the average velocity, z, along CB is equal to half the terminal velocity, v, at B. Thus,

$$z = \tfrac{1}{2}v\sqrt{2gh} = \sqrt{gh/2}$$

Then the time taken to roll the distance d along CB is:

$$t = d/z = d/\sqrt{gh/2} = d\sqrt{2/gh}$$

But,

$$h = d \sin\theta$$

So,

$$t = d\sqrt{2/gd}\sin\theta = \sqrt{2d/g}\sin\theta$$

And,

$$t^2 = 2d/g \sin\theta$$

That is,

$$t^2 \propto d$$

From the figure we can see that ACB is a right angle as is ABE. It follows that the angle CAB is also θ, and that:

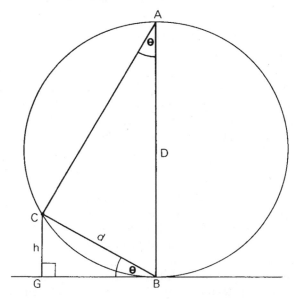

Figure 102. Modern Construction of Galileo's Chord Law (Pentz, 1970, p. 50).

$$\sin \theta = CB/AB = d/D$$

Substituting for $\sin \theta$ in the above, we see that:

$$t^2 = 2D/g$$

Thus, as g is constant, t has the same value for any chord of the circle of diameter D. This is Galileo's claim to del Monte in his historic 1602 letter.

If $t = 1$, then from the above equation we see that:

$$D = g/2$$

Thus the diameter of a "one-second" circle should be about 4.9 m, since $g = 9.8$ ms^{-2}.

With ingenuity, and simple technology, this experiment can be done for various angles of tilt. Time measurements can be taken using a form of Galileo's water weighing device, as well as more modern means. This data can be tabulated and graphed. The results will *confound*, not *confirm*, Galileo's law and the above mathematical proof. Is the discrepancy due to Galileo's usual suspects: accidents and impediments? Or is it due to more fundamental mechanisms at work in the world? This is a useful question for students to grapple with. We now know that whatever the contribution of experimental error, the fact of the ball *rolling* contributes to the discrepancy between the world on paper and the real world. If students are taught something about rotational kinetic energy and moments of inertia, they can calculate how much of the discrepancy can be accounted for. Students can be left the task of proving Galileo's addendum to the chords' law, namely the time of descent along subchords beginning at the origin and finishing at the terminus is *quicker* than along the direct and shorter chord.

The above "Law of Chords" construction can also with a slight addition be used to demonstrate Galileo's claim that the velocity (speed) of a given pendulum at its lowest point is proportional to the length of the chord through which it has been displaced prior to release.

In Figure 103 imagine a pendulum suspended from O whose bob is at C, and whose length is l (where $l = D/2$). The bob is released and swings through the arc from C to B. It remains to prove that the pendulum's velocity, v, at B is proportional to the length d of the chord CB.

From the conservation of energy law we know that:

$$\tfrac{1}{2}mv^2 = mgh$$

so,

$$v = \sqrt{2gh}$$

and,

$$h = d^2/2l$$

thus,

$$v = \sqrt{2gd^2/2l}$$
$$= d\sqrt{g/l}$$

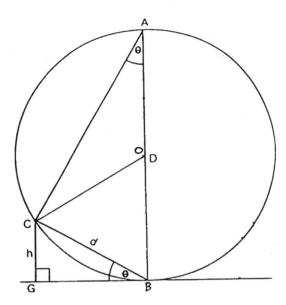

Figure 103. Velocity varies as chord length

Hence for any given pendulum, because g and l are constant, the velocity at the nadir of its swing, depends solely upon d, the length of the chord by which it is displaced from the vertical.

This result also establishes that the pendulum's momentum (mv) is also a linear function of the length of the displacement chord. Thus the pendulum is an ideal instrument for demonstrating conservation of momentum principles; the momentum of a pendulum can be doubled, tripled, halved merely by making adjustments to the length of the displacement chord—d in the above (Gauld, 1998a,b).

A simple bifilar pendulum arrangement can be used to demonstrate "in the world" the foregoing "on paper" proofs, as Galileo might have put it. One pendulum when colliding with another causes it to swing out a certain measurable distance. If the first is taken out twice as far (by chord distance), the second is moved twice as far; if the mass of the first is halved yet taken out twice as far its momentum at the nadir will remain the same and so the second will swing out to the original position. If modeling clay is put on the two pendulums, thus converting the collisions into perfectly inelastic collisions, momentum is not conserved and neither the first nor the second pendulum move upon collision. And so on. All of this can be revealed to the students through a demonstration or through the process of discovery (Lattery, 2001).

Measuring Time in Junior High School

The National Science Resources Center's *Measuring Time* (NSRC, 1994) has a nice example of an elementary school or junior high-school approach to teaching about timekeeping. This well illustrated, large format, 180-page book contains six sections on "Keeping Time with the Sun and the Moon," and a ten on "Investigating Invented Clocks." Among the first six sections are: "Before Clocks," "Making a Record of Shadows," "Observing the Phases of the Moon." Among the ten sections are: "Using Water to Measure Time," "Investigating Pendulums," "Experimenting with Pendulums," "Constructing a Clock Escapement," "Building a One-Minute Timer." It is certainly "hands-on" science, but it is also more than just hands-on. Simple historical and cultural material is introduced—not much, but nevertheless sufficient for students to appreciate that the ancient Greeks did not wear digital wristwatches, and that time is in some way connected to certain natural motions.

For instance the section on "Constructing a Clock Escapement" involves students in making a primitive escapement mechanism from Lego-like materials (Figure 104).[12] The introduction to the section says:

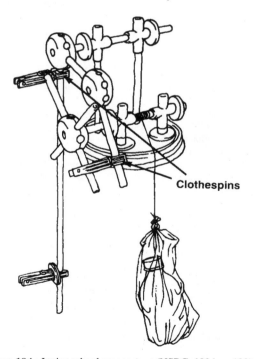

Clothespins

Figure 104. Junior school escapement (NSRC, 1994, p. 132).

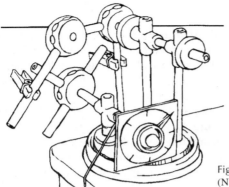

Figure 105. Clockface added to escapement
(NSRC, 1994, p. 142).

> Modern clocks constructed from pendulums, springs, or vibrating quartz crystals
> operated by microelectronic circuits have largely eliminated people's daily reliance
> on the sun and moon to keep track of time. But this transition did not occur overnight.
> Beginning with pendulum clocks driven by falling weights, scientists and engineers
> made gradual improvements to timekeeping devices in order to make them more
> accurate. The escapement mechanism that students work with in the next three lessons
> is an example of one such improvement.

It continues to explain:

> An escapement is a device that delivers the energy to drive a mechanical clock. In the
> case of the pendulum clock, the escapement delivers the regular pushes that keep the
> pendulum swinging. An escapement also regulates the energy that keeps the clock
> going. It is the design of the escapement that controls *when* the gears push on the
> pendulum and *how hard* the gears push.
>   How does an escapement work? An essential part of the mechanism is the toothed
> wheel, or driving wheel, which is turned by a string attached to a heavy weight. As
> the pendulum swings back and forth, the wheel "escapes" for just a moment, allowing
> it to turn one tooth at a time. The moving wheel gives the pendulum enough of a push
> to keep it swinging. (NSRC, 1994, p. 123)

These lessons combine some basic history with enough basic technology to
give students an appreciation of the craft skill necessary in pursuing science. These
are combined with basic scientific skills such as observation, eliminating experimental errors, and so on.

Students can also be introduced to elementary, as well as sophisticated forms
of sundials, and be tutored in observing shadow effects, inclinations, directions,
and so on. Lancelot Hogben, in his 1936 very popular book on mathematics, suggested that his readers make a sundial during their holidays, and commented, in a
way seldom seen in modern mathematics texts, that:

> If at the same time you remember that the theory was worked out by the coloured
> conquerors of Spain when the people of Britain and Germany were barbarians, living

in mud huts, ruled by ignorant priests and robber barons, it will also clear your mind of the sort of cant chanted by the white settlers of Kenya, General Hertzog, the apostles of selective immigration restrictions, and the pundits of German national socialism. (Hogben, 1936, p. 381)

High School Investigations of Time and Pendulum Motion

Ronneberg (1967) has a more sophisticated set of laboratory activities suitable for senior high school students. The laboratory course he designed:

- Is interdisciplinary.
- Follows the historical development of science.
- Is largely based upon projects, each 3 to 5 weeks in length.
- Requires quantitative solutions.
- Requires a knowledge of the precision of all experimental data.
- Requires simple equipment.
- Stresses graphical representation of data with analytical interpretations.
- Permits students to follow the methodology of science.
- Stresses the important conceptual schemes of science in considerable depth. (Ronneberg, 1967, p. 154)

His second project involved the "determination of latitude and longitude with a gnomon," the third involved "kinds of time," the fourth involved "freely falling bodies," and the fifth involved "simple harmonic motion and the motion of a simple pendulum."

In Project 4, he has students construct the apparatus and then experiment with a version of Mersenne's method for determining the length of fall in one second (see pp. 116–117). Unfortunately he does not mention Mersenne, or the place of Mersenne's determination in the history of early modern science. Knowledgable teachers can rectify this. Students plot results for h (height of fall) against t (time of fall) and see that there is no simple relationship, but that if they plot h against t^2 they do get a straight line graph, and so $h = kt^2$ (if the line goes through the origin). It is then shown that k (the slope of the line) is equal to half of g (the acceleration of gravity), and thus from their graphs that $g \cong 980 \text{cm/sec}^2$.

In Project 5, Ronneberg has students experiment with a simple pendulum. They "satisfy themselves that the period does not depend on the material in the bob, or the mass, or the amplitude—if this is small—but only on the length." Students use pendulums of different lengths and plot T against l, again ascertaining that there is no simple relationship until T is plotted against $\sqrt{l}$. Then the line has a constant slope, given by $T = k\sqrt{l}$, and yields the formula, $T = 2\pi\sqrt{l/g}$. From their results they are required to derive a value for g and then to compare it to the value derived from the previous "Mersenne-like" freefall experiment.

Ronneberg commented as follows on the strengths of his program:

> The orientation to each project makes it possible to bring considerable history of science into the course. This is highly desirable. The steps involved, for example, in the synthesis of our concepts of force, and of energy were momentous steps in building science. The steps leading to these achievements were often slow and uncertain. The results are still subject to change. These steps provide example after example of the nature, use and triumphs of the scientific method. . . . The merit of a project method in the laboratory is that the student experiences science and ceases to serve as a sponge to merely absorb scientific information. (Ronneberg, 1967, p. 161)

OPPORTUNITIES LOST: THE PENDULUM IN THE UNITED STATES NATIONAL SCIENCE EDUCATION STANDARDS

The U.S. *National Science Education Standards* (NRC, 1996) are the outcome of an extensive writing and consultative process that began in 1991.[13] Forty thousand copies of the 1994 *Draft* were printed. Eighteen thousand of these went to individuals, while 250 interest groups received multiple copies. All of the major contributors to American science education had the opportunity to be involved in the development and revision of the standards.

Throughout the document there are laudable statements that fit very well with the liberal aims of science education defended in this book. The *Standards,* for instance, maintain that science students should learn how:

- Science contributes to culture (NRC, 1996, p. 21);
- Technology and science are closely related. A single problem has both scientific and technological aspects (NRC, 1996, p. 24);
- Curriculum will often integrate topics from different subject-matter areas . . . and from different school subjects—such as science and mathematics, science and language arts, or science and history (NRC, 1996, p. 23);
- Scientific literacy also includes understanding the nature of science, the scientific enterprise, and the role of science in society and personal life (NRC, 1996, p. 21);
- Effective teachers of science possess broad knowledge of all disciplines and a deep understanding of the disciplines they teach (NRC, 1996, p. 60);
- Tracing the history of science can show how difficult it was for scientific innovators to break through the accepted ideas of their time to reach conclusions that we currently take for granted (NRC, 1996, p. 171);
- Progress in science and technology can be affected by social issues and challenges (NRC, 1996, p. 199);
- If teachers of mathematics use scientific examples and methods, understanding in both disciplines will be enhanced (NRC, 1996, p. 218).

Two pages are devoted to the pendulum (NRC, 1996, pp. 146–147). These pages deal with a set of model lessons for teaching about properties of the pendulum. Students are asked to make pendula of different length, weight, and amplitude. The class is divided into small groups, which go through the usual exercises, including plotting of results: and "After considerable discussion, the students conclude that the number of swings in a fixed time increases in a regular manner as the length of the string gets shorter" (NRC, 1996, p. 147). Students are asked to use their graphs to make a pendulum that will swing an exact number of times. It is concluded that "Students have described, explained, and predicted a natural phenomenon and learned about position and motion and about gathering, analyzing, and presenting data" (NRC, 1996, p. 147). The document says "assessment, constructing a pendulum that swings at six swings per second, is embedded in the activity" (NRC, 1996, p. 146).

There is nothing wrong with any of this. It is however a shame that with the smallest amount of historical knowledge this model lesson could have been transformed into something where not only some of the properties of the pendulum are learned, but nearly all of the wider aims that the *Standards* outline for science education could also have been achieved. In the two pages on pendula, no mention is made of timekeeping. There is no mention of a pendulum clock, much less the longitude problem and European expansion, trade and colonization. No historical figures are mentioned: There is no Galileo, no Huygens, no Newton, no Harrison. There is no mention of isochronism. There are graphs, but no properties of the circle, nothing on vectors, and no calculus.

In general, and certainly for the pendulum, the standards shy away from specifics; they leave open just what degree of scientific understanding is to be required of students at different stages in their schooling. The standards are written in "soft-focus" style. This is unfortunate. As one teacher remarked: "The irony is that the NSES assumes students are sophisticated enough to develop their own models of scientific knowledge, but are apparently incapable of understanding specific and accepted scientific theories" (Shiland, 1998, p. 616).

The *Standards* do not claim to be a curriculum, and it is maintained that teachers and school districts have to make their own curricula that embody the *Standards* recommendations and spirit. This is a particular problem when the "spirit" is not explicitly spelled out and defended (Rodriquez, 1997, Shiland, 1998). There is a commitment to a sort of personal constructivist account of science and of the learning and teaching of science, but this is no where made explicit or defended. Many teachers, with good reason, do not share this "spirit."

Even so the pendulum section might have provided more guidance to teachers, more content about the historical and cultural impact of the pendulum. If knowledge of history, of science's relationship to other disciplines, and of science's interaction with culture are important enough to be mentioned in the *Standards* as goals of science education, and as components of scientific literacy, it should be indicated when topics so obviously manifest these dimensions. There is not much point in hand-waving about liberal, or more expansive goals, for science education if nothing is done to assist teachers, or curriculum authorities, in realizing them. As one Australian teacher said during an inservice course, "Teachers are hungry for this knowledge."

The *Standards'* suggested assessment task is to make a pendulum with a period of six swings per second. This is a routine science laboratory task that most students do, and then forget. It does not do any harm, and may do some educational good. However, had the students been asked to make a pendulum with a period of half-a-swing per second (a period of two seconds), then things could have been very different. This would be a seconds pendulum, and students would see that its length is one meter. With just a few questions, or a little bit of research, this result could lead to the proposal of the seconds pendulum as the original standard of length, the original meter. This connects length standards to time standards.

With some more questions, Huygens' name might come up, and the topic of variability of length of the seconds pendulum could be broached. Richer's voyage to Cayenne might be mentioned, as might the role of the pendulum in ascertaining the shape of the earth. Once a seconds pendulum is created then the technological challenge of converting it to a clock might be raised. The crucial interdependence of hand and brain might be appreciated. All of this connects the learning and activities of contemporary students to experiments and achievements in the 17th century. Thus the sense of participation in a tradition could be engended. But there is none of this in the *Standards*. Repeatedly, model lessons are outlined that have no historical or cultural component.

In brief, there are lost opportunities in this section of the *Standards*. This is surprising because the drafts were subject to so much examination. They were read by thousands. One wonders why the foregoing simple suggestions about the pendulum were, seemingly, not mentioned during the review process. A likely explanation is that there is simply widespread ignorance of the history of science among the science education community.

Although regrettable, this ignorance is understandable. The history of science rarely figures at any stage in the professional development of scientists, science teachers, or the science educators who prepare science teachers. The history of science tends to fall between the cracks: The science faculties do not deal with it because it is deemed irrelevant; the humanities faculties do not deal with it because it is deemed too technical. The result is that the pendulum, when it appears in the *National Science Education Standards*, is a shadow of its historical self. The opportunity is, unfortunately, lost for American students to learn about the nature of science, the place of science in cultural development, the interrelationship of science with other disciplines and technology, and the importance of tradition and history in forming current scientific understanding and social practice.

The *Standards* have no monopoly on lost opportunity when it comes to teaching and learning about time and the pendulum. The American Association for the Advancement of Science, with its Project 2061,[14] is endeavouring to achieve some degree of curricular coordination between science, mathematics, technology and social studies. As it says, "the students" experiences should be designed to help them see the relationships among science, mathematics, and technology and between them and other human endeavors (AAAS, 1993 p. 320). The pendulum, which can so beautifully do all of this, is not mentioned in AAAS' *Benchmarks*. Time is mentioned, but there is a jump from the age of the earth to time in relativity theory. There is nothing in between. The simple, but challenging, conceptual issue of a time standard is not dealt with; nor is time measurement and its history mentioned.

The U.S. National Science Teachers Association's program *Scope, Sequence and Coordination* is trying to bring some scope, sequencing and coordination to the completely disjointed school science scene (NSTA, 1992). Its former director, Bill Aldridge, uses the pendulum example to illustrate what can be achieved in the way of scope and coordination, however nothing about the history or philosophy of pendulum motion appears in the NSTA document, or in Aldridge's elaboration of the pendulum case. (Aldridge, 1992). This is another sad comment on the gulf between the science education community, and the history and philosophy of science communities.[15]

In the early 1990s in the United States, a collaborative project of the Biological Sciences Curriculum Study (BSCS) and the Social Science Education Consortium (SSEC) stressed the importance of a broad approach to science instruction that encompassed the social, technological and cultural dimensions of science, and also recommended a widening of the social science curriculum to encompass scientific and technological aspects of history and social science (Bybee et al., 1992). The joint project listed six goals of this broadened, crossdisciplinary education: Science and technology influence the social, cultural, and environmental contexts within which they occur in the United States; science and technology are inextricably related to human values and activities; and so on (Bybee et al., 1992, p. 26). The pendulum is perhaps the ideal vehicle for realizing these laudable aims, yet in the two volume report it is unfortunately not mentioned. Another missed opportunity.

What the BSCS/SSEC report does recognize is that a significant barrier to achieving the project's goals is that "the preparation of teachers is inadequate" (Bybee et al., 1992, p. xiii). This important matter will be further discussed in Chapter 13.

CROSS-DISCIPLINARY TEACHING ABOUT TIME AND THE PENDULUM

This book has indicated a little of the cultural, historical, technological, religious and philosophical dimensions of the analysis of pendulum motion and of timekeeping. It is unrealistic to think that all this can be covered in a science course. But it is not unrealistic to hope that some coordination between subject areas can be achieved in a school or college, and thus for teachers in related fields to work together on the "big picture" of timekeeping. This is precisely what the BSCS/ SSEC project recommended.

Such coordination is of course almost unheard of in school systems. History, science, mathematics, music, social studies, literature, philosophy all go their own way, with barely a passing curricular nod to each other. It is notorious that physics classes teach material requiring a level of mathematics that the mathematics program has not covered, and that the mathematics is taught without reference to the physics. From the students' point of view, and even from the teacher's, knowledge is truly fragmented. However, well chosen themes that are heuristically rich can organize a curriculum to maximize the degree to which the interdependence of knowledge becomes more transparent. It may be, minimally, a matter of looking at existing independently generated curricula and simply pulling the related parts together and arranging for some coordination and crossreferencing. But it can be more than this. Timekeeping, navigation, and the longitude problem, for instance, provide rich materials for multicultural science teaching and comparative anthropology. We know that the Polynesians made great sea journeys (Akerblom, 1968). Although it needs be admitted that we know little about the failures of their journeys, nevertheless comparing their navigation methods with the marine chronometer methods of Europe is instructive.

A praiseworthy example and potential model of coordination between disciplines occurs at the Oberstufen Kolleg of the University of Bielefeld, Germany. The college utilizes a "Historical-Genetical Approach to Science Teaching," which draws its inspiration from work of the German science educators Martin Wagenschein (Wagenschein, 1962) and Jens Pukies (Pukies, 1979).[16] At the Oberstufen Kolleg:

> . . . there is attention given to the historical, social and philosophical dimension of science. Frequently, historical examples are presented in a rather anecdotal fashion in courses of science, in order to motivate students for the "real thing," the scientific content. History and philosophy are merely instrumentalized and serve to "sell the product." Our intention differs: we consider the historical and philosophical dimension to be an essential part of science and of instruction in science, that aims to present science in a social and historical context. (Misgeld, Ohly, and Strobl, 2000)

And,

> Working with case studies, particularly taken from times of historical upheaval, seems to us the way to ask questions about the process of the development of science: What they are, how they work, why they work. This basically endless search for internal and external influences on the formulation of scientific theory as well as for the consequences, social and economical, the search for social context, the relation to other human endeavours and cultural advances leads to the realization, that science is

part of social development and therefore itself capable of change. (Misgeld, Ohly, and Strobl, 2000)

What is particularly noteworthy about instruction at the college is that:

> About 10 to 15 teachers of the Oberstufen-Kolleg, representing the disciplines Biology, Chemistry, Physics, Mathematics, Economy, Philosophy, Sociology, Arts, Music, Ecological Sciences, Slavic Studies, and Greek—emulates the "Lunar Society," and meets about monthly at one of the members for supper and exchange of thought.
>
> We discuss newly read books, experiences in the course of instruction, current and historical questions about science that have come to attention. Classroom materials are exchanged as well as information about views and approaches of other colleagues, which might be useful. Regularly a previously planned and prepared topic is discussed, filling the greater part of the evening. This way, courses are developed, planned, criticised. (Misgeld, Ohly, and Strobl, 2000)

But even without this formal commitment to cross-disciplinary teaching, the pendulum can provide the occasion for useful collaboration between different academic staff in schools and universities. Some possible areas are represented in Figure 1 (p.14). Technology classes are one clear field for cooperation. The making of a simple pendulum clock is a wonderful opportunity to appreciate the combination of brain and hand that is so essential to science (pp. 311–313). Likewise the making of water clocks, candle clocks, sundials and so on. Perhaps the degree of accuracy of Galileo's waterflow measurements of time (pp. 108–109) might be discussed and held as a standard to emulate. The making of a sundial is the occasion for the world of astronomy to be opened up for students—the path of the earth, solstices, seasons, the names of the months, and so on (pp. 297–300). Questions such as whether or not the school flag pole might function as a sundial can be investigated. These technical activities can be combined with some basic historical study of ancient societies, and simple anthropology of contemporary societies that are not awash with clocks and watches. More sophisticated questions about timekeeping, wage labor, time-related cultural mores such as punctuality, sporting records, and so on, can also be raised.

Mathematics is another obvious area for collaboration. Geometrical depiction of the pendulum, and the progressive working out of its 'geometrical' properties, can give an oft-missing, applied focus, to basic geometry. How geometrical depiction allows pendulum motion to be related to freefall, and to inclined-plane motion, can be investigated (Figure 44, p. 97, Figure 101, p. 305). More advanced classes can follow some way in Huygens geometrical footsteps (pp. 125–129). Geometry allows names such as Euclid, Galileo and Huygens to be introduced and some appropriate historical background to be given to their life, times and achievements. Alternative mathematical depictions using the resolution of forces, calculus and limit theory can also be investigated (pp. 130–133). Focussing on the pendulum gives a concrete case study of the interrelationship of science and mathematics, the limitations of different forms of mathematical depiction of physical events, and of the relationship of applied mathematics to physics.

History classes can wonderfully complement science, technology and mathematics classes on pendulum and time-related topics. That Dava Sobel's *Longitude* book (Sobel 1995) shot to the top of best-seller lists around the world, indicates that the seemingly arcane problem of longitude is capable of generating great interest. It is tied up with navigation, European expansion and trade, colonization and numerous topics with which good history courses can deal. Hopefully history curricula might be encouraged to include such hitherto off-limits topics as the scientific revolution. Collaboration on the pendulum is certainly an occasion for bringing some aspects of the scientific revolution into history classes—methodology, measurement, technology, timekeeping and so on can be introduced, along with great names such as Galileo, Huygens and Newton. Del Monte is a nice and relatively accessible figure with which to make concrete the Aristotelian opposition to the new science (pp. 100–108).

Music classes can be fruitfully enlisted in pendulum based, collaborative efforts. It is usually eye-opening for music students to learn that their craft played any role in the scientific revolution, let alone an important role (pp. 276–277, 376). The metronome is familiar to these students, and provides a tangible way into the physics and history of pendulum motion. Classes can usefully investigate the distinction between timekeeping and time measurement, and practical work on the reliability of rhythmic beat as a measure of time can be undertaken. Musicology classes can profitably deal with 'Music and Science in the Age of Galileo', to use the title of just one recent anthology on the function of music in the history of science (Coelho 1992).

And much more is possible: religion classes can deal with the checkered history of the design argument (pp. 219–228); literature classes can study the rich genre of prose and poetry occasioned by the scientific revolution (pp. 228–231), and of course write their own verse on time-related themes; philosophy classes can take up methodological matters raised by the study of pendum motion, they can deal with the relationship of theory to evidence, and of experiment to both theory and evidence (pp. 243–267); perhaps drama classes could conduct a dramatic reenactment of the Galileo–del Monte relationship, which would involve historical, philosophical and scientific studies, as well as the full range of dramatic competencies.

CHAPTER 13

Science Education, Teacher Education, and Culture

John Dewey said of the cultural value of science education that:

> Our predilection for premature acceptance and assertion, our aversion to suspended judgment, are signs that we tend naturally to cut short the process of testing. We are satisfied with superficial and immediate short-visioned applications. . . . Science represents the safeguard of the race against these natural propensities and the evils which flow from them. . . . It is artificial (an acquired art), not spontaneous; learned, not native. To this fact is due the unique, the invaluable place of science in education. (Dewey, 1916, p. 189)

One might hesitate to say that science is *the* safeguard of the race against our natural propensities to superficial and self-serving judgments, but most agree that science is *one* such safeguard, provided that something about the spirit of science is internalized while science is being learned. This is often stated as the need to engender appreciation of, and commitment to, the nature of science among those learning science.

LEARNING ABOUT THE NATURE OF SCIENCE

The nature of science has long been of concern to science teachers and curriculum developers. Since the early 19th century, when science first won its place in the curriculum of some schools, it has been hoped that science teaching would have a beneficial impact on the quality of culture and personal life in virtue of students not only knowing science, but also internalizing something of the scientific spirit. Clearly these longstanding aspirations for science education depend upon some understanding by teachers and curriculum developers of the methodological and epistemological aspects of science. That is, they depend on some knowledge of the nature of science.

One reviewer of current curricular developments said:

> Although varying ideas exist for what components constitute a scientifically literate
> citizenry, two attributes that have been consistently identified in the literature are an
> understanding of the nature of science and an understanding of the nature of scientific
> knowledge. (Meichtry, 1993, p. 429)

Another reviewer stated: "For more than half a century there has been an overwhelming consensus of science education literature and science organizations of instructing science teachers and/or their students in the nature of science" (Alters, 1997, p. 39). Peter Rubba and colleagues, who advocate a science–technology–society orientation to science education, pointed out that:

> with the rise of Science–Technology–Society (STS) within the K–6 science curriculum
> during the 1980s helping students develop an understanding of the nature of science
> and technology, and their interactions in society, has become a complementary goal.
> (Rubba et al., 1996, p. 387)

These writers are the latest in a long line of scientists and educators who believed science education should impart some knowledge about the nature of science as well as imparting scientific understanding.[1] In Germany, at the end of the 19th century, Ernst Mach argued that history and philosophy should be a natural part of all school and university science instruction.[2] In the United Kingdom the influential Thomson Report of 1918 said of science teaching that

> it is desirable . . . to introduce into the teaching some account of the main achievements
> of science and of the methods by which they have been obtained. There should be
> more of the spirit, and less of the valley of dry bones . . . One way of doing this is by
> lessons on the history of science. (Brock, 1989, p. 31)

The report added: "some knowledge of the history and philosophy of science should form part of the intellectual equipment of every science teacher in a secondary school."

In the century since Mach, Dewey, and the Thomson Report, many educators have articulated the cultural, educational and scientific benefits of teaching about the nature of science, and the benefits of infusing epistemological considerations into science programs and curricula. This tradition of work, albeit it a minority one, includes Joseph Schwab's writings in the 1940s and 1950s (Schwab, 1945, 1958), the books of Leo Klopfer and James Robinson in the 1960s (Klopfer, 1969; Robinson, 1968), and the publications of Jim Rutherford, Gerald Holton, and Michael Martin in the 1970s (Rutherford, 1972; Holton, 1978; Martin, 1972, 1974). In the last two decades, many others have contributed. There has been a rapprochement between science education and the history, philosophy, and sociology of science.[3]

Understanding the nature of science is not the monopoly of liberal approaches to science education. Many who advocate a more narrow *professional,* or *technical*, approach to the subject nevertheless follow Ernst Mach in maintaining that this competence itself requires an understanding of the nature of science. For instance, the physicist Frederick Reif said:

All too often introductory physics courses "cover" numerous topics, but the knowledge actually acquired by students is often nominal rather than functional. If students are to acquire basic physics knowledge . . . it is necessary to understand better the requisite thought processes and to teach these more explicitly . . . if one wants to improve significantly students' learning of physics . . . It is also necessary to modify students' naive notions about the nature of science. (Reif, 1995, p. 281)

Certainly where more liberal or contextual approaches to science teaching have prevailed, there has been more of a concern with teaching about the nature of science, and using history and philosophy to illustrate the issues (Harvard Project Physics, for example). Where more professional approaches have prevailed, then nature of science considerations have generally been marginalized—something announced at the beginning of a book, or course, and then forgotten. The difference can be seen when liberal texts such as Hogben's *Science for the Citizen* (Hogben, 1940), Taylor's *Physics: The Pioneer Science* (Taylor, 1941), Holton's *Introduction to Concepts and Theories in Physical Science* (Holton, 1952) or Aron's *Development of Concepts of Physics* (Arons, 1965) and his *The Various Language, An Inquiry Approach to the Physical Sciences* (Arons 1977) are contrasted with any of the thousands of professional texts in physics. The liberal texts conveyed the concepts of science and additionally, a sense of the life and times of scientists, of the social circumstances that called forth the scientific developments, of the difficult birth of new scientific concepts and debate over their legitimacy. These texts give a sense of science as a part of culture, and usually there is something of a story line. Professional texts lack a story line: concepts, definitions, refinements, model problems and end-of-chapter exercises are the staple.

In the 1960s, the science educator James T. Robinson recognized that the United States was, post-Sputnik, awash with "science kits with packaged laboratory exercises," and "keeping students actively engaged in laboratory activities and well supplied with reading materials . . . is no longer a problem" (Robinson, 1965, p. 37).[4] The point of departure of his Ph.D. was the question: "Have the students, as a result of doing the activities, reading, and discussing what they have read, indeed increased their 'understanding of science'?" And further, "Can aspects of the nature of science be identified and so specified as to provide for guidance in the selection and organisation of elements which are to be included in science curricula?" (Robinson, 1965, p. 37).

DEBATE ABOUT THE NATURE OF SCIENCE

During the past three decades, questions about the nature of science have become both more contentious and more pressing than they were for Robinson and earlier writers. Previously there was a degree of cultural and philosophical unanimity about the nature and purpose of science. Of course there were some areas of dispute (inductivists versus falsificationists, positivists versus realists, Kuhnians versus

Popperians, empiricists versus rationalists, Bernalian state-interventionists versus Polyanian free-enterprisers, etc.), but these disputes were basically domestic ones. Robert Merton's characterization of science as open minded, universalist, disinterested, communal, and concerned with the pursuit of knowledge summed up professional and lay opinion on the matter (Merton, 1942).

In England at the end of the World War II this view was advanced by the noted historian Herbert Butterfield in his lectures to Cambridge students on the scientific revolution, where he said: "since the rise of Christianity there is no landmark in history that is worthy to be compared with this." He was voicing the accepted historical and philosophical wisdom about the scientific revolution.

Anyone who has read anything of the "science wars" that have been raging in professional journals and the popular press will know times have changed and there is no longer a consensus about the social or intellectual worth of science.[5] In the past three decades beginning with Thomas Kuhn's book *The Structure of Scientific Revolutions* (Kuhn, 1970) the assumptions that the scientific tradition is typified by the search for truth about the world, the exercise of individual intelligence, rational decision-making based upon empirical or other agreed nonempirical evidence, objectivity, ideological neutrality and cultural universality—have all been questioned by cultural historians of science, the Edinburgh School of sociologists of science, by some philosophers of science such as Paul Feyerabend, and by some feminist scholars. Science is, supposedly, mechanistic, materialistic, reductionist, decontextualized, ideological, masculine, elitist, competitive, exploitative, impersonal and violent (Aikenhead, 1997, p. 220). Cultural historians of science "feel the need to defenestrate science, or at least take it off its pedestal" (Pumfrey et al., 1991, p. 3). They believe:

> knowledge is no transcendental force for progress. Historically understood, it is local, it is plural, it embodies interests, it mobilizes the claims of groups and classes, and, above all, it is recruited, willy-nilly, on all sides in wars of truth. (Pumfrey et al., 1991, p. 3)

As do adherents of the strong program in the sociology of science, cultural historians believe the development of science cannot be explained in terms of epistemic factors. The growth of Newtonianism, for instance, has to be explained in political, cultural, or economic terms. On the other hand, the decline in practice of astrology among scientists cannot be explained in terms of its failure to grasp the workings of the world but, as astrology and Newtonianism are epistemically equivalent, its demise has also to be sought in external factors.

These writers are not without their critics.[6] Some of the better known are: Mario Bunge, who described much of the work in the field as "a grotesque cartoon of scientific research" (Bunge, 1991, 1992); David Stove, who said that the strong program was a "stupid and discreditable business," whose authors were "beneath philosophical notice and unlikely to benefit from it" (Stove, 1991); Larry Laudan, who said that the program's relativism was "the most prominent and pernicious manifestation of anti-intellectualism in our time" (Laudan, 1981b, 1990); Michael Devitt, who labeled constructivism "the most dangerous contemporary intellectual tendency," and said that it "attacks the immune system that saves us from silliness" (Devitt, 1991, p. ix); and Paul Gross and Norman Levitt, who in their best selling book, *Higher Superstition,* said cultural critics of science standardly exhibit "fundamental weaknesses of fact and logic" (Gross and Levitt, 1994, p. 8).

Times certainly have changed since the days of consensus about science. And science teachers, are caught up in the war. Teachers, as usually happens when cultural and social wars erupt, suffer collateral damage.

Previously the science teacher had to master his or her subject matter and the techniques of making it interesting and intelligible to students (Shulman's pedagogical content knowledge). Good teachers tried to master something of the orthodox history and philosophy of their subject. All of this was demanding enough. But now, in addition, teachers need to understand and evaluate the postmodern challenges to orthodoxy. This is a hard call, but it cannot be avoided. University students, and increasingly school students as well, encounter challenges to the legitimacy and objectivity of science. Such challenges are recurrent in their Sociology 101, Cultural Studies 102, Text Studies (formerly Literature) 103, Feminism 104, and Postmodernism 105 courses. Science teachers should be able to say something informed and intelligent about these challenges. At least they should be able to say what the strong, weak and disputed points are, and how they might bear upon the curriculum topics.

That these debates about the nature of science are not educationally idle has been dramatically shown in decisions by a string of major U.S. cities to adopt the *Portland Baseline Essays* in their science programs (Martel, 1991). The science component of these essays lists a number of commitments that are flatly at odds with a Western-scientific view of the world (see critical discussion in Good and Demastes, 1995; and Montellano, 1996). The claim is made that the Western view is just one of a number of equally valid scientific views, and thus the African science contained in the *Portland Baseline Essays* should be included in the curriculum. The education of all children in these American cities hinge on decisions regarding the historical and epistemological claims made in the *Baseline Essays.* By most standards, the claims are nonsense. Of the *Essays,* one critic has said:

> People who are genuinely concerned with improving science education in the schools and with increasing the number of minority scientists should vigorously oppose the inclusion of material into the curriculum that makes unsupported claims, introduces religion under the guise of science, and claims that the paranormal exists. The critical need to increase the supply of minority scientists requires that they be taught science at its best rather than a parody. (Montellano, 1996, p. 569)

The creation science debates and trials in the United States also pivot on considerations about the nature of science. This was explicitly the case in the

1981 Little Rock trial over the constitutionality of Arkansas's Act 590 requiring equal time for the teaching of creation science and evolutionary science.[7] In Malaysia, and other Islamic countries, Islamic science is mandated in schools, whilst Darwinian evolutionary theory is prohibited. In India, Hindu science, along with its attendant casteism, sexism and mythology, is rising on the tide of right-wing nationalism. In all these cases a great deal hinges on the difference between science and pseudoscience.

Epistemological questions are also at the heart of many feminist critiques of contemporary science and of feminist proposals for the reform of science curricula. Feminism has long ceased to be just a political movement, it now encompasses a variety of epistemological positions, some of which are in conflict with an orthodox understanding of science (see, for instance, contributions to Nelson and Nelson (1996)). Sandra Harding's "standpoint" feminist epistemology is an example of one such position that has had some influence in academic science education (Harding, 1986, 1991). It needs to be recognized that there are many critics of feminist epistemology and of feminist analyses of particular historical episodes in science. These critics cannot be dismissed as male apologists, their arguments need to be engaged with.[8] For instance, Noretta Koertge, a feminist and one of the first philosophers of science to write on science education (Koertge, 1969) said, in a piece critical of feminist philosophy of science:

> If it really could be shown that patriarchal thinking not only played a crucial role in the Scientific Revolution but is also necessary for carrying out scientific inquiry as we now know it, that would constitute the strongest argument for patriarchy that I can think of. (Koertge, 1981, p. 354)

whilst Radcliffe Richards contends:

> If the whole idea of feminist epistemology rests on a mistake, that in itself bodes ill for the details, and enough has already been said in passing—for instance, about confusions between epistemological and first-order standards, knowledge and knowledge claims, and the politics of knowledge and the politics of epistemology— to suggest a range of serious problems. (Richards, 1996, p. 404)

Epistemological questions are also prompted by much environmentalist, "deep ecology," and "new science" writing to which students are increasingly exposed. It is a rare science teacher who has not in the recent past been asked: "Do you believe in the Gaia hypothesis?"or, "Does the Big Bang prove God's existence?" or, "Do whales have rights?" All of these matters—multiculturalism, religion, feminism, environmentalism—give rise to serious epistemological and, of course, political and ethical questions. What needs to be recognised, as Radcliffe Richards pointed out, is the political folly of adopting "sophisticated, oppression-loaded, all-purpose rhetoric that actually obstructs any serious attempt at analysis" (Richards, 1996, p. 407).

The ability to distinguish good science from parodies and pseudoscience depends on a grasp of the nature of science. More generally, the widespread debates occurring over multicultural science hinge upon answers to the question: What is science? There is an epistemological question at issue in the debate, which unfortunately often gets confused with ethical and political questions. People too frequently move from the premise that colonialism and imperialism, with their destruction of indigenous worldviews and customs, were ethically bad, and po- litically exploitative, to the conclusion that Western science must be epistemo- logically bad. Conversely, it is thought that if a person affirms the epistemological superiority of Western science, then they must be in favor of discredited ethical and political practices. This is not so.[9]

CONSTRUCTIVISM AND THE NATURE OF SCIENCE

Perhaps more than anything else, what has pushed epistemological considerations to the forefront of contemporary science education discussion, is the prominence that constructivism has gained in the science education community. Peter Fensham said: "The most conspicuous psychological influence on curriculum thinking in science since 1980 has been the constructivist view of learning" (Fensham, 1992, p. 801). Russell Yeany observed: "A unification of thinking, research, curriculum development, and teacher education appears to now be occurring under the theme of constructivism" (Yeany, 1991, p. 1). Ken Tobin has remarked that: "constructivism has become increasingly popular . . . it represents a paradigm change in science education" (Tobin, 1993, p. ix). Speaking of recent American education reforms, Catherine Twomey Fosnot has commented that "Most recent reforms advocated by national professional groups are based on constructivism. For example the National Council for Teachers of Mathematics . . . and . . . the National Science Teachers Association" (Fosnot, 1996, p. x).

Constructivism is a significant influence in the *National Science Education Standards* (NRC, 1996). The 1992 *Draft Standards* recognized that the history, philosophy, and sociology of science ought to contribute to the formation of the science curriculum. But when the contribution of philosophy of science was elabo- rated in an appendix, it turned out to be *constructivist* philosophy of science. After dismissing a caricature of logical empiricism, the document endorses "A more contemporary approach, often called postmodernism [which] questions the objectivity of observation and the truth of scientific knowledge." It proceeds to state that "science is a mental representation constructed by the individual," and concludes, in case there has been any doubt, that "The National Science Educa- tion Standards are based on the postmodernist view of the nature of science." Not surprisingly these endorsements caused some eyebrows to be raised.[10]

The revised 1994 *Draft* emerged *sans* the Appendix, but its constructivist content was not rejected, merely relocated (NRC, 1994). Learning science was still identified with "constructing personal meaning." And the history of science was seen in terms of the "changing commitments of scientists [which] forge change commonly referred to as advances in science." As one commentator, sympathetic to constructivism, remarked

> even though the term *constructivism* is not used even once in the NSES, it is clear that individual constructivism . . . is the driving theory of teaching and learning throughout the document . . . the theoretical underpinning of the document is made to be invisible. (Rodriguez, 1997, p. 30)

Constructivism is fundamentally, as it was with Piaget, an epistemological doctrine. It is normally coupled with commitments to certain postpositivist, postmodernist, antirealist, and instrumentalist views about the nature of science. For instance in 1986, Rosalind Driver and Beverley Bell wrote that: "Rather than viewing truth as the fit between sense impressions and the real world, for a constructivist it is the fit of our sense impressions with our conceptions: the authority for truth lies with each of us" (Driver and Bell, 1986, p. 452). The same year, George Bodner said: "The constructivist model is an instrumentalist view of knowledge. Knowledge is good if and when it works, if and when it allows us to achieve our goals" (Bodner, 1986, p. 874). In 1989, Ernst von Glasersfeld, claimed that: "The word 'knowledge' refers to a commodity that is radically different from the objective representation of an observer-independent world which the mainstream of the Western philosophical tradition has been looking for. Instead 'knowledge' refers to conceptual structures that epistemic agents, given the range of present experience within their tradition of thought and language, consider *viable*'" (Glasersfeld, 1989, p. 124). In 1996, Catherine Twomey Fosnot wrote that "constructivism . . . describes knowledge as temporary, developmental, nonobjective, internally constructed, and socially and culturally mediated" (Fosnot, 1996, p. ix). There is no difficulty in finding other examples in the science education community of commitment to constructivist epistemology and constructivist understanding of the nature of science.

The Driver and Bell claim that "the authority for truth lies with each of us" is truly breathtaking in its educational and cultural implications. It is no wonder that so many constructivists see themselves as revolutionizing the practice of education. Ernst von Glasersfeld understates the matter when he says that: "If the theory of knowing that constructivism builds up . . . were adopted as a working hypothesis, it could bring about some rather profound changes in the general practice of education" (Glasersfeld, 1989, p. 135). These more personal, Piagetian-tradition, constructivist views undoubtedly do have, if true, profound consequences for education, for understanding the role of the teacher, for assessment, and so on. So also do the more social constructivist views such as the Edinburgh Strong Programme. Of the latter, Peter Slezak commented:

> There could scarcely be a more fundamental contribution to science education than the one offered by constructivist sociological theories, since they purport to overturn

"the very idea" of science as a distinctive intellectual enterprise with its special values. (Slezak, 1994a, p. 291)

It needs to be recognised that constructivist epistemology has been challenged by scholars inside and outside the education community.[11] Professional opinion is, to put it mildly, divided on its philosophical merits. For instance, Wallis Suchting commented on Ernst von Glasersfeld's version of personal constructivism.

First, much of the doctrine known as "constructivism" . . . is simply unintelligible. Second, to the extent that it is intelligible . . . it is simply confused. Third, there is a complete absence of any argument for whatever positions can be made out. . . . In general, far from being what it is claimed to be, namely, the New Age in philosophy of science, an even slightly perceptive ear can detect the familiar voice of a really quite primitive, traditional subjectivistic empiricism with some overtones of diverse provenance like Piaget and Kuhn. (Suchting, 1992, p. 247)

And, on the more sociological varieties of constructivism, Michael Devitt remarked:

I have a candidate for *the* most dangerous contemporary intellectual tendency, it is ... constructivism. Constructivism is a combination of two Kantian ideas with twentieth-century relativism. The two Kantian ideas are, first, that we make the known world by imposing concepts, and, second, that the independent world is (at most) a mere 'thing-in-itself' forever beyond our ken. . . . [considering] its role in France, in the social sciences, in literature departments, and in some largely well-meaning, but confused, political movements [it] has led to a veritable epidemic of "worldmaking." Constructivism attacks the immune system that saves us from silliness. (Devitt, 1991, p. ix)

I do not want in this chapter to argue the pros and cons of constructivist epistemology as I have done that elsewhere.[12] It is however worth pointing out that frequently the debate is misplaced because each side attributes to the other, positions that the other does not hold. For instance most realists are sophisticated about science and its history. They recognize that science is a human creation, that it is bound by historical circumstances, that it changes over time, that its theories are underdetermined by empirical evidence, that its knowledge claims are not absolute, that its methods and methodology change over time, that it necessarily deals in abstractions and idealizations, that it involves certain metaphysical positions, that its research agendas are affected by social interests and ideology, that its learning requires that children be attentive and intellectually engaged, and so on. These are shared positions that both sides can happily agree about, and can encourage students to appreciate. If these positions collectively amount to constructivism, then we are all constructivist—although, as Ernst von Glasersfeld remarked on one occasion, this is *trivial* constructivism.

There certainly are nonverbal differences between realists and serious constructivists. Realists believe that science aims to tell us about reality, not about our experiences; that its knowledge claims are evaluated by reference to the world, not by reference to personal, social, or national utility or viability; that scientific methodology is normative, and consequently distinctions can be made between good and bad science; that science is objective in the sense of being different from personal, inner experience; that science tries to identify and minimize the impact of noncognitive interests (political, religious, gender, class) in its development; that decision-making in science has a central cognitive element and is not reducible to mere sociological considerations; and so on.[13]

A MODEST PROPOSAL

It is unrealistic to expect students or preservice teachers to become competent historians, sociologists, or philosophers of science. We should have limited aims in introducing questions about the nature of science in the classroom: a more complex understanding of science, not a total, or even a very complex, understanding. Karen Dawkins and Allan Glatthorn rightly stressed this point:

> Since it takes historians and philosophers of science decades of study to derive generalizations about the nature of science from the historical evidence, it is unrealistic to expect teachers or students to develop profound understandings from a necessarily limited exposure to the world of science. (Dawkins and Glatthorn, 1998, p. 164)

Of course the questions strictly do not have to be introduced: they are already there, they only have to be recognized. Philosophy is not far below the surface in any science classroom. At a most basic level any text or scientific discussion will contain terms such as "law," "theory," "model," "explanation," "cause," "truth," "knowledge," "hypothesis," "confirmation," "observation," "evidence," "idealization," "time," "space," "fields," "species." Every science teacher has had a student who puts a hand up and asks: "Excuse me, but if no one has ever seen atoms, why are we drawing pictures of them?" Similarly history (minimally in the form of names such as Galileo, Newton, Boyle, Darwin, Mendel, Faraday, Volta, Dalton, Bohr, Einstein, and so on) is unavoidable. A professional teacher should be able to elaborate a little on these matters: A music teacher is certainly expected to be able to say something about Beethoven, Mozart, Chopin and the other major, and more contemporary, contributors to our musical heritage.

Philosophy begins when students and teachers slow down the science lesson and ask what "law," "model," "confirmation," and so on, mean and what the conditions are for their correct use. All of these concepts contribute to, and in part arise from, philosophical deliberation on issues of epistemology and metaphysics: questions about what things can be known and how we can know them, and about what things actually exist in the world and the relations possible between them. Students and teachers can be encouraged to ask the philosopher's standard questions: What do you mean by . . . ? and, How do you know . . . ? of all these concepts. This kind of philosophical analysis can begin in elementary school.[14] Such introductory philosophical analysis allows greater appreciation of the distinct empirical and conceptual issues involved when for instance Boyle's Law, Dalton's model, or Darwin's theory is discussed. It also promotes critical and reflective thinking.

There is no need to overwhelm students with "cutting-edge" questions. They have to crawl before they can walk, and walk before they can run. This is no more than commonsensical pedagogical practice: simple pendulums are dealt with before compound pendulums, addition and substraction before multiplication and division, Euclidean geometry before non-Euclidean geometry, and so on. There are numerous low-level questions that can be used to engage students: What is a scientific explanation? What is a controlled experiment? What is a crucial experiment? How do models function in science? How much confirmation does a hypothesis require before it is established? Are there ways of evaluating the worth of competing research programs? Did Newton's religious belief affect his science? Was Darwin's "damaged book" analogy a competent reply to critics who pointed to all the evidence that contradicted his evolutionary theory? Was Planck culpable for remaining in Nazi Germany and continuing his scientific research during the war? And so on.

Studies show that all the standard logical fallacies, known since Aristotle's day, are routinely committed by science students (Jungwirth, 1987; Zeidler, 1997). Given that being able to reason clearly is an obvious component of being scientific, then one of the low-level nature of science objectives might be to teach some elementary formal and informal logic. This can surely do no harm, and does a great deal of good.

Science educators should be modest when urging substantive positions in the history and philosophy of science, or in epistemology. James Robinson in the 1960s tied understanding the nature of science to *agreement* with 85 logical–empiricist propositions about science (Robinson 1968). It would have been wiser to aim for *understanding* the propositions, or *evaluating* the propositions, rather than being in agreement with them.

Teachers can of course have strong opinions on various substantive matters in the philosophy of science. They can be fierce partisans of particular viewpoints; they may believe that theoretical terms are mere conveniences for linking observational statements, that all theoretical progress in science is the result of change in external social conditions, that there is no scientific method, that Newton sent Western science down the wrong epistemological track, and so on. Some science teachers might strongly believe in the 3Rs: realism, rationality, and reason. Modesty does not entail vapid fence-sitting, but it does entail the recognition that there are usually two, if not more, sides to most serious intellectual questions. This recognition needs to be intelligently and sensitively translated into classroom practice.

Bertrand Russell, an outspoken and fierce partisan of numerous philosophical and social causes, recognized the antieducational effect of indoctrination. In his essay *On Education*, written during World War I, he pointed out:

> If the children themselves were considered, education would not aim at making them
> belong to this party or that, but at enabling them to choose intelligently between the
> parties; it would aim at making them able to think, not at making them think what
> their teachers think . . . we should educate them so as to give them the knowledge and
> the mental habits required for forming independent opinions . . . (Egner and Denonn,
> 1961, p. 401)

A case study approach allows nature of science matters to be dealt with inductively, tentatively and not didactically. The philosophical issues can arise from questions about, and discussion of, episodes in the history of science, from appropriate biographies of scientists, from laboratory exercises including ones that replicate historical experiments, from textbook illustrations, from popular writings, or from science-related social issues.[15] This book's case study of the pendulum has, hopefully, demonstrated the benefits of an inductive approach to teaching about the nature of science.

TEACHER EDUCATION

The vitality of the scientific tradition depends upon teachers introducing children to its achievements, methods and thought processes. Without teachers there are neither scientists nor scientifically literate citizens. This fact is often overlooked. In a recent book dealing with the life and achievements of illustrious members of the pre-World War II Hungarian intellectual dispora, the Nobel Laureate Eugene Wigner, remarked:

> I wish to say at this occasion a few words on a subject about which we think little
> when we are young, but which we appreciate increasingly when we reflect on our
> intellectual development. I mean our gratitude to our teachers. . . . Our gymnasium
> teachers had a vital presence. To kindle interest and spread knowledge among the
> young—this is what they truly loved. They were preoccupied with teaching and they
> impressed us all, not only with their array of facts, but with the intense and loving
> attitude they held toward knowledge. (Marx, 1994, p. 175)

Contextual teaching about the pendulum requires teachers who have knowledge about the history and philosophy of science (HPS). Unfortunately HPS is rarely included in science teacher education programs, these being dominated by educational psychology and teaching methodology. This situation will have to change, if for no other reason, because the curricula and national and state standards being implemented in the United States, Canada, Britain, Denmark, Spain, Japan, Australia and other countries explicitly include episodes in the history of science, and questions about the nature, or philosophy, of science. And, apart from curricular considerations, science teachers are increasingly caught up in controversies that demand philosophical and historical acumen: feminist critiques of science, multicultural claims about the status of Western science, fundamentalist Christian and Islamic religious claims about the scientific worldview and, specifically, evolution and its place in the curriculum.

History and Philosophy of Science in Teacher Education

It has long been argued that the history and philosophy of science should be a staple part of the education of science teachers. As long ago as 1918, the British *Thomson Report* said "some knowledge of the history and philosophy of science should form part of the intellectual equipment of every science teacher in a secondary school" (Thomson, 1918, p. 3). More recently, the Science Council of Canada, after advocating increased attention to historical and philosophical matters in the science curriculum, said: "Although Council does not expect children or adolescents to be trained in the philosophy of science, it does expect science educators to be trained in this area" (SCC, 1984, p. 37). A 1981 review of the place of philosophy of science in British science-teacher education said:

> This more philosophical background which is being advocated for teachers would, it is believed, enable them to handle their science teaching in a more informed and versatile manner and to be in a more effective position to help their pupils build up the coherent picture of science—appropriate to age and ability—which is so often lacking. (Manuel, 1981, p. 771)

Michael Polanyi suggested the history and philosophy of science should be as much a part of science teacher education as literary and musical criticism is part of literary and musical education. Leaving aside at the moment the debate about the "canon" in English and Music, most English teachers feel a professional obligation to read and know something about the major contributors to English literature (or French and German teachers, to the major French and German writers, and so on). Good music teachers are assumed to have some knowledge and appreciation of the great musicians. So also a science teacher should have some knowledge of the major contributors to science, and the "big episodes" in the history of science.[16]

Thirty years ago Israel Scheffler argued for the inclusion of philosophy of science courses in the preparation of science teachers. It was part of his wider argument for the inclusion of courses in the philosophy of the discipline in programs that are preparing people to teach that discipline. His suggestion was that: "philosophies-of constitute a desirable additional input in teacher preparation beyond subject-matter competence, practice in teaching, and educational methodology" (Scheffler, 1970, p. 40). He summarized his argument as follows:

> I have outlined four main efforts through which philosophies-of might contribute to education: (1) the analytic description of forms of thought represented by teaching subjects; (2) the evaluation and criticism of such forms of thought; (3) the analysis of specific materials so as to systematise and exhibit them as exemplifications of forms

of thought; and (4) the interpretation of particular exemplifications in terms accessible to the novice. (Scheffler, 1970, p. 40)

Scheffler's suggestion fell on deaf ears, as witnessed to by the title of a 1985 paper: "Science Education and Philosophy of Science: Twenty-Five Years of Mutually Exclusive Development" (Duschl, 1985).

Teachers, as professionals, should have historical and philosophical knowledge of their subject matter quite independently of whether this knowledge is directly used in classrooms: teachers ought to know more about their subject than they are required to teach. Teachers have a professional responsibility to see beyond the school fence. Science teachers, in particular, are introducing children to a tradition that is complex, rich, influential and of great cultural significance. In contrast to drill sergeants, political commissars, or coaching staff, educators should have a historical, philosophical, and cultural perspective on their discipline. Some degree of this perspective is what distinguishes educators from mere trainers. This is the case for educators in literature, music, philosophy, art, history, and religion. Good philosophers, historians, and theologians, for instance, take it for granted that they have to know something of the history of their disciplines, the major epistemological tendencies that have characterized their disciplines, and exhibit something of a methodological competence in appraising these tendencies. This lack of a sense of tradition, and of the historicity of cognition, is typical of fundamentalisms of all kinds, including scientific fundamentalism or scientism.

The Stanford-based, Carnegie-funded, National Teacher Assessment Project, directed by Lee Shulman, is one influential project that stresses the importance of teachers having a broad professional knowledge base. For Shulman:

> Teachers must not only be capable of defining for students the accepted truths in a domain. They must also be able to explain why a particular proposition is deemed warranted, why it is worth knowing, and how it relates to other propositions, both within the discipline and without, both in theory and in practice. (Shulman, 1986, p. 9)

Shulman's ideas are reflected in the United States National Board for Professional Teaching Standards—*What Teachers Should Know and Be Able To Do* (1989). An evaluation package for biology teachers that has been developed as part of the Carnegie project, tries to assess teachers' grasp of the nature of science, its processes, and determinants. In their words, "Do teachers hold a rich conception of the scientific enterprise as an interaction of the facts, laws and theories of a domain, mastering the skills to construct such knowledge, and recognizing that this knowledge is influenced by and has influence on human society?" (Collins, 1989, p. 64).

The foregoing are all varieties of *inprinciple* arguments for the inclusion of HPS in science teacher education, but these have now been superceded by *inpractice* arguments because so many countries have included HPS elements in their national standards and curriculum. The United States National Science Education

Standards have previously been mentioned (see Good and Shymansky, 2001; Rutherford, 2001). They require that United States students completing a program of science teaching should, among other things, know that:

> Science contributes to culture (NRC, 1996, p. 21);
>
> Technology and science are closely related. A single problem has both scientific and technological aspects (NRC, 1996, p. 24);
>
> Scientific literacy also includes understanding the nature of science, the scientific enterprise, and the role of science in society and personal life (NRC, 1996, p. 21);
>
> Tracing the history of science can show how difficult it was for scientific innovators to break through the accepted ideas of their time to reach conclusions that we currently take for granted (NRC,1996, p. 171);

A position paper of the United States Association for the Education of Teachers in Science, the professional association of those who prepare science teachers, has recognized something of the foregoing concerns, and has recommended to the association that:

> *Standard 1d*: The beginning science teacher educator should possess levels of understanding of the philosophy, sociology, and history of science exceeding that specified in the [US] reform documents. (Lederman et al., 1997, p. 236)

In the Australian state of New South Wales, the new *Draft Stages 4–5 (Grades 7–10) Science Syllabus* (BOS, 1998) includes "History of Science" and "The Nature and Practice of Science" as two of its five focus areas. The syllabus says of the Focus Areas that:

> Attention to the Prescribed Focus Areas will assist students to develop an understanding that scientific activity has become an integral part of the culture in which we live and, as such, contributes a distinctive view of the world. The Prescribed Focus Areas increase students' knowledge and understanding of science as an ever-developing body of knowledge, the provisional nature of scientific explanations, the complex relationship between evidence and ideas, and the impact of science on society. (BOS, 1998, p. 5)

Concerning the history of science, the syllabus says:

> Knowledge of the historical background is important for an adequate understanding of science. Students should develop an understanding of: the developmental nature of scientific knowledge; the part that science has played in shaping society; how science has been influenced and constrained by societies. (BOS, 1998, p. 9)

Concerning the nature and practice of science, the syllabus says

> students should develop an understanding of the nature and practice of science, including the importance of creativity, intuition, logic and objectivity. Students should develop an understanding of the nature of scientific explanations, their provisional character, the development of ideal cases from raw phenomena and the complex

relationship between: existing scientific views and the evidence supporting these; the processes and methods of exploring, generating, testing and relating ideas; and the stimulation provided by technological advances and constraints imposed by the limitations of technology. (BOS, 1998, p. 9)

In the United Kingdom, a group of prominent science educators, reflecting on Britain's National Curriculum and the most appropriate form of science education for the new millennium, wrote a report with ten recommendations, the sixth of which said that: "The science curriculum should provide young people with an understanding of some key-ideas-about science, that is, ideas about the ways in which reliable knowledge of the natural world has been, and is being, obtained" (Millar and Osborne, 1998, p. 20). In elaborating this recommendation, the writers say:

Pupils should also become familiar with stories about the development of important ideas in science which illustrate the following general ideas:

- that scientific explanations "go beyond" the available data and do not simply "emerge" from it but involve creative insights (e.g. Lavoisier and Priestley's efforts to understand combustion);
- that many scientific explanations are in the form of "models" of what we think may be happening, on a level which is not directly observable;
- that new ideas often meet opposition from other individuals and groups, sometimes because of wider social, political or religious commitments (e.g. Copernicus and Galileo and the solar system);
- that any reported scientific findings, or proposed explanations, must withstand critical scrutiny by other scientists working in the same field, before being accepted as scientific knowledge (e.g. Pasteur's work on immunization). (Millar and Osborne, 1998, pps. 21–22)

In all these cases, HPS is not simply another item of subject matter added to the science syllabus; what is proposed is the more general incorporation of HPS themes into the content of curricula. As the American Association for the Advancement of Science said:

Science courses should place science in its historical perspective. Liberally educated students—the science major and the nonmajor alike—should complete their science courses with an appreciation of science as part of an intellectual, social, and cultural tradition. . . . Science courses must convey these aspects of science by stressing its ethical, social, economic, and political dimensions. (AAAS, 1990, p. 24)

Clearly the realization of the laudable liberal goals of all these curricula is only possible if teachers themselves have some knowledge of how science and culture have been, and are, related as well as some appreciation of the epistemological status of scientific knowledge. In a number of countries, HPS has moved from a "wish" list to an "examinable" list in the school science programme. This makes it necessary, not just desirable, that appropriate HPS courses be included in teacher education programs.

Unfortunately teacher education programs have lagged significantly behind these curricular developments. In Australia, less than ten percent of teacher education programs have a significant HPS component. The United States proportion is much the same (Loving, 1991). Most such programs do not adequately prepare teachers to cope with, let alone inspire enthusiasm for, the broader cultural dimensions of the new curricula. The joint BSCS–SSEC document recognized this when, after developing its curriculum framework for a more historical and philosophical approach to the teaching of natural and social science, it said that the first barrier to implementing such a curriculum was that "the preparation of teachers is inadequate" (Bybee, et al. 1992, p. xiii).

For HPS to contribute to science teacher education much greater cooperation between the HPS and science education communities will be required. Ideally HPS departments should be encouraged to offer "HPS in Science Teaching" courses. Simply including a HPS requirement in the teacher education course is not ideal. Student teachers should be able to see the relevance of what they are studying to their own future occupation. Alternatively education departments should hire staff who have formal qualifications in HPS, and have these staff teach such "HPS in Science Teaching" courses. The latter recommendation was made 20 years ago by the Australian researchers Ted Cawthron and Jack Rowell. They wrote a comprehensive paper on the subject of "Epistemology and Science Education" concluding that:

> Clearly the teacher training institutions need to look more closely at these matters and to employ professional staff well versed in both the methodology and the philosophy of science. If possible persons who also have worked in scientific research with distinction for several years. Progress is not going to be very marked, as long as science education is left simply to practising teachers or administrators with no training in the philosophy and epistemological foundations of science. Length of classroom service is not an appropriate qualification for critical analysis on the epistemological level. (Cawthron and Rowell, 1978, p. 51)

Constructivism and Science Teacher Education

Constructivist theories of teaching and learning are a significant influence in contemporary science teacher education programs. As Catherine Twomey Fosnot observed, "Most recent reforms advocated by national professional groups are based on constructivism" (Fosnot, 1996, p. x). Concerning the place of history and philosophy of science in teacher education, constructivism has had a mixed impact. Some constructivists have championed the cause of HPS in teacher education as a way of demonstrating the contingent, culturally-bound and constructivist nature of science. Others, often embracing some form of the Hegelian–Piagetian "Recapitulation Thesis," have turned to the history of science as a way to understand difficulties children have with learning scientific concepts, and hence to facilitate children's learning (Gauld, 1991; Nussbaum, 1983, 1998; Wandersee, 1985, 1990; Wandersee and Roach, 1998). But more commonly, where constructivism is embraced, teacher education programs place more emphasis on the "child as sense-maker," the classroom as a "knowledge construction zone," and "empowerment" of the teacher so they will become a "facilitator" of children's learning. The subject matter being taught, its conceptual structure, and its history and philosophy, are marginalised.[17] Joan Solomon rightly said that constructivism has always "skirted around" the learning of the content of science.[18] Self-awareness, student-awareness, and metacognition are the priorities.

For example, one teacher who participated in a large New Zealand constructivist-inspired teacher development program spoke of the ideal of good teaching that she had prior to undertaking the program:

> . . . those teachers I thought were effective at school were those who seemed . . . very knowledgeable and put it across in the form that I could understand. And so I always strived in my teaching to try and put things across in a nicely structured form that made sense and had easily followed steps. I thought that was what a good teacher was. (Bell, 1992, p. 7)

She was encouraged, "facilitated," to change her ideas. The program director had written that "teachers need to understand what is in the mind of students— their ideas, explanations and understandings, questions, concerns, values, prior experiences and interests" (Bell, 1991, p. 160). This is an undertaking that does not leave much time for other things, such as learning some history or philosophy of the subject matter being taught. Especially if one is teaching 100 high school students, or 1,000 university students in Physics 101. The reason for giving this advice about understanding the mind of the student is that constructivists hold that learning is a matter of individual mental construction, knowledge is what makes sense to the individual, and teaching is facilitating the transformation of preexisting ideas and concepts. Hence the individual focus. Hence the stress on process skills rather than content mastery.[19] There are serious practical and theoretical, to say nothing of ethical, problems with this advice.

The practical flaws had been pointed out by Brian Simon, who wrote masterful critiques of class-based English education policy and practice, when he criticized the British Plowden Report of 1967. He commented that "by focusing on the individual child...the Plowden Committee created a situation from which it was impossible to derive an effective pedagogy" (Simon, 1981, p. 141). And noted that "research has shown that primary school teachers who have taken the priority of individualisation to heart [create]...complex management problems . . . [which are] far too complex and time-consuming...for the teacher to perform" (Simon, 1981, p. 141).

Simon's objections are not just to the obvious practical problems of child-centered pedagogy. With small classes, and an abundance of time, these may be overcome. He also had theoretical objections:

> ... to start from the standpoint of individual differences is to start from the wrong position. To develop effective pedagogy means starting from the standpoint of what children have in common as members of the human species; to establish the general principles of teaching and, in the light of them, to determine what modification of practice are necessary to meet specific individual needs. If all children are to be assisted to learn, to master increasingly complex tasks, to develop increasingly complex skills and abilities or mental operations, then this is an objective that schools must have in common; their task becomes the deliberate development of such skills and abilities in all their children. And this involves imparting a definite structure into the teaching, and so into the learning experiences provided for the pupils. (Simon, 1981, p. 141)

Simon's approach to teaching is opposite to that advocated by most constructivists. He is a radical, a humanitarian, and a staunch opponent of schools' role in perpetrating an unjust British social system, and yet he is clearly directive in his pedagogical advice:

> This approach, I am arguing, is the opposite of basing the educational process on the child, on his immediate interests and spontaneous activity, and providing, in theory, for a total differentiation of the learning process in the case of each individual child. This latter approach is not only undesirable in principle, it is impossible of achievement in practice. . . . new pedagogy requires carefully defined goals, structure, and adult guidance. Without this a high proportion of children, whose concepts are formed as a result of their everyday experiences, and, as a result, are often distorted and incorrectly reflect reality, will never even reach the stage where the development of higher cognitive forms of activity becomes a possibility. (Simon, 1981, pp. 142, 143)

Simon's critique of child-centered pedagogy, in as much as this pedagogy minimizes the importance of subject-matter competence, echoed similar critiques by Antonio Gramsci, the Italian Marxist killed in the 1920s by Mussolini's facists, who regarded progressive education as a vehicle for enslaving rather than empowering the working class (Gramsci, 1971).[20]

Teaching Literacy

The personal and cultural stakes are high here. Bad theories in teacher education do harm. Fosnot, for instance, after commenting on the dominance of constructivist theory in science education, observed:

> In literacy, much in-service work is going on under the rubic of whole language/ writing process. The psychological theory behind all of these reforms is constructivism. (Fosnot, 1996, p. x)

She is right to identify constructivism as, minimally, supportive of, if not responsible for, whole-language approaches to the teaching of reading, of which the best known recent example is New Zealand's Reading Recovery program instituted in the 1970s by Marie Clay. After dominating reading instruction in New

Zealand, Reading Recovery was exported to most Western countries in the 1970s and 1980s. It quickly became, as Fosnot noted, the norm in teacher education and inservice fields. Children are read to individually or in small groups; they are surrounded by writing and books, and somehow they construct their meaning for the whole words they see on paper. The alternative phonic approach is decidedly nonconstructivist: It depends on teaching children some or all of the 44 phonemes of the English language, and then teaching them how to break words down to their constituent phonic sounds and reconstitute the sound of the word. Phonic instruction is tied to a transmission model of learning. But once the basics are there, the hope is that children will be free of the teacher and able to read by themselves. The debate between whole-language and phonics has been intense enough to be labeled "The Reading Wars" (Lemann, 1997).[21]

Fosnot, along with numerous others, clearly believe constructivism is a good theory to support science and literacy education. Others, a minority, have disagreed. Thomas Nicholson, for example, has over a long period conducted appraisals of Reading Recovery in New Zealand (Nicholson, 1993, 1997, 1998, 2000).[22] He believed the program was a disaster, especially for low-income kids and kids from homes that do not support education. Of the new whole-language based English curriculum introduced in New Zealand in 1993, he wrote:

> In the proposed new English curriculum, however, the present system [whole language] will be well and truly engraved in every teacher's reading programme . . . It's an appalling prospect, because the theory behind the curriculum statement (i.e., constructivism theory) is wrong. There are dozens of research studies which are critical of this theory. But the curriculum-makers in the Education Ministry are immune to such criticism. (Nicholson, 1994)

Nicholson's appraisal has been echoed by a major welfare group in Australia, the Smith Family, who conducted a national study of reading attainment of the children in its welfare programs. Their report said:

> the way children have been taught to read may have been making it harder than ever for disadvantaged kids to catch on. . . . [there] are serious doubts about using the "whole word" approach to teaching reading in the early school years. The dumping of the phonetic approach—sounding the word out—means children have no structure to hang onto. . . . this [whole language] may not work for kids whose lives already lack structure and whose parents, coping with unemployment, desertion, violence or illness, may not feel like a cosy read at night. (*Sydney Morning Herald*, February 4, 1995, p. 5)

In Australia, Macquarie University researchers commented adversely on the effect of whole-language instruction in New South Wales primary schools:

> There are obvious dangers in the widespread use of a programme . . . which has not received the benefits of stringent empirically based evaluations. This is not unusual for educational innovations, which are famous for their cycle of early enthusiasm, widespread dissemination, subsequent disappointment and eventual decline—the classic swing of the pendulum. (Center, Wheldall, and Freeman, 1992, p. 263)

Following stringent empirical evaluations, which demonstrated a 20-year-study of the decline of reading ability in all school grades, the Macquarie professor wrote:

> The decline in systematic reading instruction in favour of the less intensive whole language philosophy (in Britain, the United States and Australia) presents as the most likely cause. Whole language teaching methods may be adequate to assist high- and average progress readers to learn to read, but low-progress readers are likely to be even more disadvantaged. (Wheldall, 1999, p. 17)

The contentious effects of constructivism on the teaching of literacy have been mentioned because the same debate is repeated in science education. Constructivism is in the ascendency in colleges of education and in professional associations, yet its philosophical and psychological underpinnings have been questioned by philosophers and cognitive scientists (Matthews, 1998), and its pedagogical usefulness is still to be demonstrated (Matthews, 1994b, Chap. 7, 2000).

Constructivist Language and Educational Realities

Constructivism introduced some new words and meanings. It has borrowed terminology from progressive education traditions, and it appropriated concepts from

CONSTRUCTIVIST LANGUAGE	ORTHODOX LANGUAGE
perturbation	anomaly
viability	confirmation
construction of knowledge	learning
facilitating cognitive transformation	teaching
scheme	theory
conceptual ecology	ideas
accommodation	theory change
negotiation of meaning	student discussion
dialogical interactive processes	talking with each other
student engagement	paying attention
off-task behavior	not paying attention
community of discourse	group
distinctive discoursive communities	different groups
personal construction of meaning	understanding
discourse	writing
verbal discourse	speaking
discoursive resources	concepts
habitus	cultural environment
symbolic violence	causing offense
mediational tools	graphs
conversational artifacts	diagrams
inscription devices	drawings, diagrams, graphs
cognitive apprenticeship	education

postmodernist sources. However it is not clear that new realities have been identified, or that old realities are better explained. Nor is it clear that longstanding problems of epistemology have been avoided, transcended, or solved. Translations, such as the foregoing, can easily be made from constructivist language to standard English and orthodox philosophy of science.

Using such a translation manual, the following constructivist passages can be rewritten in simple everyday terms.

CONSTRUCTIVIST EXPRESSION	PLAIN EXPRESSION
Since coparticipation involves the negotiation of a shared language, the focus is on sustaining a dynamic system in which discursive resources are evolving in a direction that is constrained by the values of the majority culture while demonstrating respect for the habitus of participants from minority cultures, all the time guarding against the debilitation of symbolic violence. (Tobin, 1998, p. 212)	Teach in a way that is sensitive to cultural values.
. . . through our presence as facilitators and mentors, we can provide settings that are constrained and have minimal complexity so that students can construct conceptual and procedural knowledge with low risks of failure. (Roth, 1993, p. 168)	If students are taught simple things first, they are more likely to learn.
[Constructivism] suggests a commonality amongst school science students and research scientists as they struggle to make sense of perturbations in their respective experiential realities. (Taylor, 1998, p. 1114)	Students and scientists consider adjusting their theories when confronted with anomalies.
Making meaning is thus a dialogic process involving persons-in-conversation, and learning is seen as the process by which individuals are introduced to a culture by more skilled members. As this happens they "appropriate" the cultural tools through their involvement in the activities of this culture. (Driver et al., 1994, p. 7)	Students need the assistance of teachers when learning new concepts.
If students are to learn science as a form of discourse, then it is necessary for them to adapt their language resources as they practice science in settings in which those who know s cience assist them to learn by engaging activities in which coparticipation occurs. (Tobin, McRobbie, and Anderson 1997, p. 493)	Students need the assistance of teachers to learn the new concepts and vocabulary required for proficiency in science.
Our microanalytical view of the learning processes in one group showed how much the evolution of students' activities depended on features of the physical context, discourse contributions from individual group members, material actions on and with instructional artifacts, contingent interpretations, and the past history of the activity itself. (Duit et al., 1998, p. 1070)	Learning is affected by peers and by the availability of educational resources.

There is no inprinciple problem with specialized vocabularies and theoretical terms: natural science is full of them. But whereas natural science uses theoretical terms to simplify complex matters, social science, at least in the above examples, is using theoretical terms to make simple matters complex. This is especially odd as so often constructivists in science education accuse orthodox science teachers of committing "symbolic violence," and of confusing students with arcane terminology and concepts.

George Elliot gave good advice when she wrote: 'We have got to exert ourselves a little to keep sane and call things by the names other people call them by'. If this advice is not heeded, then science education research runs the risk of being further ignored by teachers and administrators, and of becoming the province of an isolated in-group whose membership is restricted to those who know the code.

Education is notoriously subject to fashions and fads, especially ones imported from psychology and philosophy. They usually pass if they do not sustain the core educational business of teaching, learning and, hopefully, growing in wisdom. It is perhaps too early to judge constructivism, but there are clear signs that it too will eventually be seen as a passing fad—at least the excesses of the position will be so seen. A great deal of constructivism is laudable—the emphasis on teaching for understanding, on student engagement in learning, on cooperative learning, on being cogniscant of the ideas and prior understanding of students, on developing epistemological sophistication, and so forth. However all of these laudable aspects are really ancient educational verities—most can be found in Socrates' dialogues. They have for centuries been part of the liberal educational tradition. As frequently happens in education, the pedagogical wheel is being reinvented by contemporary constructivists—although some are inventing a square wheel, and others a wheel without any tread.

Philosophy of Education

These considerations raise questions in the philosophy of education. Questions of curriculum construction and content, forms and purposes of assessment, even textbook choice and classroom management depend on being clear about the purposes and aims of education. Unless we know where we should be going, it is difficult to ascertain how well we are getting there. Teachers and administrators need to ask questions such as: What as a society do we want our schools to achieve? How should the study of science be organized so as to advance the educational goals that we identify for schools? Education is a normative enterprise: it seeks to create better societies and better people. Unfortunately normative questions are less and less attended to. Philosophy of education, in any rigorous, systematic and examined sense, is waning in education circles. Which, predictably, means that philosophy of education in non-rigorous, non-systematic and unexamined senses— that is wooly thinking—is waxing.

The contextual, intellectualist, crossdisciplinary proposals of this book find their natural home in the liberal education tradition, whose core commitment is that education is concerned with the development of a range of knowledge and a depth of understanding, and with the cultivation of intellectual and moral virtues.[23] Although this is traditionalist education, it is by no means conservative, elitist, nor tied to authoritarianism or didacticism. To emphasize the role of the teacher as someone who understands the subject matter and can explain it appropriately and engagingly to students is not to endorse any particular teaching style. Straightforward lecturing and instructing might be involved, as might student projects, experiments, investigations, essays, debates or whatever else. The subject matter and the tradition of scientific investigation need to be respected, and so do students. However, respect for students is consistent with telling them they are wrong, that they have misunderstood the meaning of certain key concepts, and that they have to work and think harder to master certain subject matters. A capable teacher can do all of this in a way that is not disrespectful, that is consistent with student's intellectual attainment and capacity, and is not counterproductive to learning. What is less than respectful is to occupy students with tasks that, even if good fun, contribute little to intellectual growth.

This conception of good science teaching is hardly novel. Seventy years ago, F.M. Westaway, an English inspector of schools, who had training in philosophy of science and who wrote a book on the teaching of scientific method (Westaway, 1919), published a popular education text used in science teacher education programs. He said that a good science teacher is one who:

> knows his own subject . . . is widely read in other branches of science . . . knows how to teach...is able to express himself lucidly . . . is skilful in manipulation . . . is a logician . . . is something of a philosopher . . . is so far an historian that he can sit down with a crowd of [students] and talk to them about the personal equations, the lives, and the work of such geniuses as Galileo, Newton, Faraday and Darwin. More than all this, he is an enthusiast, full of faith in his own particular work. (Westaway, 1929, p. 3)

Westaway extolled the idea of teachers as educators, as people who have a rich understanding of their subject matter and are enthusiasts for it. Mortimer Adler long ago argued for the same conception of the teacher, and warned against reducing teacher competence to mere "know how" or pedagogical competence. He wrote:

> For the most part, the members of the teaching profession are overtrained and undereducated. Teaching is an art and a teacher must be trained, but since the technique is one of communicating knowledge and inculcating discipline, it is not educational psychology and courses in method and pedagogy that train a teacher, but the liberal arts . . . Further, a teacher should have a cultivated mind, generally cultivated regardless of his field of special interest, for he must be a visible and moving representative of the cultural tradition to his students. But how can he be this if he has no acquaintance with the cultural heritage, if he cannot read well, and if he is not well-read? (Adler, 1939/1988, p. 79)

These visions or ideals of education are paramount importance for teachers; they set goals and provide guidance to teachers across the spectrum of their engagements with students, with administrations, with parents and with colleagues. Psychology alone simply does not provide this orientation; philosophy is needed.

Some constructivists oppose the inclusion of historical and philosophical considerations in the science curriculum because they are hostile to the idea of a curriculum. Or at least they are hostile to a curriculum that is not student-, or student and teacher-, generated. Michael O'Loughlin, a social constructivist, warned that even in progressive classrooms children still "learn that the voice of authority, whether teacher or text, is privileged and authoritative" (O'Loughlin, 1992, p. 807). Against traditional understanding of education as the mastery of a body of knowledge created by people who are authorities in the subject matter, O'Loughlin said knowing is "a process of examining current reality critically and constructing critical visions of present reality and of other possible realities so that one may become empowered to envisage and enact social transformation" (O'Loughlin, 1992, p. 799).

Measured against these revolutionary epistemological standards, Galileo's diligent efforts in understanding the pendulum, or Huygens' patient determination of the isochronous property of the cycloid curve, hardly rate as knowledge. The historical, philosophical, and cultural study of the pendulum recommended in this book fits uneasily with the teacher's role as a facilitator and empowerer of the construction of critical visions of present reality and other possible realities. The book's assumption is that it is best to understand current reality, including one's intellectual and cultural tradition, before constructing critical visions of alternatives. This surely is the lesson from, for instance, Karl Marx's patient and diligent efforts to understand the capitalist mode of production.

Although he or she may not be helping construct other possible realities and critical visions, a teacher who introduces students to something of the rich story of the pendulum, its connection with timekeeping, its place in cultural and social history, and who helps students understand the science of pendulum motion can nevertheless be confident that they are making a contribution to their pupil's education and the development of their mind. This surely is the prerequisite for intelligent and realistic "critical visions." If more such modest contributions were made by teachers, then the appreciation of science and its role in culture would be more widespread.

The historian Alastair Crombie gave expression to this view when he wrote:

> ... enlightenment ... can be gained only by looking beneath the surface of immediate scientific results, by seeking to identify the intellectual and technical conditions that made certain discoveries possible and explanations acceptable to a particular generation or group, others not, and the same not to others. By displaying the science in historical relation to the philosophical assumptions, technical equipment, and social context of its designers and discoverers, they [historians] showed how the history of science could illuminate the nature both of European culture and of scientific thinking. (Crombie, 1981, p. 271)

CULTURE AND EDUCATION

Certainly the world has moved on since Westaway's characterization of good teachers. It is now the norm that scientists know only their own discipline, and indeed only a small part of that. Knowing something about science outside of one's speciality is uncommon. The bulk of science training is carried on without historical and philosophical considerations—students spend a lot of time learning scientific procedure, laws and theories, but hardly any time is devoted to the meaning of laws and theories, or how they relate to empirical and other evidence. With rare exceptions, scientists are not interested in the tradition of scientific thought. Such interest is seen as a luxury when the struggle to keep up-to-date is so demanding. The number of teachers, who are interested in, as Westaway puts it, the "personal equations, the lives, and the work of . . . Galileo, Newton, Faraday and Darwin" is negligible. Education has also moved on. History, and the cultivation of an historical sense or appreciation, has largely disappeared from education. And school history programs have frequently substituted the learning of "historical method" for the learning of history. Most children leave school barely knowing anything about the 20th century, let alone about earlier centuries. The sense of continuity with, dependence upon, and being able to learn from, the past has largely gone.

The bleak funds- and equipment-starved schools portrayed in Jonathan Kozol's *Savage Inequalities* (Kozol, 1991) defy the best efforts of teachers. But the Western educational crisis is deeper than funds and equipment. Learning and education has flourished in circumstances blighted by desperate poverty. Frank McCourt, in his poignant best seller *Angela's Ashes*, describes how in Ireland during the long centuries of English occupation, "Catholic children met in hedge schools in the depths of the country and learned English, Irish, Latin and Greek. The people loved learning." (McCourt, 1996, p. 209). In Central Europe after World War II schools had few books, but children learned writing and arithmetic often by tracing letters and numbers in sandboxes. Indo-chinese refugee children attend the most crowded and poorly funded Australian and U.S. schools, yet their educational performance is overall far better than children in more privileged schools. In the 70 years since Westaway wrote, the social support for education, as something to be valued in itself, has evaporated. Despite the massive increase in numbers of schools, and in the years of attendance at school, Western culture is becoming increasingly hostile to education. There is a concern with the wrapping, but not the content.

Westaway's teachers could assume their students and their student's parents valued education and the work of teachers. The English working class, for instance, took education very seriously. It was valued. Thomas Huxley's 1868 address at the opening of the South London Working Men's College, titled "A Liberal Education and Where to Find It" well encapsulated the aspirations and assumptions that were once widespread concerning education (Huxley, 1868).

Lánczos Kornél, one of many in the between-wars Hungarian dispora, wrote:

> I was born [1902] in a town of about thirty thousand inhabitants, a town of remarkable vigour and intellectual interest. People lived a life which was rich in many respects. Not rich in material values, but rich in intellectual values. . . . We enjoyed the privilege of intellectual endeavours, in music, in poetry, in science. We grew up in the late Victorian era as one may characterize this Middle-European scene. . . . We appreciated the fact that we could study, therefore it was our duty to work hard. So it was very different from the kind of permissive society we have today [1970] in America. (Marx 1994, p. 138)

The assumption that education is valued can no longer be made. The world, and the home, is geared for entertainment. The staggering amount of television watched is both a cause and an effect of this antieducational culture. Recent research indicates that the average 18-year-old in the United States has spent 20,000 hours watching TV compared to 11,000 in school, and that families spend, on average, 7.5 hours per day watching television (Hamburg, 1992). The silver medal for television watching is held by New Zealand, whose television programs are considered "brain dead" by experienced commentator. Neil Postman, in his recent critique of television culture, had this warning:

> When a population becomes distracted by trivia, when cultural life is redefined as a perpetual round of entertainments, when serious public conversation becomes a form of baby-talk, when, in short, a people become an audience and their public business a vaudeville act, then a nation finds itself at risk; culture-death is a clear possibility. (Postman, 1985, p. 161)

School is just one competitor in the entertainment market, and compared to video games, chat lines, TV soaps, and Friday night basketball—it is not such a great competitor. Westaway could assume a seriousness of intellectual purpose among his students and his fellow teachers. This can no longer be assumed. There might be a seriousness about grades, in as much as these allow progression to higher salaries, or a higher SAT score, but the notion of understanding ideas and being interested in them, the heart of a liberal education, is evaporating. Courses are more and more fragmented, and covered in smaller and smaller time units; TV "sound bites" are replicated in "curriculum bites." Progressive assessment—assessment "bites"—standardly makes anything previously assessed irrelevant, and best forgotten in order to create cognitive space for new material.

The confusion of entertainment with education is widespread. In universities, student assessment of teachers frequently reward the entertainer rather than the teacher who is serious about his or her subject. Universities, through their "publish or perish" policies conspire against genuine scholarship. If newly appointed staff have not published prodigiously in their first five years, irrespective of what half-baked garbage is published, they are fired. Promotion and funding regimes reward "salami" publication, and pretty inadequate salami at that. The tradition of university staff being encouraged to master their subject, its history, and its philosophical implications—to spend time in the library—is under threat. In Australia, at least, there has never been more print expended on the need for excellence in education. The amount expended is inversely related to concern with scholarship and its encouragement.

The case of the late Wallis Suchting is illustrative. He took seven years (1954–1961) to obtain his Ph.D. because he felt he needed to master Latin, Greek and six modern European languages to do justice to his topic: "The Criterion of Empirical Verifibility." He published on average two works each three years. His scholarship, and command of philosophical literature was prodigious. He had an exacting care with words, a devotion to good writing, and a patient analytic ability that was remarkable. As one colleague said on his death, "he was a man of titanic intellect whose appetite for learning knew no bounds." Yet if he were restarting his university career, he would not be appointed in the first place (seven years for a Ph.D. and no conference presentations); and if he were appointed, he would not be given tenure (too few publications, too few students electing to take his courses, and no research grants). That one of the most learned philosophers Australasia has produced would be so treated, is a sad commentary on the direction of Australian university education. But, as one well-placed critic put the matter:

> Unfortunately, Australian university education in the 1990s has become a hotchpotch of academic teller machines, dispensing degrees galore of questionable value to thousands of students, many of whom haven't been well taught, and some of whom shouldn't have been encouraged to enrol. . . . Behind the expensive public relations facade, and with help from tawdy commercial advertising, universities have been involved in a quality-ignoring chase for student numbers. That's led to falling standards of admission and assessment; the fragmenting of knowledge into smaller and smaller areas of specialty; a multitude of undergraduate and postgraduate programs; the abandonment of several unfashionable disciplines; and chronic degree inflation. (Crowley, 1998, p. 1)

The critic, Frank Crowley, an Australian professor of history and dean, did not draw parallels with higher education in the United States, United Kingdom and elsewhere. Others can.

Another increasingly important competitor for schools, especially United States schools, is work. Twenty years ago 70% of all 16- and 17-year-old United States students were also in the labor force flipping hamburgers, taking orders, passing bar-codes over checkout counter scanners, instructing customers to "have a nice day", and so on (Graham,1998, p. 232). The participation rate has increased since then. A New Hampshire study in 1990 showed 84% of students in grades 10–12 had jobs outside of school; and nearly half of these worked more than 20 hours per week (Graham, 1998, p. 231). Most studies confirm that it is "greed not need" which drives this participation. Those who most need the money are the ones who do not get the jobs—to those who already own Doc Martens, Discmen, or some other ephemeral alternative, more will be given. Revealingly, most United

States studies show that working up to 20 hours per week has *no* adverse effect on school results.[24] This is a sad comment on the base level of school performance, and the amount of study required to achieve that performance. But then:

> American schools don't consider academic achievement their principal goal. Social development, emotional well-being, the instillation of democratic values, and athletics all take precedence over academics . . . the whole system seems designed to keep the students busy with anything but studying. (Cromer, 1993, p. 199)

One review of United States education reached this sobering conclusion[25]:

> A large number of recent reports and studies point to the relatively low level of academic achievement registered by contemporary American students. These writers explicitly or implicitly blame a wide variety of factors for this problem: undereducated and underskilled teachers; distracted, spoiled, and unmotivated students; an educational organization clogged with politics, bureaucracy, and unionism; and an unchallenging, watered-down curriculum. But I am suggesting that it is more valid to point the finger at a powerful purpose for schooling that is at core anti-educational. By structuring schooling around the goal of social mobility, Americans have succeeded in producing students who are well schooled and poorly educated. The system teaches them to master the forms and not the content. (Labaree, 1997a, p. 68)

This dismal cultural and educational background makes it more important to strive for good education, particularly good science education.[26] Science education is important not just for instrumental reasons, such as having more engineers, mechanics and nurses—or "knowing where to put the soufflé," as two sociologists of science have recently opined (Collins and Pinch, 1992, p. 150)—it is also important for cultural reasons. Science has a vital part to play in the development of culture, and in people's deep-rooted understanding of their place in the wider scheme of things. Since the scientific revolution of the 17th century science has been a dominant influence in the development of world views, of philosophy, of social science, of large parts of literature and poetry, and even of religious understanding. It is a truism that we live in a scientific age. In order to understand this age—its achievements and limitations—we need some understanding of science.

Science affects styles of thinking and reasoning in society. At its best, science in the Enlightenment tradition has been a bulwark against superstition and self-centered interpretations of the world. For all its faults, the scientific tradition has promoted rationality, critical thinking and objectivity. It instills a concern for evidence, and for judging ideas not by personal or social interest, but by how the world is. Science contributes to a sense of "cosmic piety," as Bertrand Russell once called it.

These values are under attack both inside and outside universities. When the objectivity and rationality of science are eroded, then the power of science to counteract narcissistic personal and ideological tendencies is diminished. We have centuries of evidence to show that natural thinking is neither rational nor scientific. Scientific thinking has to be cultivated and nurtured. It is the result of education. Pseudoscientific and irrational world views have a long history and a seductive appeal.

The flight from science in the Western world needs to be arrested. Gerald Holton, who has for decades been championing the cause of good and enriched science education, observed:

> . . . history has shown repeatedly that a disaffection with science and its view of the world can turn into a rage that links up with far more sinister movements . . . the record from Ancient Greece to Fascist Germany and Stalin's USSR to our day shows that movements to delegitimate conventional science are ever present and ready to put themselves at the service of other forces that wish to bend the course of civilisation their way—for example by the glorification of populism, folk belief, and violence. (Holton, 1993, pp. 148, 184)

CONCLUSION

This book has dealt with a very small portion of the scientific picture: the pendulum and its history, philosophy, physics, and utilization. It is hoped that the book has provided information, ideas and questions that can assist teachers in their task of "kindling interest in, and spreading knowledge about," the larger scientific picture.[27]

This larger picture is coloured with creativity, perserverance, genius, intelligence and emotion; it is painted with technological and mathematical 'brushes'; it is framed by cultural, social and philosophical commitments. It is a human picture, but one whose worth is judged not just by humans, but by the world as well: Nature judges the adequacy of scientific pictures. By contextual teaching of the pendulum, children can experience something of the making of the scientific picture; they can re-enact some of the picture's elements and derive satisfaction from so doing; they can feel some of the curiosity and frustration that drive the great scientists, and perhaps share the insights of Galileo, Huygens, Newton and others. The science curriculum and classroom can give students a feel for, and appreciation of, the scientific tradition. But not if history is excluded, and the opportunity to relive some of the great moments of scientific achievement, is denied. The pendulum story is tailor made for rich and engaging teaching, for displaying the human face of science. The book's argument echoes Arthur Koestler who, in lamenting the "anti-humanism" and "boredom" of science classrooms wrote:

> To derive pleasure from the art of discovery . . . the student must be made to re-live, to some extent, the creative process. In other words he must be induced, with proper aid and guidance, to make some of the fundamental discoveries of science himself, to experience in his own mind some of those flashes of insight which have lightened its path. This means that the history of science ought be made an essential part of the curriculum. (Koestler, 1964, p. 268)

While World War II raged, Karl Popper defended, in a footnote to his *The Open Society and Its Enemies*, a liberal approach to the teaching of science. His words are a fitting conclusion to this book as, although times have changed, the social importance of good science education remains the same:

> There can be no history of man which excludes a history of his intellectual struggles and achievements; and there can be no history of ideas which excludes the history of scientific ideas . . . Only if the student experiences how easy it is to err, and how hard to make even a small advance in the field of knowledge, only then can he obtain a feeling for the standards of intellectual honesty, a respect for truth, and a disregard of authority and bumptiousness. But nothing is more necessary to-day than the spread of these modest intellectual virtues. (Popper, 1945, p. 283)

APPENDIX

Some Significant Dates

ca. 2000 BC	Sundials used for timekeeping in the Tigris and Euphrates valleys.
circa 1500 BC	Sundials widely used in Egypt.
circa 200 BC	Eratosthenes' calculation of the circumference of the earth as equivalent of 39,690 kms (compared to modern 40,000 kms value).
circa 250 BC	Ctesibius' refined water clocks in Alexandria.
circa 335 BC	Aristotle proposes a reductionist account of time whereby time is identified with movement.
400	Saint Augustine in his *Confessions* raises modern philosophical questions about time and proposes a nonreductionist account whereby movements occur in time
1092	Su Sung's heavenly clockwork built in China
1283	A mechanical clock, probably the first in Europe, is built at Dunstable Priory in England
circa 1300	Foliot and verge escapement first utilized in mechanical clocks.
1350	Giovanni de Dondi builds his masterpiece clock at Padua.
1352	Strasbourg cathedral clock built.
1360	Charles V of France installs Henry de Vick's clock in his Paris palace and requires all other clocks including church clocks to keep to its time.
1377	Nicole Oresme writes on the pendulum in his *On the Book of the Heavens*, describes the world a clock and uses this metaphor in his version of the design argument.
1419	Prince Henry establishes the Sagres Institute for navigational improvement.
circa 1490	Leonardo da Vinci sketches pendulums and writes on mechanical clocks.
1492	Columbus's voyage to the Americas.
1494	Treaty of Tordesillas, dividing Spanish and Portuguese spheres of influence.

1530	Gemma Frisius, Flemish astronomer, writes of timekeeping as the solution to longitude determination.
1590	Jesuit priests, Father Ricci and companion, take European mechanical clocks to China.
1598	King Phillip III of Spain establishes a Longitude Prize.
1602	Galileo writes to Guidobaldo del Monte, describing his analysis of pendulum motion.
1610–1620	Galileo corresponds with Spanish Court over the Longitude Prize.
1636	Galileo writes to States-General of Holland claiming his proposed pendulum clock will solve the longitude problem.
1638	Galileo's *Two New Sciences* is published, containing his mathematical proof of the Laws of Pendulum Motion including the mistaken claim that circular pendular motion is isochronous.
1639	Galileo publishes, in Paris, a work on the theory and construction of a pendulum clock.
1657	Christiaan Huygens builds his first pendulum clock, with cycloidal pendula motion, accurate to 10 seconds per day.
1658	Huygens publishes his first book on clocks, the *Horologium*.
1662	British Royal Society established, with ascertaining longitude being one of its preoccupations.
1664	Huygens sends pendulum clocks to sea with Captain Holmes to determine longitude during the voyage.
1666	Louis XIV of France establishes the *Académie Royal des Sciences* with Huygens as president. It is charged with solving the longitude problem.
1672	Jean Richer's voyage to Cayenne, and his discovery that pendulum clocks beat slower in equatorial regions.
1673	Huygens publishes his major horological work, *Horologium Oscillatorium*, proves that the cycloid is the isochronous curve, describes construction of pendulum clock, and proposes the length of a seconds pendulum (0.9935m) as a universal standard of length.
1675	Huygens creates the balance wheel regulator for watches. Greenwich Observatory established in order to solve the longitude problem.
1676	Robert Hooke creates the anchor escapement for clocks.
circa 1678	Thomas Tompion and George Graham create a dead-beat escapement for watches and clocks.
1687	Isaac Newton's *Principia* published with its extensive utilization of pendulum properties.
1714	British government passes the Longitude Act and establishes a £20,000 longitude prize.

1715	William Derham publishes his clockwork version of the Design Argument.
1717	Newton, in the Scholium to the second edition of his *Principia*, articulates his absolutist, nonreductionist account of time.
1759	John Harrison builds his H4 marine chronometer and, after a sea trial to the West Indies, claims the British Longitude Board's prize.
circa 1790	First wrist watch.
1795	French Revolutionary Assembly rejects Huygen's proposal for an international standard of length and adopts the geodesic unit of one ten-millionth of the quarter meridian of the earth as the standard meter.
1798	An estimated 50,000 watches produced in England.
1802	William Paley opens his *Natural Theology* treatise with the clock work metaphor.
1851	Léon Foucault utilizes a "Foucault Pendulum" to establish the rotation of the earth.
1883	Ernst Mach, in his *Science of Mechanics*, criticises Newton's absolutist view of time as being metaphysical and nonscientific.
1905	Albert Einstein's Special Theory of Relativity published, claims that simultaneity cannot have an absolute meaning. First radio time signals broadcast.
1922	William Shortt's free pendulum clock built, with an accuracy of 3/100 of a second per day.
1929	First quartz crystal clock built, with an accuracy of two milliseconds per day.
1946	Jean Piaget's *The Child's Conception of Time* published.
1955	Caesium clock built.
1956	Second defined as 1/31,556,925.9747 part of the tropical year 1900.
1967	Second defined as time elaspsed during 9,192,631,770 oscillations of an undisturbed cesium atom.
1996	United States National Science Education Standards released, with two pages devoted to the pendulum. There is no mention of the longitude problem, timekeeping, pendulum clocks, Galileo, Huygens, the seconds pendulum as a proposed international standard of length, Harrison or any other methodological, historical, or cultural aspects of the pendulum.

Endnotes

CHAPTER 1

1. Dating the scientific revolution is, understandably, a bit more complex than here indicated. I have taken fairly conventional dates (the publication of Galileo's *Dialogue* and Newton's *Principia*) as boundaries. But they are fairly porous boundaries; and they are idealized. Clearly all revolutionary episodes are prepared for, and they rarely culminate in a discrete moment or achievement. Irrespective of timing, there has not been unanimity about just what were the revolutionary aspects of the scientific revolution. On the historiography of the scientific revolution, see Cohen (1994). For recent essays reappraising its methodological and cultural achievements, see Lindberg and Westman (1990).
2. On Newton's philosophy and methodology, see contributions to Bricker and Hughes (1990). On the influence of Newtonianism, see contributions to Butts and Davis (1970), and Stayer (1988).
3. The ideological function of science during Europe's colonial expansion has been much researched. An introduction to the literature can be found in Pyenson (1993). It is important, though, to separate cognitive value from ideological function. Even so, this still leaves serious ethical and political questions about the imposition of Western science on cultures where it is incompatible with, if not destructive of, indigenous science, worldviews, and social structure. On this see Matthews (1994a, Chap. 9) and Siegel (1997).
4. When, immediately after its publication in 1638, Descartes bought a copy of the *Discourse on Two New Sciences,* he wrote to Mersenne that "I also have the book by Galileo and I spent two hours leafing through it, but I find so little to fill the margins that I believe I can put all my comments in a very small letter. Hence it is not worth that I should send you the book" (Shea, 1978, p. 148).
5. That such a huge and well funded body as NSTA could, in the early 1990s, write a national science curriculum document without involving the participation of historians or philosophers is a sad commentary on the gulf between science education and the HPS communities.
6. For discussion of the extent that non-European observers recognized or utilized the properties of pendulum motion, see Hall, 1978, pp. 443–447.
7. The basic documents are the biennial *Science and Engineering Indicators* published by the National Science Board. Jon D. Miller has conducted a series of large-scale studies on scientific literacy in the United States (Miller, 1983, 1987, 1992). On the basis of ability to say something intelligible about concepts such as "molecule," "atom," "byte," in 1985 he judged only three per cent of high-school graduates, 12 percent of college graduates, and 18 percent of college doctoral graduates to be scientifically literate. See also Jones (1988), Darling-Hammond and Hudson (1990) and Shamos (1995). The United States' performance on the 1996 Third

International Mathematics and Science Study (TIMSS) is analyzed in Schmidt, McKnight, and Raizen (1997).

8. The interpretation of United Kingdom figures is made more difficult by the effects of introducing, in 1988, the National Curriculum in England and Wales. Some maintain that the "balanced science" approach of the national curriculum has meant that 16-year-olds are less prepared for, and hence less inclined to choose, specialist discipline courses for their A-levels. See Conroy (1994) and Hughes (1993).

9. The study used scales refined over a number of years by Jon Miller in the United States (Miller, 1992).

10. For discussion of the antiscience phenomena, see Passmore (1978), Holton (1993), Gross and Levitt (1994), Grove (1989), and Levitt (1998).

11. The largest American projects are Project 2061 of the American Association for the Advancement of Science (Rutherford and Ahlgren (1990), AAAS (1993)), the Scope, Sequence and Coordination project of the National Science Teachers Association (Aldridge (1992), NSTA (1992), Pearsall (1992)), and the National Research Council's National Science Education Standards (NRC 1994). Aspects of these are discussed in Eisenhart, Finkel and Marion (1996).

12. See especially the Science Council of Canada's 1984 report, *Science for Every Student: Educating Canadians for Tomorrow's World* (SCC, 1984).

13. See Nielsen and Thomson (1990).

14. See the RMCER (1994) curriculum document which, incidentally, is perhaps the most beautifully illustrated school curriculum document yet produced.

15. The British National Curriculum is the main initiative (NCC, 1988, 1991).

16. This test, and the more general situation of New Zealand science education, is discussed in *Challenging New Zealand Science Education* (Matthews, 1995a).

17. As one might expect, there are political and economic dimensions to this educational crisis and government responses. See Stedman (1987).

18. Douglas Roberts (1982) had provided an account of the range of curricular emphases embraced in the history of science pedagogy.

19. In the United States, the NSTA Yearbooks of 1982 (NSTA 1982) and 1985 (Bybee 1985) deal with the rationale and content of STS programs. The NSTA publication *The Science, Technology, Society Movement* (Yager, 1993) reviewed their implementation. In the United Kingdom, the Science in a Social Context (SISCON), and Science and Technology in Society (SATIS) programs have been the most influential. Joan Solomon and Glen Aikenhead edited an anthology on STS education (Solomon and Aikenhead, 1994).

20. This tradition—encompassing Holton, Schwab, Klopfer, Conant, Mach, Whitehead and others – is discussed in Matthews (1994a). Robert Carson provided a nice contemporary statement of the place of science in a liberal education Carson (1997).

21. These influences are discussed in Matthews (1994a, pp. 54–56). For a description of the rationale and workings of the Amherst College course, which was inspired by Conant and began in 1947, see Arons (1959). This Amherst course led to Arons text, *The Development of Concepts in Physics* (Arons, 1965).

22. The idea for this visual representation of the argument comes from Gerald Holton's AAAS lecture, which was subsequently published (Holton, 1995).

23. Author of *Physics as a Liberal Art* (Trefil, 1978).

24. I have discussed some of the literature in Matthews (1994a, pp. 31–33). See also the review by Paul DeHart Hurd (Hurd, 1998).

25. I discussed problems with Hirst's epistemology in *The Marxist Theory of Schooling* (Matthews, 1980, pp. 167–174). Hirst subsequently backed away from many of the positions stated in his landmark early "Forms of Knowledge" paper (Hirst, 1991). A recent supportive interpretation of Hirst's thesis is by Mackenzie (1998).

CHAPTER 2

1. This method of determining the solstice was common in ancient civilizations for ascertaining the date for religious observances. When the noon sun was reflected directly off the water in a deep well, the sun was directly overhead, and thus it was the solstice.
2. French (1965, pp. 259–261) elaborates Eratosthenes's argument. There is scholarly debate about whether Eratosthenes corrected this figure to 252,000 stadia to get a figure divisible by 60, and thus giving a round figure of 700 stadia to the degree. For the modern equivalents of these ancient measures, see Berriman (1953, pp. 15–19).
3. A century after Eratosthenes, Poseidonius, a geographer at Rhodes, made another calculation of the circumference of the earth using star sights. His figure was about one-sixth less than Eratosthenes, and thus one-sixth less accurate than modern values. Poseidonius shrunk the earth. His longitude degree was about 57 miles. This smaller, and inaccurate figure, was used by many subsequent navigators, Columbus among them. On the determination of the length of a longitude degree see Heath (1913, p. 339), Waters (1958, pp. 64–65), Williams (1992, pp. 8–9), Hogben (1973, pp. 30–35, and Harley and Woodward (1987).
4. The astrolabe was known to Ptolemy in the second century AD; it was refined over centuries by Islamic astronomers. For an account of the development and workings of medieval and renaissance European navigation instruments, see Howse (1980, pp. 57–60), Maddison (1969), Andrewes (1998c) and Taylor and Richey (1962).
5. For accounts of medieval and renaissance navigation see May (1973), Taylor (1971), Waters (1958), and Williams (1992, Chaps. 1–3).
6. The pole star does rotate slightly so navigators were cautioned to measure its altitude only when its 'Guard' stars were in a certain position.
7. For accounts of Portuguese navigational science, the achievements of the Sagres institute, the life of Henry the Navigator, see Mathew (1988), Waters (1967, 1968), Mendelssohn (1976), Russell (1984), Newitt (1986), and Goodman (1991).
8. Accounts of the rise of the Portuguese seaborne empire can be read in Boxer (1969, 1984, 1985), Diffie and Winius (1977) and Chaunu (1979). John Law purports to offer a "nonreductionist sociological perspective on technological change" (Law, 1987, p. 227) using Portuguese expansion as a test case.
9. Something of the human cost of these voyages of exploration can be appreciated in Gama's account of his meeting with an Arab dhow off the Malabar coast in 1502: "We took a Mecca ship on board of which were 380 men and many women and children, and we took from it fully 12,000 ducats, and goods worth at least another 10,000. And we burned the ship and all the people on board with gunpowder, on the first day of October" (Boorstin, 1983, p. 177). A few weeks later, off Calicut on the southern shores of India, Gama instructed that negotiators sent to him by the Muslim prince be cut into small pieces and returned with a note suggesting that the prince make his curry from them. Gama was a great navigator and sailor, perhaps one of the best of the age. He also embodied the standard Christian mores of his time. For more on his life and times, see Hart (1950); on this navigational methods, see Morison (1974).
10. For an account of the magnetic deviation method of determining longitude, see Gould (1923, p. 4). Bennett describes its early investigation by the Royal Society (Bennett 1985). Jonkers discusses its utilization by early Portuguese and Dutch navigators (Jonkers, 1996).
11. On the papal treaties of 1493 and 1494 see Mathew (1988, pp. 94–96).
12. On the Badojoz Conference, including full references to Spanish and Portuguese sources, see Randles (1985, pp. 236–237).
13. Part of Columbus' problem was his acceptance of Ptolemy's reduced dimensions for the size of the earth. These were in error by a factor of about one-third. His longitude error in the 1492 voyage was 22°; on the 1504 voyage, it was 38°. On the navigation of Columbus, see Morison

(1942, pp. 183–196) and Maddison (1991).

14. Dava Sobel provides a readable, account of the British efforts to solve the problem of longitude (ultimately done in the 1770s by John Harrison) (Sobel, 1995). See also May (1976) and Quill (1966) for an account of Harrison's achievement. The problem of finding longitude at sea is dealt with in many places, but see especially Gould (1923, pp. 1–17) and Williams (1992, Chap. 6). See also the contributions to Andrewes (1998a), these being the papers presented at the 1993 Harvard Longitude Symposium on which Sobel's book is based.

15. For the history of the Royal Observatory, especially its foundation and early years, see Howse (1980, Chap. 2).

16. One unnamed seaman was apparently not convinced of the navigational competence of his superiors and had been keeping his own reckoning of longitude. He approached Sir Clowdisley saying the navigators were in error. For this he was promptly hanged by the admiral, as such navigational attentions were expressly forbidden to ordinary seamen. Shortly later the ships struck the Scilly Isles (Sobel, 1995, p. 12).

17. A 1714 petition to parliament of merchants and ship captains, implored the following:

> The Discovery of the Longitude is of such Consequence to Great Britain, for the Safety of the Navy, the Merchant Ships, as well as Improvement of Trade, that for want thereof, many ships had been retarded in their Voyages, and many lost, but if due Encouragement were proposed by the Publick for such as shall discover the same, some Persons would offer themselves to prove the same, before the most proper Judges. (Williams, 1992, p. 80)

18. The act is included as an appendix to Humphrey Quill's *John Harrison: The Man Who Found Longitude* (Quill, 1966). Details of the Act's speedy 70 days passage through Parliament are given in Turner (1998, p. 116).

19. Just what "time" is that is being measured, is a long story. It is the duration between *regularly* recurring recurrent events, with the year (the time between the earth doing one revolution of the sun, 365 days), the month (the time between successive new moons, 29.5 days), and the day (the time between two risings of the sun, 24 hours) being the standard durations of *regular* events. There is a difference between the solar day and the sidereal day. The solar day is the duration between consecutive noons. Because of the earth's elliptical orbit, and its tilted axis of rotation, this varies by about 15 minutes from winter to summer. The stellar or sidereal day is the time between two star "noons"; that is, the time between a given star being directly overhead. Because this is not affected by the elliptic orbit, or the inclination of the earth's axis of rotation, its length does not vary with the time of year or seasons. But because the earth moves about one degree in its solar orbit while we wait for the following star noon, the sidereal day will be about four minutes shorter than the mean solar day. (See Jespersen and Fitz-Randolph, 1982, p. 61)

20. The politics of choosing London as the international zero meridian is described in Williams (1992, Chap. 5) and Howse (1980; 1985).

CHAPTER 3

1. Milham (1945) contains an excellent account of the theory and use of sundials and clepsydra. Green (1926), Rohr (1965), and Waugh (1973) deal extensively and exclusively with sundials. The historiography of dialling is discussed in Turner (1989). Although sundials are discussed here as timekeeping devices, it is useful to note what De Solla Price said of the function of sundials in antiquity:

> It would be a mistake to suppose that . . . sundials . . .had the primary utilitarian purpose

of telling the time. Doubtless they were on occasion made to serve this practical end, but on the whole their design and intention seems to have been the aesthetic or religious satisfaction derived from making a device to simulate the heavens. Greek and Roman sundials, for example, seldom have their hour-lines numbered. (Price, 1964, quoted in Waugh, 1973, p. 5)

2. Brearley (1919, p. 41) attributes this breakthrough to the Chaldean historian Berosus in 250 BC.
3. Looking at the shadow direction at noon as told by a watch or clock will only approximate the meridian, because clock time is mean solar time. The actual length of days varies (see Chapter 11) by plus or minus 15 minutes through the year. Thus the shadow at clock noon will fall either side of the true noon shadow (the meridian) at different times of the year.
4. This formula and its application is discussed in Waugh (1973), and Hogben (1936, pp. 378–385).
5. For an account of Ctesibius' clocks, see Bond (1948, pp. 58–61). An ancient description is given by Virtuvius in the first century BC (see Vitruvius, 1960). See also, Sleeswyk and Hulden (1990). On water clocks more generally, see Turner (1985).
6. The monastery's practice is inspired by the psalmist who, in the Old Testament, says: "Seven times a day do I praise thee." The Church's "office hours" were: matins (midnight), lauds (dawn), prime (first hour, 6 am), terce (third hour, 9 A.M.), Sext (sixth hour, noon), none (ninth hour, 3 P.M.), vespers (when the Vesper star appears, 5 P.M.), and compline (night prayers when all duties and occupations come to an end, 8 P.M.). On the role of medieval monasteries in the development of timekeeping, see Mumford (1934, pp. 12–18), North (1975), Rossum (1996, pp. 33–45), and Drover (1954). For a more general treatment of medieval ideas of time, see Higgins (1989).
7. The choice of 24 hours for a day goes back at least to the ancient Egyptians, who divided their day into 24 unequal hours. This fitted in with their sexagesimal number system, which gave twelve months each of 30 days for the year, with five days added on at the end of each year to bring their calendar into line with the sun's transit.
8. The word deriving from the fact that only at the spring (March 21) and autumn (September 23) equinoxes do day and night have the same duration, and hence all hours in the day are equal.
9. Examples are on display in the Deutsches Museum in Munich.
10. There are numerous good books on the history of timekeeping and clockmaking. See, for instance: Brearley (1919), Robertson (1931), Jespersen and Fitz-Randolph (1977), Milham (1945), Whitrow (1988), Lloyd (1970), Dale (1992), Bruton (1979), Barnett (1998), Mesange (1969), and Williams (1992).
11. See Rossum (1996) and Cipolla (1967) for the cultural history of the spread of town clocks. The public clocks were large and very expensive. The iron mechanism of the clock installed at Montpellier in 1410 weighed 2,000 pounds (909 kg) as did its bell (Cipolla 1967, p. 121). They took up to ten years to build, and required a full-time paid attendant.
12. For the dating of this spread of town clocks see Bell and Bell (1963, pp. 56–60), Cipolla (1967, pp. 39–47), Macey (1980, p. 18), Lloyd (1958, 1970), and Rossum (1996, Chap. 5). The Salisbury clock, built in 1386, is the world's oldest still-functioning, clock (Maltin, 1993).
13. It is interesting to note that this clock, as described, struck "Italian time." That is, the new day began at dusk. In Italy, Italian time lasted for centuries, with some records of its use well into the 19th century.
14. Dondi's documentation is so exact that replicas of his clock have been made by the Smithsonian Institute in Washington, D.C. and the Science Museum in London. For elaboration of Dondi's achievement, see Bedini and Maddison (1966).
15. Initially the term "clock" was used for devices that merely sounded bells, "horologe" was the term for devices that additionally indicated the hours on a dial.
16. See, for instance, Haber (1975).

17. For a fuller account of de Vick's clock see Milham 1945, pp. 81–85, and Robertson 1931, pp. 49–66.

18. A faint echo of this can be seen when the philosopher and former priest Anthony Kenny remarked in his autobiography, *The Path from Rome*, that his English seminary refused to move to daylight-saving time. The seminary principal said that such time disturbed God's natural order.

19. Jost Burgi, at the end of the 16th century, tried to make clocks that gave the accuracy Tycho Brahe demanded. Burgi's crystal clock—so called because of its rock crystal dials—provided separate dials for hours, minutes, and seconds. Macey noted this was one of the first recorded uses of the seconds hand (Macey, 1980, p. 26). Burgi's clocks are described in Maurice (1980).

20. The definitive study of this subject is Joseph Needham, Wang Ling, and Derek de Solla Price's *Heavenly Clockwork: The Great Astronomical Clocks of Medieval China* (Needham, Ling, and Solla Price 1960). The core of this book is a translation of the long-lost 1090 book by Su Sung titled *New Design for a Mechanised Armillary Sphere and Celestial Globe*. See also Needham and Ling (1954–1965), and Cipolla (1967, Part II).

21. For an account of Ricci's life, see Louis Gallagher's translation of Ricci's journals (Gallagher 1961), Vincent Cronin's *The Wise Man from the West* (Cronin, 1961) and J.D. Spence's *The Memory Palace of Matteo Ricci* (Spence 1984). Boorstin discusses the episode (Boorstin, 1983, Chap. 7).

22. An excellent study of the cultural factors that inhibited China's scientific development is Huff (1993, Chaps. 7, 8). See also Rosenberg and Birdzell (1990), and Sivan (1984).

CHAPTER 4

1. See White (1966) for a discussion of this tendency. White specifically comments on Alexandre Koyré's influential 1939 treatment of Galileo, saying "Koyré's Galileo is a bit like Jacques Maritain's Aquinas—a vast intellect palpitating in a realm of pure ideas unsullied by....practical considerations" (White, 1966, p. 98).

2. On Galileo's contribution to philosophy, see especially Finocchiaro (1980, pp. 150–157).

3. Galileo's lasting contribution to theology was, of course, to the field of exegesis (the principles of interpretation of biblical texts). This was a large part of the Galileo Affair. His core contribution to exegesis is found in his 1615 *Letter to the Grand Duchess Christina* (Galileo, 1615/1957). For more on Galileo and theology, see Fantoli (1994, Chap. 3) and Russo (1987).

4. On Galileo's contribution to the development of scientific instrumentation, see Bedini (1967, 1986).

5. The most focused history of the medieval pendulum is that of Bert Hall (1978). Lynn White (1962, pp. 117, 172–173, and 1966, p. 108) discusses some aspects of the topic.

6. Much has been written on medieval impetus theory. Good introductory accounts can be found in Murdoch and Sylla (1978), Grant (1971, Chap. 4), Moody (1966), and Weisheipl (1959, pp. 69–81; 1985, Chaps. 2, 4). More developed accounts can be found in Moody (1951), Grant (1964), and Maier (1964, Chaps. 4, 5).

7. Bodies around us do seem to require a mover in order to keep in motion; we even feel impelled to postulate "gravity" as the cause for stones falling and, in unguarded moments, even "lightness" (*levitas*) as a reason for balloons and the like rising. The modern Aristotelian, Mortimer J. Adler, made this observation in his *Introduction to Aristotle*:

> In an effort to understand nature, society, and man, Aristotle began where everyone should begin—with what he already knew in the light of his ordinary, commonplace experience. Beginning there, his thinking used notions that all of us possess, not because we are taught them in school, but because they are the common stock of human thought about anything and everything. (Adler 1978, p. xi)

Thomas Kuhn, in his *The Copernican Revolution*, wrote:

> Aristotle was able to express in an abstract and consistent manner many spontaneous perceptions of the universe that had existed for centuries before he gave them a logical verbal rationale. In many cases these are just the perceptions that, since the seventeenth century, elementary scientific education has increasingly banished from the adult Western mind. . . .but the opinions of children, of the members of primitive tribes, and of many non-western peoples do parallel his with surprising frequency. (Kuhn, 1957, p. 96)

8. On this important, but often overlooked distinction, see Weisheipl (1965).
9. Aristotle overlooked whether the natural motion of heavy bodies was *constant* or *accelerated* motion. If the former, as the medievals would show, and as Galileo would grapple with at Pisa, then accelerated motion would require a mover; if the latter, then everyday accelerated motion would require no mover.
10. *Physics* Bk. IV, Ch. 8, 215a 1–8; *Physics* Bk. VIII, Ch. 10, 266b27– 267a22; *De Caelo* Bk. III, Chap. 2, 301b, 17–33. Sachs (1995) is a resonable guided introduction to the text of Aristotle's *Physics*.
11. On the engaging and culturally significant history of this argument for God's existence from the fact of motion in the world, see Buckley (1971).
12. Norwood Russell Hanson argues that this view is not entirely improbable, and that Aristotle's recognition that the air (more generally the medium) has something to do with the motion of a body through it, anticipates the physical core of later aerodynamical theory: "The philosopher today who seriously asks 'how can birds, and aeroplanes, fly' cannot but be struck with the omnipresence of *Physics* IV in most of the modern aerodynamic answers to that question" (Hanson, 1965a, p. 145).
13. For Philoponus' arguments, translations of some of his texts, and his influence on subsequent impetus theory, see especially Sorabji (1987, 1988, Chap. 14) and Wolff (1987).
14. This situation of the whole of a systematic intellectual tapestry unraveling when one string is pulled might be seen in debates, for instance, about female ordination to the priesthood. Initially it appears to be a singular matter, but conservatives are perhaps right to insist that if the string is pulled too hard, then other things will begin unraveling: the teaching authority of the church, the truth of biblical assertions–including perhaps assertions of more fundamental kinds, and so on.
15. On the means whereby Philoponus influenced the later medievals, and on the translation history of his works, see Moody (1975, pp. 234–235), Koyré (1968, pp. 29, 68), and contributions to Sorabji (1987).
16. For an account of Marchia's impetus theory, and why he thought it to be naturally decaying, see Maier (1982, p. 85).
17. Translations of Buridan's key scientific texts (*Questions on the Eight Books of the 'Physics' of Aristotle, Questions on the Four Books on the Heavens and the World of Aristotle*) are contained in Clagett (1959).
18. There is an understandable tendency to see, as Pierre Duhem did in his landmark studies at the beginning of the century (Duhem, 1913), Buridan's concept of undiminished impetus as a precursor to Galileo's principle of circular inertia and Newton's theory of linear inertia. For Buridan, a body set in motion, and experiencing no resistance, just keeps on going. But this thesis overlooks the fact that Buridan's, and all other medieval, impetus theory was rooted in Aristotelian ontology: Impetus was the "mover" required by Aristotle's thesis that all violent motion required an external mover. Impetus theory was an elaborate rescue operation for Aristotelian physics: Galilean and Newtonian physics signalled the death-knell of this physics and its associated ontology and cosmology.

The interpretation of Buridan's impetus theory is more complex than is here being made out. He does for instance believe that bodies have a natural tendency to rest and to oppose motion (an *inclinatio contraria*). Thus the resistance experienced by projectiles arises not only from the medium through which they pass, but there is also an inherent resistance to motion. For these reasons, Maier believes that in so far as Buridan has a law of inertia, it only applies to heavenly bodies, not to terrestrial bodies (Maier, 1982, pp. 89–92).

19. Galileo repeats this thought experiment in his 1633 *Dialogues Concerning the Two Chief World Systems.*)

20. Of course one needs to be aware here, as elsewhere, that although the *words* may be the same, the *ideas* may be different. This is a complex matter, but there are good grounds for saying that the idea of impetus that Galileo has at Pisa is that of Marchia's self-decaying motor force. At Padua he has the idea of Buridan's self-conserving motor force, and in his mature mechanics he no longer has the idea of the Aristotelian tradition's motor forces, although he retains the word "impetus."

21. About 40–50 oscillations before stopping is characteristic of most casually swung, heavy-body simple pendulums. Much less for light weight (cork, polystyrene) simple pendulums.

22. It has been pointed out that the Italian *mille*, the word Galileo uses to denote the number of pendulum swings, although technically meaning "thousand," also is often used as "several" or "many." Galileo may have been making the latter, more modest, claim.

23. Science deals with statements, not observations. It is not what is seen (a private psychological matter), but what is reported, that is basic to science. Reports have to be expressed in a language that includes the relevant theory thought to be appropriate to the situation. This is the correct sense of the much written upon theory dependence of observation thesis. For accounts of this thesis, see Hanson (1958) and Brown (1977).

24. For Leonardo's use of the pendulum and for his contribution to clockmaking, see Bedini and Reti (1974), Edwardes (1977, pp. 12–16), and Hall (1976).

25. It is a moot point whether Galileo in the 1580s recognized these properties of the pendulum that underlie the behavior of the *pulsilogium*. He does mention in his 1590 *De Motu* that heavy and light pendulums swing in synchrony, but he clearly does not state his pendulum laws until his 1602 letter to del Monte.

26. For an account of the "priority debate" see Robertson 1931, Chapter vi.

CHAPTER 5

1. This same process would later be repeated with Darwin's revolutionary claims in biology: Darwin's new science could not be accepted until there were changes in what constituted scientific evidence and scientific proof. In the case of both Galileo and Darwin there is much dispute about whether the new science drove the change in philosophy, or whether the change in philosophy drove the new science.

2. This work is discussed in Drake (1976), Wisan (1974), Fredette (1972) and Damerow, Freudenthal, McLaughlin & Renn (1992, pp. 126–149). Dijksterhius described it as "thoroughly medieval" in its approach (Dijksterhius, 1961/1986, p. 334). Maurice Clavelin provided the foundational analysis of Galileo's early mechanics, including *De Motu* and *Le Mecaniche* in his *The Natural Philosophy of Galileo* (Clavelin, 1974, Chap. 3).

3. See Drake (1967, pp. 313–318), Clavelin (1974, pp. 165–174), and Wisan (1978, 1980) for discussion of the fecundity of these geometrical depictions of the balance situation, and the insightfulness of Galileo's concentration on vertical displacement as a common measure of the movement involved on different planes and balances.

4. Del Monte's *Mechanics* is translated in Drake and Drabkin (1969) who describe del Monte as the greatest mechanician of the 16th century. Dugas (1988, pp. 100–101) discusses sympathetically del Monte's contribution to mechanics. See also the pen-picture of del Monte in Drake (1978, p. 459); the more extensive picture in Rose (1980); the accounts of del Monte's patronage of Galileo in Biagioli (1993, pp. 30–31) and Sharratt (1994, p. 43–44); and del Monte's role in the late-16th century Archimedean revival in Bertoloni Meli (1992). An extensive discussion of del Monte as representative of the newly emerging engineer-scientist class is in Renn et al. (1998, pp. 36–40). These latter authors also translate a number of key letters between del Monte and Galileo, and provide documentation for archival sources.

   The historian Jügen Renn, who among other things has studied and translated a 245 page notebook of del Monte's found in the Bibliothèque Nationale de Paris, is reluctant to call del Monte an "Aristotelian," nor even a "progressive Aristotelian" (personal communication). He believes that del Monte is firmly in the Archimedean-inspired, engineer–scientist camp, and that its members were distinct from the Aristotelian tradition. I am content with the Aristotelian label in as much as del Monte shared the Aristotelian conviction that scientific (natural philosophical) claims had to accord with, and be judged against, experience. To what extent he was an Aristotelian beyond this point is tangential to the argument of this book.

5. The letter was written in October 1602 (*Opere*, Edizione Nazionale, Florence 1934, vol. 10, pp. 97–100), and a translation has been provided by Stillman Drake (Drake, 1978, pp. 69–71). Arguably a better, but less accessible, translation is in Renn et al. (1998, pp. 104–106). Ronald Naylor (1980, pp. 367–371) and W.C. Humphreys (Humphreys 1967, pp. 232–234) discuss the letter in the context of Galileo's work on the law of fall.

6. In the last two decades there has been much research on social patronage and science. For studies of Galileo's engagement in the system, see Biagioli (1993) and Westfall (1985).

7. The topic of Galileo's use of rhetoric has also been researched. See, among others, Finocchiaro (1980, Chaps. 1–3) and Moss (1986; 1993).

8. There are three different properties of motion on a curve that need to be separated:

   Tautochronous, meaning that a body falling freely on the curve reaches the lowest point at the same time, regardless of where on the curve it was released.
   Brachistochronous, meaning that a body freely falling (but not in a straight line) will reach its lowest point fastest by travelling on this curve.
   Isochronous, meaning that each successive oscillation on a curve takes the same period of time.

   Tautochronous and Isochronous motion, though conceptually different, are in reality much the same thing.

9. For Galileo's length and time units, see Drake (1990, p. 7).

10. Discussion of Galileo's inclined plane investigations can be read in Humphreys (1967) and Costabel (1975).

11. Stillman Drake discusses these measurements in his *Galileo: Pioneer Scientist* (Drake, 1990, p. 23–25). So also does James MacLachlan, in his *Galileo Galilei* (MacLachlan, 1997, pp. 114–117).

12. William Wallace has translated and discussed Galileo's earliest Pisan notebooks and lecture materials (Wallace, 1977). These are relevant to the question of how he did, and did not, fit into the tradition of Italian natural philosophy in the late 16th century. On this question see also Wallace (1981a; 1981b; 1984).

13. Some biographies of Galileo are Drake (1978;1980), Geymonat (1957/1965), Shapere (1974), and Sharratt (1994).

14. The historian A. Rupert Hall called it a great year for astronomy, but an unfortunate year for physics. This because it delayed by 25 years the publication of Galileo's physics, the essentials

of which had been worked out by the time he turned his telescope to the skies. Thus Galileo's physics never influenced the work of Mersenne, Descartes, Beeckman, and Roberval who were all working at this same period (Hall, 1965, p. 185).

15. James MacLachlan (1976) conveniently itemizes the 11 passages in Galileo's *Dialogue* (1633) and *Discourse* (1638) where he deals with the pendulum.

16. Galileo believed his account of tidal motion (false as it turns out) was the definitive argument for Copernicus's theory of the rotating earth. On this see Brown (1976), Mertz (1982), and Shea (1970).

17. This is a mistake. It is possible that there are other curves, the descent along whose arcs are also faster than along their corresponding chords, and also faster than along the arc of a corresponding circle. This is the mistake that will be pointed out by Huygens, who showed mathematically that it is the cycloid, not the circle that is the brachistochrone. And also that it is cycloidal, not circular, motion that is isochronic.

18. Stillman Drake discusses these *Discourse* writings and is at pains to point to Galileo's experimental support for them (Drake, 1990, pp. 209–213).

19. Some key texts that deal with the opposition to Galileo's pendulum claims are Koyré (1953), MacLachlan (1976), Segre (1991, pp. 41–44), and Settle (1961).

20. Mersenne was a notable French mathematician, philosopher, scientist, theologian, and Catholic priest. He is largely forgotten, but it is worth recording what one historian has said of him: "Mersenne was possibly the most selfless servant the scientific community had ever had. His disarming generosity, his tireless eagerness to learn and to communicate, earned him the trust, respect, and gratitude of most of those who played a part in ushering in . . .the century of genius" (Jaki, 1974, p. 284). The range of Mersenne's work, and its relationship to its time, is discussed in Dear (1988).

21. Crombie lists this correspondence in Crombie (1994, p. 1458, n. 197).

22. Now, armed with the formula $s = \frac{1}{2}\,\mathbf{g}t^2$, we can see that if t = 1 second, then $\mathbf{g} = 2s$, or double the distance a freefalling body covers in one second.

23. A seconds pendulum was one whose bob took one second to swing from one side to the other. As the period of a pendulum is the time to swing from one side to the other, *and then back to the commencement point*, the period of a seconds pendulum is two seconds. A pendulum clock's escapement device moves one notch each half oscillation (one side-to-side swing), thus for a seconds pendulum, the escapement moves each second.

24. See Yoder (1988, pp. 11–14), Dear (1988, pp. 160–169), and Koyré (1953) for discussion of Mersenne's experiments.

25. On the 17th century's, and more specifically Galileo's, understanding of science requiring certainty, see McMullin (1978, 1990) and Wisan (1978).

26. Riccioli's work is described in detail in Koyré (1968, pp. 104–107).

27. I discuss this in Matthews (1994a, pp. 123–124).

CHAPTER 6

1. He sent these proofs to his father for possible publication, but almost immediately he obtained and read a copy of Galileo's *Discourse* wherein was Galileo's original proofs. He wrote to his father retracting the request for publication, saying "I did not wish to write the *Iliad* after Homer had" (Yoder, 1988, p. 9).

2. This was an early manifestation of the view that science should be above, or at least beyond, politics. During the Napoleonic Wars the English scientist Davy would be allowed to travel freely in France conducting his research; during World War I, the French scientist and historian, Pierre Duhem, defended the achievements of German science (Duhem, 1916/1991); during

World War II, the German scientist Max Planck appealed to such a separation to justify his remaining as director of the Kaiser Wilhelm Institute in Nazi Germany (Heilbron, 1986).

3. The life and achievements of Huygens are written up in Bell (1947), Bos (1980a), and in the entry in *The Dictionary of Scientific Biography*.

4. See William A. Wallace's *Galileo's Early Notebooks: The Physical Questions* (Wallace, 1977). James Franklin provides an informative account of how "medieval and Renaissance reasoning with diagrams, both physical and mental, trained Europeans to think adequately to do science" (Franklin, 1999, p. 53).

5. John Locke, Newton's "underlabourer," complained that his mathematical competence was not adequate enough to allow him to critically read Newton's work.

6. For Galileo's influence on Huygens horological investigations, see Dobson (1985) and Leopold (1998).

7. A translation of this work is appended to Edwardes (1977).

8. The cycloid curve had been studied by Galileo and his pupil Torricelli. Huygens recognized that, for large amplitudes the effective length of the pendulum had to be shortened with respect to the amplitude; a cycloid enabled this to happen.

9. The relevant manuscripts are analyzed in Blay (1998, pp. 18–27).

10. For comment on this derivation, see Mahoney (1980, 1990 pp. 476–485), Yoder (1988, pp. 50 ff), and Blay (1998, pp. 21–27).

11. For Huygens' commitment to geometrical analysis and his appreciation of the calculus, see Bos (1980b).

12. Newton was 55 years old and managing the English Mint when he received the problem from Bernoulli in 1697. His niece, Catherine Conduitt, wrote:

> Sir I.N. was in the midst of the hurry of the great recoinage [when he received the problem] and did not come home till four from the Tower very much tired, but did not sleep till he had solved it, which was by four in the morning. (Dunham, 1990, p. 201)

On this brachistochrone contest, see Landes (1983, p. 420), Gjertsen (1986, p. 146), and Dunham (1990, pp. 199–202).

13. Huygens was appreciative of patronage. He concluded his dedication with "Since you are very busy and must attend to the highest affairs, which is to be expected in one who stands at the summit of things, I realize that you hardly have the liberty to attend to these contemplations, even though you are capable of all things…if these writings can prove to posterity that the sciences and the arts flourished at this time, they will also teach them that above all else this was due to your virtue and your magnanimous spirit" (Huygens, 1673/1986. pp. 8–9).

14. For discussion of attempts at using Huygens' pendulum for determination of longitude at sea, see Leopold (1998), Mahoney (1980), Robertson (1931, Chap. 9), and Schliesser and Smith (1996).

15. A detailed discussion of Bruce's efforts can be read in Yoder (1988, pp. 151–154) and Leopold (1998, p. 104).

16. Nothing came of Wren's proposal, and the English persisted with their artificial yard standard. In 1758 it was found that even the two standard yards preserved at the exchequer were not the same length (Heilbron, 1989, p. 990)! An Act of 1824 did specify that if the imperial standard yard were to be destroyed, then it should be replaced by the natural unit of a seconds pendulum. Such a catastrophe did happen in 1834 when a fire in the House of Parliament destroyed the standard yard. However by then it was realized that the length of a seconds pendulum was not as invariant as originally thought; and so the English, and their colonies, persisted with the artificial yard standard (Scarr, 1967, pp. 2–5).

17. He had calculated that the radius of the earth would have to be hundreds of times larger than it is for the centrifugal force to equal a body's weight, and hence for the body to be thrown off the

rotating earth. Galileo made this point in his arguments with those who claimed a rotating earth would hurl bodies and buildings into space.

18. Full details of Richer's voyage, its aims and accomplishments, can be found in Olmsted (1942).
19. The contemporary estimate is 92,800,000 miles.
20. Richer had previously not excelled as a careful experimentalist (Mahoney, 1980, p. 253).
21. Accounts of the development of the standard meter can be found in Alder (1995), Thompson (1967), Kline (1988, Chap. 9), Kula (1986, Chaps. 21–23), Heilbron (1989), Petry (1993, pp. 302–303) and Berriman (1953, Chap. XI). Kula provides a most extensive guide to French historical documents. Both he and Heilbron discuss the politics of the metric system. Additionally, see *L'Epopé du Metre Historie Systeme Decimel*, Ministry of Weights and Measures, Bicentennary of the French Revolution Documentation, Paris, September 1989.
22. On this greatly covered subject see at least Martins (1993) and Stein (1990).
23. Newton himself, of course, was never happy about action at a distance; in one place he wrote to his friend Bentley saying "do not ascribe such an unphilosophical notion to me." The classic treatment of the scientific and philosophical reactions to Newton's postulation of "attraction at a distance" is Mary Hesse's *Forces and Fields* (Hesse, 1961). See also Koyré (1965, pp. 115–138). Recently a philosopher of science noted:

> But modern science faced a fundamental intellectual problem right at its very beginning with its failure to indicate a mechanism for gravitation – a problem it has still been unable to solve not only as regards gravitation but all four of the fundamental forces. Dilworth, 1996, p. 205)

24. The demonstration is described in Dijksterhuis (1961/1986, p. 462).
25. For discussion of Huygens' reasoning and calculations about the effect of latitude on g and hence on determination of longitude, see Schliesser and Smith (1996).
26. In fact, at the poles g is equal to 32.257 ft/sec^2, at the equator it is 32.089 ft/sec^2.
27. A classic work on the history of geodesy is Todhunter (1873/1962). See also Chapin (1994) and Greenberg (1995).
28. See Macey (1980, p. 29).

CHAPTER 7

1. In Newton's December 10, 1692 letter to Bentley, he said:

> When I wrote my treatise about our system, I had an eye upon such principles as might work with considering men for the belief of a Deity; and nothing can rejoice me more than to find it useful for that purpose. (Thayer and Randall, 1953, p. 46)

2. The society's formal name was *The Royal Society of London for Improving Natural Knowledge*. The first history of the Royal Society was written by Sprat (1667); Gerrylynn Roberts provides a very readable account of the origins of the Royal Society and of its French and Italian equivalents (Roberts 1991); see also Emerson (1990). David Lux provides a historiographical review of 20th-century writings on these scientific societies (Lux, 1991).
3. Details of Hooke's life can be found in the entry in the *Dictionary of Scientific Biography*, and Hooke's incomplete autobiography which was published as "The Life of Dr. Robert Hooke" in Richard Waller's *Posthumous Works of Robert Hooke* (Hooke 1705/1969, pp. i–xxviii). See aslo Ellen Drake's biography of Hooke, *Restless Genius* (Drake, 1996), and contributions to Hunter and Schaffer (1989).

4. On the Royal Society's concern with the longitude problem, see Bennett (1985).

5. The history of escapement technology, and illustrative drawings, are provided in Landes (1983, Appendix A).

6. Rosenberger, a 19th-century commentator, said: "Since Hooke, like Borelli, was a very clever physicist but no mathematician, he too could only be the percursor of a greater discoverer in whose brightness the lesser sun of his fame faded from view more than it should" (in Jourdain, 1913, p. 361).

7. Robertson thought that the lecture, on account of its intemperate language, was probably never delivered. But Hooke's diary shows that he read it to the Royal Society on June 26, 1673 (Patterson, 1952, p. 286).

8. The best accounts of this controversy are Robertson (1931, Chap. 10), Patterson (1952), Hall (1978), Leopold (1993), and Jardine (1999, pp. 144–157).

9. Harrison had long languished in undeserved obscurity, known only to historians of timekeeping and of navigation, plus a limited circle of scientists. All this changed in 1995 with the publication of Dava Sobel's bestselling biography, *Longitude. The True Story of a Lone Genius Who Solved the Greatest Scientific Problem of His Time* (Sobel 1995). For more scholarly work see Andrewes (1998b), Quill (1966), May (1976), King (1998) and Gould (1923, Chaps. 3, 4). See also other contributions to Andrewes (1998a), being the symposium held at Harvard University in 1993 that inspired, and provided much material for, Sobel's book.

10. A seconds pendulum, that is one with a length of one meter, or about 39 inches, will gain or lose one second per day for each 0.001 inch it is lengthened or shortened. This change is caused by a temperature change of 4° F.

11. *Journals of the House of Commons*, June, 11, 1714, p. 677. Cited in Quill (1966, p. 4). See also Andrewes (1998b).

12. For an account of Harrison's H1 see Andrewes (1998b, pp. 199–207). For Harrison's other clocks, see Andrewes (1998b), Gould (1923, Chaps. 3, 4), King (1998) and Randall (1998).

13. Details of the struggle between Harrison and the Longitude Board can be read in Gould (1923, Chap. 4) and Sobel (1995, Chaps. 11, 12).

14. Cook was killed in Hawaii in 1779. K1 was brought back to the Greenwich Observatory where it was put on display. The horologist Rupert Gould reported that "this machine, which had not been subsequently cleaned or adjusted in any way [since 1878], was set going on the occasion of the Annual Visitation of the Royal Observatory in 1922, and that its error on G.M.T. between 11 a.m. and 7 p.m. amounted to 0.4 seconds only" (Gould, 1923, p. 50). For the life and achievements of Cook, see Hough (1994).

15. On the utilization of the marine chronometer, see Waters (1980), Davies (1978), Gould (1923), and Sadler (1968). For its late adoption in Spain, see Lafuente and Sellé (1985).

16. Books such as *The Collector's Guide to Clocks* (Roberts, 1992) give an indication of the artistry of 17th and 18th century clockmakers.

17. This is an early example of the division and deskilling of labour occasioned by the ascendent capitalist mode of production. The classic work, apart from Marx on the deskilling and fragmentation of labor under capitalism, is Harry Braverman's *Labour and Monopoly Capital* (Braverman, 1974).

18. For an account of the life of Babbage, his involvement with British industry, and his championing of technical education, see Hyman (1984) and Collier and MacLachlan (1998). For a more general account of the historical dependence of computing machines on the craft of clockwork, see Pratt (1987). The latter work describes Schickard's 1624 "calculating clock," and Leibniz' 1687 calculating machine.

CHAPTER 8

1. Recent excellent biographies of Newton include the definitive Westfall (1980) and its shortened version (Westfall, 1993), Hall (1992), and Christianson (1996).
2. Hessen's thesis is that "The Marxist analysis of Newton's activity...will consist first and foremost in understanding Newton, his work and his world outlook as the product of [his] period" (Hessen, 1933, p. 32). It is the harbinger of modern social constructivist analyses of science. Interestingly, for this theme of this book, Hessen explicitly draws attention to the Longitude Act of 1713, and Newton's pendulum studies designed to perfect timekeeping, as support for his thesis. For a review and restatement of Hessen's thesis, see Freudenthal (1988).
3. On the Newtonian method, see the contributions to Bricker and Hughes (1990), the contributions to Stayer (1988), Harper and Smith (1995), and Suchting (1995). On the influence of Newtonianism, see Cohen (1980), Butts and Davis (1970), and Guerlac (1981).
4. By no means was everybody envious. Bishop Berkeley was a trenchant critic of Newton, and he wrote his *De Motu* (Berkeley, 1721/1901) to refute Newton's physics of force. The British empiricist, Hutchinson, said of the geometrical constructions of the *Principia* that they were just "cobwebs of circles and lines to catch flies in" (Cantor, 1991, p. 219). Huygens and Leibniz could never accept what they regarded as the anti-mechanical, verging on occult, claims about attraction at a distance. Elsewhere on the continent, and indeed in England, Aristotelianism lingered long in university "natural philosophy" departments. And within Newtonism there was internal tension between those who modeled their work on the more rational-mathematical methods of the *Principia*, and those who looked to the more descriptive–experimental methods of the *Opticks*. On the cultural and philosophical factors affecting the diffusion of Newtonianism, see Guerlac (1981).
5. A translation is in Edwardes (1977).
6. For the impact of Copernicus on views of time, see Blumenberg (1987, part IV).
7. The qualification is called for, as ancient atomists merely explained macro phenomena by postulating invisible atoms that had the observed property: diamond consisted of very hard atoms, clay consisted of soft atoms. This got rid of Aristotelian forms, of diamondness and clayness, as the explanation of the macro phenomena; but not much advance was being made, as the explanations merely pushed what was to be explained down to another, micro, level. Seventeenth-century atomists were made of sterner stuff. They thought that the ancients had the right ideas—explanation of the macro in term of the micro, and limiting the real to atoms and, for some, a void—but wanted more rigour and more genuine explanations.
8. Although in 1277 Bishop Etienne Tempier of Paris issued an edict, the *Rotula*, listing 219 philosophical and theological errors of Aristotelianism and forbidding the masters of Paris University to teach any of them. A similar list was issued by Archbishop Robert Kilwardby of Canterbury binding on Oxford masters. Nevertheless, Aristotle's matter–form distinction did provide a metaphysical framework that allowed sense to be made of the central Catholic doctrine of transsubstantion, concerned with the real presence when the host is consecrated at Mass. With atomism, the doctrine became a complete mystery.
9. Rom Harré (Harré, 1964, part 2) provides a nice introduction to the tenets and structure of corpuscularian philosophy. See also Westfall (1971, Chap. 2), and Einstein and Infeld (1938, chaps. 1, 2).
10. Alexandre Koyré aptly remarked that:

> The Cartesian Universe . . .is built with very meagre materials. Matter and motion; or rather (since Cartesian matter, homogeneous and uniform, is nothing but extension) extension and motion; or better still, since Cartesian extension is strictly geometrical, space and motion. The Cartesian universe is the realisation of geometry. (Koyré, 1939/1978, p. 252)

11. Seventeenth-century atomism was not monolithic. Gassendi, for instance, maintained that matter was not infinitely divisible, that there was a lower limit, the atom, that could not be further divided. His ontology consisted of atoms moving in a void. On the other hand, Descartes for instance, believed that even atoms could be further divided, indeed divided to infinity. His ontology consisted of notional atoms moving in a plenum, an aether. On these matters, see Lennon (1989), Melsen (1952, Chap. 3), and Pyle (1997).

12. Newtonians believed in the reality of attractive forces, while Cartesians and Leibnizians denied this. For the latter groups, Newton, by admitting forces into his fundamental ontology, had placed himself outside the fold of mechanical philosophers. For a discussion of force and natural motion in Newton, see McGuire (1994), Westfall (1971), and Hunt and Suchting (1969).

13. Alan Chalmers, in recent publications, makes the useful point that, in discussion of 17th-century atomism, we need to distinguish the mechanical philosophers' philosophy (or perhaps rhetoric) from their science. He claims, for instance, that Boyle's good science owed nothing to Boyle's atomism, and that the latter was an impediment to the former (Chalmers, 1993; 1998). He agrees with Stephen Shapin's claim that Boyle never in his life provided a mechanical explanation: he waved his hands in their direction, but never followed through with any explantaion of his own (Shapin 1994, p. 334).

14. For the history of ideas about gravitation and action at a distance, see Jammer (1957) and Hesse (1961).

15. The excellent introductory college physics book by A.P. French, *Newtonian Mechanics* gives the modern working through of this example (French, 1965, pp. 256–259). French was an advocate for bringing history of science into science programs, and his book is an admirable witness for the marriage. Elsewhere he has addressed some of the problems with this approach (French 1983; 1989). It is interesting to note that many introductory college physics texts do not even list "moon" in their index, let alone describe Newton's experiment.

16. For Newton's concept of mass, see Jammer (1961, Chap. 5).

17. There is a conceptual problem in saying that gravitational attraction is an *essential* property of matter. Such attractive forces, logically, depend upon other bodies being present. But essential properties are ones that must be attributed to bodies all bodies, even in the circumstance of there only being one of them. Can a single body be intelligently said to be exerting or exercising gravitational attraction?

18. Colin Gauld (1998a) has described some of the 17-century's investigation of the physics of collisions and the role of the pendulum in illustrating the phenomena.

19. See Clarke's exchanges with Liebniz in Alexander (1956).

20. The debate, and its attendant separation of the concepts of momentum and energy, will be further discussed in Chapter 9. A classic work on the "discovery" of the conservation laws is Mach (1872/1911). See also Elkana (1974b) and Hiebert (1962).

21. I discuss this paper in Matthews (1994a, pp. 101–103).

22. Some classic papers on the verificationist criterion in empiricist philosophy are Hempel (1965, Chap. 4), Popper (1963, Chap. 11) and Quine (1951).

23. d'Alembert's *Traité de dynamique* was first published in 1743 then revised and corrected in 1758. The work is discussed in Dugas (1988, pp. 244–254). He thought that the dispute in large measure stemmed from both Newton and Leibniz reifying the essentially *relational* qualities they were talking about. Thus he said:

> Rather than understanding by the word "force" some sort of being which resides in the body, one should use it only as an abridged way of expressing a fact—in the same way as, when one says that one body has twice the velocity of another instead of saying that in equal times it traverses a distance twice as great, one does not mean by the word "velocity" something inherent in the body. (Alexander, 1956, p. xxxii)

d'Alembert was a "precursor" of Mach and the positivists: He proposed an account of motion that was devoid of motive forces, things that he thought were "obscure and metaphysical."

24. On this matter see Alexander (1956, p. xxix–xxxi), and Taylor (1941, p. 217–218).
25. Coriolis introduced the factor ½ into *vis viva* in 1829. He thus formulated the contemporary definition of kinetic energy as $KE = \frac{1}{2}mv^2$.
26. On this, see Scott (1960).
27. Henri Poincaré, at the turn of the century, wrote on these a priori aspects of science, maintaining that:

> As we cannot give a general definition of energy, the principle of conservation of energy simply signifies that there is *something* which remains constant. Whatever fresh notions of the world may be given us by future experiments, we are certain beforehand that there is something which remains constant, and which may be called *energy*. (Poincaré, 1905/1952, p. 166)

28. On this important subject of the a priori in physical theory, see Pap (1945) and Mittelstrass (1977) among others.
29. His experiments, using a pendulum to beat in time with the echo heard in a 208 foot corridor at Trinity College, and making adjustments for winter (denser air) and summer (thinner air) are described in Book II of the *Principia* (Newton, 1739/1934, pp. 382–384). They are discussed in Westfall (1980, pp. 734–736).
30. See Heath (1913) for an account of Aristarchus' life, times, and astronomical speculations.
31. See an account in Mattila (1991).

CHAPTER 9

1. Among many excellent cultural histories of timekeeping see Bell and Bell (1963), Borst (1993), Cipolla (1967), Fraser (1987), Landes (1983), Macey (1980), Rossum (1996), Welch (1972), Barnett (1998), and Whitrow (1988).
2. Another popular book on the history of clocks repeats the same chandelier story, but has Leonardo da Vinci watching it and feeling his pulse! (Roberts, 1992, p. 10).
3. It is noteworthy, in passing, that Comenius does not mention accurate time measurement among the things that cause his amazement. Clocks at this stage, the mid-17th century, are still timekeepers rather than time-measurers. More or less striking the hour does not constitute time measurement. Precise time measurement is some decades away with Huygens' and Hooke's clocks.
4. On the role of the clock metaphor in the history of psychology, see McReynolds (1980). The La Mettrie and Helvétius qoutes are from this source. McReynolds wrote: "My thesis is that the example of the mechanical clock was responsible to a significant degree—I do not claim that it was the only factor—for mediating the entrance into psychology of the concept of motive" (McReynolds, 1980, p. 105).
5. Macey (1980, Chap. 4) and his entry "Watchmaker God and Argument from Design" in Macey (1994) contain good guides to the literature on this subject. See also Haber (1975).
6. For some of this history see Raven (1953) and Knight (1986, Chap. 3). For a discussion of Paul Davies' version of the design argument, see Slezak (1996).
7. Newton also articulated the design argument using, as Paley would do a century later, biological examples. His "Short Scheme for the True Religion" (Thayer, 1953, pp. 65–66) is a nice statement of the argument. There, as one example, he mentions the eye:

Whence is it that the eyes of all sorts of living creatures are transparent to the very bottom, and the only transparent members in the body, having on the outside a hard transparent skin and within transparent humors, with a crystalline lens in the middle and a pupil before the lens, all of them so finely shaped and fitted for vision that no artist can mend them? Did blind chance know that there was light and what was its refraction, and fit the eyes of all creatures after the most curious manner to make use of it? These and suchlike considerations always have and ever will prevail with mankind to believe that there is a Being who made all things and has all things in his power, and who is therefore to be feared. (Thayer, 1953, p. 65)

8. For the religious background to Newton's work, see Jacob (1976, 1986). For a brief but informative history of natural theology from the 17th to the 19th centuries see Brooke (1991, Chaps. iv, vi).

9. On Newton's philosophical method, and for philosophical issues raised by his science, see contributions to Bricker and Hughes (1990).

10. This celebrated correspondence was first published in 1717. The letters, along with some other key philosophical and theological texts of Newton, are published in Alexander (1956). Newton's hand in Clarke's replies to Leibniz was confirmed when outlines of the latter's replies were found among Newton's own papers.

11. On the physics, metaphysics, theology, and politics involved in this celebrated public debate see, among numerous others, Cassirer (1943), Laudan (1968), Perl (1969), Hall (1980), and Priestly (1970).

12. Leibniz' principle of sufficient reason takes three different forms in his work: (1), the causal principle, that nothing happens without a cause; (2) God must always have a motive for acting; and (3) that God must always act for the best (Alexander, 1956, p. xxii).

13. In his *A True Inquiry into the Vulgarly Received Notion of Nature* (1686), Boyle wrote that the medieval Aristotelian idea of self-perfecting natures detracted from "the honor of the great author and governor of the world, that men should ascribe most of the admirable things, that are to be met with in it, not to him, but to a certain nature" (Deason, 1986, p. 180).

14. Paley's *Evidences* (1802) proved so popular that 20 editions were printed in the following 18 years. As late as 1831 Charles Darwin, and everyone else studying for a Cambridge degree, had to master Paley's major works in natural theology. This requirement ensured the educated classes had some contact with natural history. M.L. Clarke's biography of Paley is informative and most readable (Clarke, 1974).

15. Olaf Pedersen pointed out that this famous argument was first presented by the Dutchman Bernard Nieuwentyt nearly a century earlier in 1716. Nieuwentyt's version is: "Would anyone think himself to be clever if, having found this watch in a lonely place, he firmly believed that it was not the work of a skilled artisan who had assembled its parts, but that there were in the universe a non-intelligent law of nature which had collected and assembled its parts and joined so many pieces together" (Pedersen, 1992, p. 72)

16. In 1829 Earl of Bridgewater left in his will a considerable sum of money for the publication of a series of monographs to be written by scientists who would advance the program of natural theology. Eight treatises were written, one by William Whewell, the illustrious anti-inductivist proponent of the hypthetico-deductive method in science (Knight, 1986, pp. 43–47).

17. For discussion of this genre, see Jones (1966, Chap. IV) and Macy (1980, Chap. 6).

18. The following paragraphs, with some revisions, are taken from *Science Teaching: The Role of History and Philosophy of Science* (Matthews, 1994, pp. 205–208). For a discussion of metaphor, models, and analogy in science see, among countless others, Black (1962), Hesse (1963), Holton (1986), Leatherdale (1974), Schon (1963, Chap. 4), and Wartofsky (1979a).

19. For something of the history of the interaction of science and philosophy in the development and decline of atomism, see Einstein and Infeld (1938/1966), Harman (1982), Chalmers (1993, 1998), Pyle (1997), and Melsen (1952).

20. For a discussion from a cognitive science perspective, of metaphorical and analogical reasoning, see contributions to Vosniadou and Ortony (1989).

CHAPTER 10

1. I have provided source material outlining the scientific engagements of the 17th century philosophers in my edited collection, *The Scientific Background to Modern Philosophy* (Matthews 1989b). Against the normal philosophical trend, one biographer of Descartes wrote:

   > The opposite approach is adopted here. I interpret the extant writings of Descartes as the output of a practising scientist who, somewhat unfortunately, wrote a few short and relatively unimportant philosophical essays. (Clark, 1982, p. 2)

2. The interconnections between science and metaphysics are discussed in Amsterdamski (1975), Buchdahl (1969), Burtt (1932), Gjertsen (1989), Popper (1963, Chap. 2), Trusted (1991), Smith (1965), Hatfield (1990), and Wartofsky (1968).
3. For the philosophical roots of behaviorism, see Smith (1986) and Mills (1998).
4. For a philosophical appraisal of constructivism, see contributions to Matthews (1998).
5. A number of philosophers have written on the laws of motion as a bridge between physics and philosophy. See especially Nagel (1961, pp. 174–202), Whitrow (1971), Hanson (1959, 1965a, b). For the study of motion as a bridge between physics and theology, see especially Michael Buckley's *Motion and Motion's God* (Buckley, 1971).
6. On Aristotle's world view or cosmology see, among others, Collingwood (1945, pp. 81-85) and Grant (1971, chap. 5).
7. On metaphysics as a heuristic for physics, see Wartofsky (1968) and Agassi (1964).
8. Richard Westfall, writing on "The Problem of Force in Galileo's Physics," said that: "motion towards a gravitating center remains for him a metaphysically privileged motion to which the adjective 'natural' testifies on virtually every page that he writes" (Westfall, 1966, p. 89). This then, in Bachelard's terms, is an "epistemological obstacle" to Galileo formulating an adequate theory of force. For discussion of the philosophical tradition of self-motion from Aristotle to Newton, see contributions to Gill and Lennox (1994).
9. For discussion of how the natural–violent motion distinction is carried over into Newton's mechanics, see Ellis (1965), Hanson (1965b), and McGuire (1994).
10. As Max Jammer, in his *Concepts of Space*, commented:

    > It is perhaps not wholly unjustified to suggest a comparison between the notion of physical space in Aristotle's cosmology and the notion of Einstein's "spherical space" as expounded in early relativisitic cosmology. . . .the idea of "geodesic lines," determined by the geometry of space and their importance for the description of the path of material particles or light rays, suggest a certain analogy to the notion of "natural places" and the paths leading to them. (Jammer, 1954, p. 20)

11. Richard Rorty (Rorty, 1963), and James Weisheipl (Weisheipl, 1985, p. 23), provide an analysis of the Aristotelian elements in Whitehead's philosophy and science.
12. A second edition appeared in 1946, and a new impression of the book appeared annually in each of the next 25 years. It was translated into 12 languages.
13. But having cleared other pretenders from the ground, Ayer did not then engage with an explication of the theory, methodology, logic, or philosophy of science. In replying to a critic, Ayer stated that:

my greater fidelity to the tradition of British empiricism was displayed by my paying more attention to the theory of knowledge than to what Carnap called the logic of science, and this has increasingly been the case in my career. Not that I have abandoned my view that the best prospect for philosophy lies in its integration with science but my own lack of a scientific background has debarred me from pursuing this end. (Ayer, 1992, p. 301)

14. Good discussion of the difficulties, but also the benefits, in the marriage of philosophy of science with the history of science can be found in Garber (1986), Wartofsky (1976), McMullin (1970, 1975), and Shapere (1977).
15. The most comprehensive study of the different interpretations of the scientific revolution is Floris Cohen's *The Scientific Revolution: A Historiographical Inquiry* (Cohen, 1994).
16. On this tendency to project one's own philosophical position onto Galileo, see Crombie's essay on "Philosophical Presuppositions and the Shifting Interpretations of Galileo" (Crombie, 1981).
17. The following are especially comprehensive accounts of Galileo's methodological innovation and its relation to other medieval and renaissance traditions: McMullin (1978, 1985, 1990), Wisan (1978, 1981), and Wallace (1981a). In terms of the epistemological lessons to be learnt, few articles are better than Mittelstrass (1972). See also a fine discussion in Suchting (1995).
18. Finocchiaro (1980, p. 149) has provided a table of 14 philosophical topics to be found in Galileo's *Dialogue on the Two Chief World Systems* (1633).
19. This opposition to the mathematizing of physics, something that Galileo mentions in his 1602 letter to del Monte, was a deeply held Aristotelian, and more generally empiricist, conviction. The British empiricist, Hutchinson, would later say of the geometrical constructions of Newton's *Principia* that they were just "cobwebs of circles and lines to catch flies in" (Cantor, 1991, p. 219).
20. On Galileo's recourse to "accidental" factors, and their crucial role in his scientific methodology, see McMullin (1985) and Koertge (1977).
21. Thomas McTighe discusses some of the issues surrounding Galileo's putative Platonism (McTighe, 1967).
22. For formula for circular error, K, is: $K = 1.65x^2$ where x is the amplitude in degrees. The formula is independent of the length of the pendulum. On this see Landes (1983, p. 420).
23. Except if the circular arc is greatly reduced, so that the difference between it an an cycloidal arc in negible. This small amplitude option was tried, but it has its own mechanical drawbacks, namely increased susceptibility to interference.
24. In 1677 Edmund Halley took a London seconds-pendulum to St. Helena (16°Lat. South) and found that it also had to be shortened (Cook, 1998, p. 87). However he attributed the effect to the height above sea level at which he took his readings on the island. That is, the greater distance from the earth's center and hence the less gravitational attraction. In 1682 Robert Hooke informed the Royal Society that a London clock beat faster, that is, its period was shorter, when taken to Tangiers. On these and other examples, see Ariotti (1972, p. 406–407).
25. In the 20th century, Carl Hempel provided the classic study of the logic of confirmation (Hempel, 1945), and of the relation between confirmation and explanation (Hempel and Oppenheim 1948).
26. Nancy Cartwright has based her analysis of the laws of physics on the fact that the laws are, literally, not true (Cartwright, 1983, 1993).
27. On the topic of experiment, and its role in science, see Franklin (1986), Galison (1987), and Hacking (1983).
28. On this, see Nancy Cartwright's discussion (Cartwright, 1993).
29. Such objectivist epistemology is articulated in Popper (1972, Chaps. 3, 4), Althusser and Balibar (1970), Baltas (1988, 1990), Chalmers (1976), Lewin (1931), Mittelstrass (1972), Sneed (1979) and Suchting (1986a).
30. Ayer's *The Problem of Knowledge* (Ayer, 1956) is a quintessential empiricist account of

epistemology. Likewise is his *The Foundations of Empirical Knowledge* (Ayer, 1955) which, characteristically, is preoccupied with the problem of perception. This is a tell-tale marker of empiricism.

31. On these matters see Matthews (1994, Chaps. 7, 8) and Matthews (1992a, 1992b).

32. Many, for instance, have remarked on how Hegelian and even theological are the structures of Marx's world view.

33. The Index entry for "objectivism" in Kenneth Tobin's *The Practice of Constructivism in Science Education* (Tobin 1993) reads "see positivism"—a position that Popper spent his life arguing against.

34. One extensive study of the logic of idealisation in science is Nowak (1980). Ronald Laymon has also addressed the issue (Laymon, 1978, 1982, 1985).

35. The distinction between scientific data and scientific phenomena has been developed at length by James Bogen and James Woodward (1988). For an interpretation of Galileo's work in terms of the data/phenomena distinction, see Hemmendinger (1984).

36. For an analysis of the meaning of "phenomena" in Newton's work, see Achinstein (1990).

37. For example, David Hume wrote: "all our simple ideas in their first appearance are deriv'd from simple impressions, which are correspondent to them, and which they exactly represent" (Hume, 1734/1888, p. 4)

38. Compare Edwin Boring's famous reductionist remark that "intelligence is what intelligence tests measure." For discussion of comparable reductionist views of time, see Ariotti (1973) and Cover (1997).

39. A comment that psychologists have often enough made about measuring people's IQ.

40. John Findlay discusses this argument of Augustine's in his "Time: A Treatment of Some Puzzles" (Findlay, 1968). Findlay believes that this puzzle, and others of Augustine, "have their origin, not in any genuine obscurity in our experience, but in our ways of thinking and talking" (Findlay, 1968, p. 37).

41. On these, see Bunge (1966).

42. See for instance Lenzen (1938, pp. 295–302) and Reichenbach (1927/1958; 1951, Chap. 9). For a review of contemporary philosophers arguments about Mach's relativist account of space and time, see Mitchell (1993).

CHAPTER 11

1. The standard source book is Marshall Clagett's *The Science of Mechanics in the Middle Ages* (Clagett, 1959).

2. See, for instance, contributions to Basalla (1968) and Lindberg and Westman (1990). See also Hall (1970) and Huff (1993).

3. See contributions to Coelho (1992). Also Crombie (1990), James (1993), Kassler (1982), and Palisca (1961).

4. Much has been written on Aristotle's understanding of perception and judgement. One book on the subject is Deborah Modrak's *Aristotle: The Power of Perception* (Modrak, 1987). See also Block (1961), Krips (1980), and Slakey (1961).

5. This is why the Aristotelian tradition had a built-in bias against experiment, and in favor of observation. It is noteworthy that a version of such Aristotelian views are recurring in contemporary feminist criticism of scientific experimentation.

6. See the contributions to Part II of Swartz (1965).

7. For an account of this transition in educational constructivism, see O'Loughlin (1992) and Kelly (1997).

8. Piaget was a philosopher but, as Richard Kitchener, in a review of philosophical responses to his work, remarks: "Unfortunately, Piaget's program of a genetic epistemology has not been widely discussed by philosophers, nor critically evaluated in a comprehensive and scholarly way" (Kitchener, 1985a, p. 4). Among the exceptions are Marx Wartofsky, David Hamlyn, and Stephen Toulmin in their contribution to Mischel (1971), and Kitchener (1986). See also Kitchener's bibliography (Kitchener, 1985b).

9. Unfortunately psychological and educational writing usually ignores the distinction between psychology and epistemology, and uses children's "ideas" and children's "knowledge" interchangeably.

10. Colin Gauld, has drawn my attention (private communication) to the thus far untranslated *Sociological Studies* of Piaget, where apparently he moves a considerable distance from the epistemological individualism that I have been describing. These *Studies* are discussed in Chapman (1986) and Mays (1982).

11. It is symptomatic of the flight from science that this major book by an award-winning educator has chapters devoted to mythic, romantic, philosophic, and ironic understanding but contains no discussion of scientific understanding.

12. In the 1930s Piaget initiated the study of the psychology of time, to which many subsequently contributed. One landmark publication was Paul Fraisse's *The Psychology of Time* (Fraisse, 1963). Twenty years later, in recognition of Fraisse's work, an important anthology, *The Developmental Psychology of Time,* was published (Friedman, 1982). One study that concentrated on refining the Piagetian tasks and questions, still concluded that: "Children do not understand the concept of time before they understand the concept of speed; in fact, the reverse is the case, with speed being mastered well in advance of time" (Siegler and Richards, 1979, p. 297)

CHAPTER 12

1. In the Australian state of New South Wales, less than 1/3 of Higher School Certificate students study history. In the state of Victoria about 10% of final year students study history, and many school history departments have been closed, meaning that no history at all is taught in the school. In New Zealand, history has been subsumed into "Social Studies," and then basically ignored. Even when history is taught, it is notorious that school and university history departments have turned their back on the history of science. All of this is a great pity, as the study of history gives students a sense of society as having a past, and a sense of institutions, ideas and ideologies as contingent and evolving.

2. See also Conant's foreword to Thomas Kuhn's *The Copernican Revolution* (Kuhn 1957), where he discussed the "growing concern" about unsatisfactory science courses for the nonspecialist, and suggests that "more benefit can be obtained by an intensive study of certain episodes in the development of physics, chemistry and biology" (Kuhn, 1957, p. xix). He commended the just-then-published *Harvard Case Studies in Experimental Science* (Conant, 1957) as an appropriate contribution to the problem. On the relationship between Conant's education program and Kuhn's view of science, see Fuller (2000).

3. Fifteen percent of United States high school physics students were following this program at its peak, and it was widely used outside the United States. The philosophy behind this program can be read in Gerald Holton (1978), and in the symposium published in *The Physics Teacher* (Holton, 1967). Other evaluations of Harvard Project Physics can be found in Aikenhead (1974), Brush (1969, 1989), Russell (1981), and Welch and Walberg (1972).

4. There are numerous good books on the history and diversity of calendars. See for instance O'Neill (1975), Richards (1998), and Duncan (1998).

5. Actually there are two months. The *sidereal* month is the time for the moon to complete an orbit of the earth with respect to a fixed star. This is 27.32166 days. The *synodic* month is the time for the moon to complete an orbit of the earth with respect to the sun. This is 29.53059 days (longer than the sidereal month because the earth has moved in its orbit around the sun).

6. The eccentricity of the earth's orbit means that the shortest shadow does not coincide with noon at the equinoxes and solstices. On these dates, apparent solar time and clock time are out by about 9.5 minutes.

7. It is of course only refined scientific *purposes* that necessitate the addition of leap seconds to different years. Most of life would go on as usual if they were not added. The very institution of leap seconds is indicative of the growth of precision in time measurement since Galileo first sketched plans for a pendulum clock.

8. Some of the extensive research literature on school practical work is gathered together in Woolnough (1991). See also the Derek Hodson edited special issue of *International Journal of Science Education* 20(6), 1998.

9. On Ernst Mach, and his place in the discipline of science education, see Matthews (1990; 1994a, pp. 95–99).

10. The 1905 New York state physics syllabus contained 260 topics which were to be covered in 120 hours—a half-hour per topic! (Mann, 1912, p. 66). Eighty years later, the same state's biology syllabus required the student to learn 1,440 scientific terms and concepts in one year (Swift, 1988). Mary Budd Rowe commented that "Science books have turned into fantastic dictionaries...the plot, the story line—the way in which ideas interact—have disappeared" (Fisher, 1992, p. 53). Science teaching is frequently like intellectual spray painting: teachers hope that some students will sit still long enough to get a reasonable coverage, and that the information will not peel off before the final exam.

11. Galileo stated it as "the forza at the bottom of the incline is equal to the forza required to lift it up the incline."

12. Another source of ideas about amateur clock making is Zubrowski (1988).

13. Some of the history of the standards is given in the early pages of the *Standards* document (NRC, 1996, pp. 13–15). Angelo Collins, one of the principal writers, provided a more detailed history (Collins, 1995), and commented on the internal and external politics of the exercise (Collins, 1998).

14. The director of Project 2061 is James Rutherford, who 20 years earlier was one of the three directors of Harvard Project Physics, and who in 1972 wrote in support of "A Humanistic Approach to Science Teaching" (Rutherford, 1972). The project's basic document is *Science for All Americans* (Rutherford and Ahlgren 1990), which is elaborated in *Benchmarks for Science Literacy* (AAAS, 1993). Some of the educational and philosophical underpinnings of the project are discussed in Cossman (1989) and Matthews (1994a, pp. 35–44).

15. Apparently the NSTA made some effort to involve United States historians of science in their project, but were not successful in finding anyone willing to participate. Philosophers were not even sought. This is a telling comment on the separation of the professional disciplines of HPS from the realities and practice of science education.

16. See also contributions to Misgeld (1994), particularly that of Rühaak (Rühaak, 1994).

CHAPTER 13

1. A collection of papers, along with extensive bibliographies, are in two special issues of *Science & Education* (**6**(4), 1997 and **7**(6), 1998) devoted to "The Nature of Science and Science Education." See also the extensive bibliography in Bell et al. (2001).

2. See particularly his essay "On Instruction in the Sciences and in the Classics" (Mach, 1886/ 1943). Mach's educational ideas are discussed in Matthews (1990).

3. See, for instance, the work of Derek Hodson (Hodson, 1986, 1988), Rick Duschl (Duschl, 1985, 1990), Norman Lederman (Lederman, 1992), Joan Solomon (Solomon, 1991), Edgar Jenkins (Jenkins, 1990, 1994, 1996) and myself (Matthews, 1992c, 1994a). See also the contributions to the journal *Science & Education* from 1992 to the present.

4. Robinson's Ph.D. at Stanford University was supervised by Paul DeHart Hurd and completed in 1964. The thesis was summarised the following year in an article in the *Journal of Research in Science Teaching* titled "Science Teaching and the Nature of Science" (Robinson, 1965). In 1968 a book version—titled *The Nature of Science and Science Teaching*—was published (Robinson, 1968). I have elsewhere written on Robinson and on some lessons for today of his 1960s engagement with nature of science questions (Matthews, 1997a).

5. For contributions to the "Science Wars" see, Passmore (1978), Holton (1993), Gross and Levitt (1994), Ross (1996), and Levitt (1998).

6. See contributions to Gross, Levitt, and Lewis (1996), and Koertge (1998).

7. For documents, and discussion of this trial, see contributions to Ruse, 1988.

8. See for instance Koertge (1981, 1996), Olson (1990), Pinnick (1994), Richards (1996), and Ruskai (1996).

9. For discussion of these matters see Matthews (1994a, Chap. 9) and Siegel (1997).

10. Gerald Holton has documented efforts made to counteract constructivist interpretations of the history and philosophy of science in the *Standards* (Holton, 1996). Also Good and Shymansky (2001).

11. See for instance Suchting (1992), Slezak (1994 a,b), Solomon (1994), Nola (1997), Phillips (1995, 1997), Osborne (1996), Gross and Levitt (1994), Cromer (1993), and Brown (1994). See also contributions to Matthews (1998).

12. See Matthews (1992a, b, 1993; 1994a, Chap.7, 1995a pp. 115–125, 1997a, 2000).

13. I have elaborated the core commitments of a "modest realism" in Matthews (1994a, pp. 176– 177).

14. For discussion of philosophical practice in elementary school, see Berlin and Gaines (1966) and Wagner and Lucas (1977).

15. I have described a number of "nature of science courses in Matthews 1994a, pp. 202–203. See also contributions to McComas (1998).

16. Others to argue the advantages of including HPS in science teacher education programs include Cawthorn and Rowell (1978), Eger (1987), Robinson (1969), Scheffler (1970) and Summers (1982).

17. For instance, in the four thousand or so articles listed in the 4th edition of Helga Pfundt and Reinders Duit's *Bibliography* of constructivist-based science education research, there are only three or four entries on learning about pendulum motion, and none on learning the history or philosophy of pendulum motion.

18. In Solomon's words:

> Constructivism has always skirted round the actual learning of an established body of knowledge...students will find that words are used in new and standardised ways: problems which were never even seen as being problems, are solved in a sense which needs to be learned and rehearsed. For a time all pupils may feel that they are on foreign land and no amount of recollection of their own remembered territory with shut eyes will help them to acclimatise. (Solomon, 1994, p. 16)

19. For instance, a former New Zealand colleague marked a group of 35 constructivist-based Bachelor of Education students' practice teaching lessons on goldfish. Of the 35 lesson plans, only two were concerned with knowledge outcomes in their classes. The rest were all pleasant enough

activities – imagine you are a goldfish, make a goldfish collage, brainstorm goldfish – but not the things for which parents primarily send their children to school, or for which governments fund education. Only two of the lessons were concerned to inform pupils about what food goldfish ate, what temperature water they survive in, and whether they are fresh or saltwater fish.

20. I have touched on some of these matters in an earlier book on *The Marxist Theory of Schooling* Matthews (1980).

21. I have discussed some aspects, especially the New Zealand case, of constructivism and literacy teaching in Matthews 1995a, pp. 49–57.

22. Apart from Nicholson's publications, see Adams (1990), Center, Wheldall and Freeman (1992) and Iverson and Tunmer (1992).

23. Some recent philosophical contributors to the liberal education tradition include Mortimer Adler (Adler, 1988), Richard Peters (Peters, 1966), Richard McKeon (McKeon, 1994), Paul Hirst (Hirst, 1974), Israel Scheffler (Scheffler, 1973), and G.H. Bantock (Bantock, 1981). Elliot Eisner, in his review of curriculum ideologies, calls this educational tradition "rational humanism" (Eisner, 1992). Liberal educational theory is a *normative* theory, not a social scientific theory about the function of schools in society. One can defend liberal educational ideals whilst being a thorough-going Marxist about the function of schools in capitalist and totalitarian societies. A theory of education is not the same as a theory of schooling, although they are frequently confused. On liberal education, see also Schneider and Shoenberg (1998) and Kimball (1986).

24. A good deal of the literature on students' work commitments and their impact is discussed in Owen (1995, Chaps. 1–3).

25. This position is elaborated in Labaree's book, *How to Succeed in School Without Really Learning: The Credentials Race in American Education* (Labaree, 1997b).

26. There is debate about just how dismal current education is, and what contributes to, or militates against, educational attainment. For the U.S., the TIMSS (Third International Mathematics and Science Study) results are related and analysed in Schmidt et al. (1997). E.D. Hirsch, for example, is one of many who paint a pessimistic picture of the U.S. educational landscape (Hirsch, 1996). Biddle and Berliner (1995) argue, unconvincingly in my opinion, that the U.S. educational crisis is illusory. Various Authors (1992) have argued a similar case in the U.K. The dismal situation of Australian university education is documnted in contributions to Coady (2000).

27. Individual parts of the story are elaborated in the references and, increasingly, on the world wide web. The web provides an abundance of information and material that previously has been beyond the capacity of individual books and teachers. For instance, one search turned up 1,700 entries for "pendulum clock," 1,000 entries for "pendulum motion," 300 entries for "Christiaan Huygens" and 20 entries for "Guidabaldo del Monte." One of numerous guides to using the web for teaching purposes is Brooks (1997).

References

Achinstein, P. (1990). Newton's corpuscular query and experimental philosophy. In P. Bricker & R.I.G. Hughes (eds.), *Philosophical Perspectives on Newtonian Science*, MIT Press, Cambridge, MA., pp. 135 –173.

Adams, M. J. (1990). *Beginning to Read. Thinking and Learning about Print,* MIT Press Cambridge, MA.

Adas, M. (1989). *Machines as the Measure of Men: Technology and Ideologies of Western Dominance,* Cornell University Press, Ithaca, NY.

Adler, M.J. (1978). *Aristotle for Everybody,* Macmillan, New York.

Adler, M.J. (1988). *Reforming Education,* (G. van Doren ed.), Macmillan, New York.

Agassi, J. (1964). The nature of scientific problems and their roots in metaphysics. In M. Bunge (ed.), *The Critical Approach,* Free Press, Glencoe, IL. Reprinted in his *Science in Flux,* Reidel, Boston, 1975 .

Aikenhead, G.S., & Solomon, J. (eds.). (1994). *STS Education: International Perspectives on Reform,* Teachers College Press, New York.

Aikenhead, G.S. (1974). Course evaluation II: Interpretation of student performance on evaluation tests. *Journal of Research in Science Teaching* 11: 23–30.

Aikenhead, G.S. (1997). Towards a first nations cross-cultural science and technology curriculum. *Science Education* 81(2): 217–238.

Akerblom, K. (1968). *Astronomy and Navigation in Polynesia and Micronesia: A Survey,* The Ethnographical Museum, Stockholm.

Alder, K. (1995). A revolution to measure: The political economy of the metric system in France. In Wise, M.N. (ed.), *The Values of Precision,* Princeton University Press, Princeton, NJ, pp. 39–71.

Aldridge, W.G. (1992). Project on scope, sequence, and coordination: A new synthesis for improving science ducation. *Journal of Science Education and Technology,* (1): 13–21.

Alexander, H.G. (ed.) (1956). *The Leibniz-Clarke Correspondence,* Manchester University Press, Manchester.

Allchin, A., Anthony, E., Bristol, J., Dean, A., Hall, D., & Lieb, C. (1999). History of science—with labs. *Science & Education* 8(6): 619–632.

Allen, R.E. (ed.), (1966). *Greek Philosophy. Thales to Aristotle,* The Free Press, New York.

Alters, B.J. (1997). Whose nature of science?, *Journal of Research in Science Teaching* 34(1): 39–55.

Althusser, L. & Balibar, E. (1970). *Reading Capital,* New Left Books, London.

American Association for the Advancement of Science (AAAS) (1989). *Project 2061: Science for All Americans,* AAAS, Washington, D.C. Also published by Oxford University Press, 1990.

American Association for the Advancement of Science (AAAS) (1990). *The Liberal Art of Science: Agenda for Action,* AAAS, Washington, D.C.

American Association for the Advancement of Science(AAAS) (1993). *Benchmarks for Science Literacy*, Oxford University Press, New York.

Amsterdamski, S. (1975). *Between Experience and Metaphysics*, Reidel, Dordrecht.

Andrewes, W.J.H. (1985). Time for the astronomer. *Vistas in Astronomy*, Murdein, P. & Beer, P. (eds.), 28: 69–86.

Andrewes, W.J.H. (ed.) (1998a). *The Quest for Longitude: The Proceedings of the Longitude Symposium, Harvard University, Cambridge, Massachusetts, November 4–6, 1993*, Collection of Historical Scientific Instruments, Harvard University, Cambridge, MA.

Andrewes, W.J.H. (1998b). Even Newton could be wrong: The story of Harrison's first three sea clocks. In Andrewes, W.J.H. (ed.) *The Quest for Longitude: The Proceedings of the Longitude Symposium, Harvard University, Cambridge, Massachusetts, November 4–6, 1993*, Collection of Historical Scientific Instruments, Harvard University, Cambridge, MA., pp. 190–234.

Andrewes, W.J.H. (1998c). Finding local time at sea and the instruments involved. In Andrewes, W.J.H. (ed.) *The Quest for Longitude: The Proceedings of the Longitude Symposium, Harvard University, Cambridge, Massachusetts, November 4–6, 1993*, Collection of Historical Scientific Instruments, Harvard University, Cambridge, MA., pp. 394–404.

Aquinas, T. (1270/1952). *Summa Theologica*, trans. English Dominican Province, Great Books, Chicago, vol. 19.

Ariotti, P.E. (1968). Galileo on the isochrony of the pendulum. *Isis* 59: 414–426.

Ariotti, P.E. (1972). Aspects of the conception and development of the pendulum in the 17th century. *Archive for History of the Exact Sciences* 8: 329–410.

Ariotti, P.E. (1973). Toward absolute time: The undermining and refutation of the Aristotelian conception of time in the 16th and 17th Centuries. *Annals of Science* 30: 31–50.

Ariotti, P.E. (1975). The concept of time in western antiquity. In Fraser, J.T. and Lawrence, N. (eds.), *The Study of Time II: Proceedings of the Second Conference of the International Society for the Study of Time*, Springer-Verlag, New York, 1975, pp. 69–80.

Armstrong, H.E. (1903). *The Teaching of Scientific Method and Other Papers on Education*, The Macmillan Co., London.

Arons, A.B. (1959). Structure, methods, and objectives of the required freshman calculus-physics course at Amherst College. *American Journal of Physics* 27(9): 658–666.

Arons, A.B. (1965). *Development of Concepts of Physics*, Addison-Wesley, Reading MA.

Arons, A.B. (1977). *The Various Language, An Inquiry Approach to the Physical Sciences*, Oxford University Press, New York.

Arons, A.B. (1989). What science should we teach? In *Curriculum Development in the Year 2000*, BSCS, Colorado Springs, CO.

Arons, A.B. (1990). *A Guide to Introductory Physics Teaching*, John Wiley, New York.

Assis, A.K.T. & Zylbersztajn, A. (2001). The Influence of Ernst Mach in the teaching of mechanics. *Science & Education* 10(1).

Atwood, S.G. (1975). The development of the pendulum as a device for regulating clocks prior to the 18th century. In Fraser, J.T. & Lawrence, N. (eds.), *The Study of Time II: Proceedings of the Second Conference of the International Society for the Study of Time*, Springer-Verlag, New York, 1975, pp. 417–450.

Augustine, St. (400/1961). *The Confessions*, (R.S. Pine-Coffin, trans.), Penguin, Harmondsworth, Middlesex.

Aveni, A.F. (1989). *Empires of Time: Calendars, Clocks, and Cultures*, Basic Books, New York.

Ayer, A.J. (1936). *Language, Truth and Logic*, Victor Gollancz, London. Second edition, 1946.

Ayer, A.J. (1955). *The Foundations of Empirical Knowledge*, Macmillan, London.

Ayer, A.J. (1956). *The Problem of Knowledge*, Penguin, Harmondsworth.

Ayer, A.J. (1992). Reply to Tscha Hung. In Hahn, L.E. (ed.), *The Philosophy of A.J. Ayer*, Open Court, New York, pp. 301–307.

Bachelard, G. (1934/1984). *The New Scientific Spirit*, Beacon Books, Boston.

Bailey, H. (1919). *Time Telling Through the Ages*, Doubleday, New York.

Baltas, A. (1988). On the structure of physics as a science. In Batens, D. & van Bendegens, J.P. (eds.), *Theory and Experiment*, Reidel, Dordrecht, pp. 207–225.

Baltas, A. (1990). Once Again on the Meaning of Physical Concepts. In P. Nicolacopoulos (ed.), *Greek Studies in the Philosophy and History of Science*, Kluwer Academic Publishers, Dordrecht, pp. 293–313.

Bantock, G.H. (1981). The idea of a liberal education. In his *The Parochialism of the Present*, Routledge & Kegan Paul, London, pp. 65–79.

Barnes, J. (ed.) (1984). *The Complete Works of Aristotle*, two volumes, Princeton University Press, Princeton, NJ.

Barnett, J.E. (1998). *Time's Pendulum: From Sundials to Atomic Clocks, the Fascinating History of Timekeeping and How Our Discoveries Changed the World*, Harcourt Brace & Co., New York.

Basalla, G. (ed.) (1968). *The Rise of Modern Science: External or Internal Factors?*,Heath, Lexington, MA.

Beardsley, T. (1992). Teaching real science, *Scientific American* October, 78–86.

Bedini, S.A. and Maddison, F. (1966). *Astrarium* of Giovanni de Dondi. *Transactions of the American Philosophical Society* 56:1–69.

Bedini, S.A. & Reti, L. (1974). Horology. In Reti, L. (ed.), *The Unknown Leonardo*, McGraw-Hill Book Company, New York.

Bedini, S.A. (1956). Chinese mechanical clocks. *Bulletin of the National Association of Water and Clock Collectors* 7:211–221.

Bedini, S.A. (1963). Galileo Galilei and time measurement. *Physis* 5, 145–165.

Bedini, S.A. (1967). The instruments of Galileo Galilei. In McMullin, E. (ed.), *Galileo Man of Science*, Scholar's Bookshelf, Princeton Junction, NJ, pp. 256–294.

Bedini, S.A. (1980). The mechanical clock and the industrial revolution. In Maurice, K. & Mayr, O. (eds.), *Clockwork Universe. German Clocks and Automata 1550–1650*, Neale Watson Academic Publications, New York.

Bedini, S.A. (1986). Galileo and scientific instrumentation. In Wallace, W.A. (ed.), *Reinterpreting Galileo*, Catholic University of America Press, Washington, pp.127–154.

Bedini, S.A. (1991). *The Pulse of Time: Galileo Galilei, the Determination of Longitude, and the Pendulum Clock*, Olschki, Florence.

Bell, A.E. (1947). *Christiaan Huygens and the Development of Science in the Seventeenth Century*, Edward Arnold, London.

Bell, B.F. (1991). A constructivist view of learning and the draft forms 1–5 science syllabus. *SAME Papers 1991:* 154–180.

Bell, R., Abd-el-Khalick, F. Lederman, N.G., McComas, W.F. & Mathews, M. R. (2001). The nature of sciene and science education: A Bibliogrpahy. Science & Education 10(3).

Bell, B.F. (ed.): 1992, *I Know About LISP But how do I put it into Practice: Draft Report*, Centre for Science and Mathematics Education Research, University of Waikato, Hamilton.

Bell, T.H. & Bell, C. (1963). *The Riddle of Time*, Viking Press, New York.

Benjamin, A. (1966). Ideas of time in the history of philosophy. In Fraser, J. (ed.), *The Voices of Time*, George Braziller, New York.

Bennett, J.A. (1985). The longitude and the new science. *Vistas in Astronomy* 28.

Bereiter, C. (1994). Constructivism, socioculturalism, and Popper's world 3. *Educational Researcher* 23(7): 21–23.

Bergmann, P.(1949). *Basic Theories of Physics*, Prentice-Hall, New York.

Berkeley, G. (1721/1901). *De Motu*. In Fraser, A.C. (ed.), *The Works of George Berkeley*, Oxford University Press, Oxford. (extracts in Armstrong, D.M. (ed.), *Berkeley's Philosophical Writings*, New York, 1965).

Berlin, B.M. & Gaines, A.M. (1966). Use philosophy to explain the scientific method. *The Science Teacher* 33(5).

Bernal, J.D. (1939). *The Social Function of Science*, Routledge & Kegan Paul, London.

Berriman, A.E. (1953). *Historical Metrology: A New Analysis of the Archaelogical and Historical Evidence Relating to Weights and Measures*, J.M. Dent & Sons, London.

Bertele, von H. (1953). Precision timekeeping in the pre-Huygens era. *Horological Journal* 95: 794–816.

Bertoloni Meli, D. (1992). Guidobaldo del Monte and the Archimedean revival. *Nuncius* 7:30–34.

Biagioli, M. (1993). *Galileo Courtier: The Practice of Science in the Culture of Absolutism*, University of Chicago Press, Chicago.

Biddle, B. & Berliner, D. (1995). *The Manufactured Crisis in American Public Schools*, Addison, Wesley & Longmans, New York.

Black, M. (1962). *Models and Metaphors*, Cornell University Press, Ithaca NY.

Blackmore, J.T. (1983). Philosophy as part of the internal history of science. *Philosophy of the Social Sciences* 13: 17–45.

Blay, M. (1998). *Reasoning with the Infinite: From the Closed World to the Mathematical Universe*, University of Chicago Press, Chicago.

Block, I. (1961). Truth and error in Aristotle's theory of sense perception. *Philosophical Quarterly* xi: 1–9.

Blumenberg, H. (1987). *The Genesis of the Copernican World*, MIT Press, Cambridge MA.

Board of Studies, NSW (BOS) (1998). *Draft Stages 4 and 5 Science Syllabus*, Board of Studies, Sydney.

Bochner, S. (1966). *The Role of Mathematics in the Rise of Science*, Princeton University Press, Princeton, NJ.

Bodner, G.M. (1986). Constructivism: A Theory of Knowledge. *Journal of Chemical Education* 63(10): 873–878.

Boeha, B. (1990). Aristotle, alive and well in Papua New Guinea science classrooms. *Physics Education* 25: 280–283.

Bogen, J. & Woodward, J. (1988). Saving the phenomena. *The Philosophical Review* XCVII(3): 303–350.

Bond, A.R. (1948). *The Story of Mechanics*, Collier & Son, New York.

Boorstin, D.J. (1983). *The Discoverers*, Random House, New York.

Booth, E.H. & Nicol, M.N. (1962). *Physics* (16th edition), Australasian Medical Publishing Company, Sydney. (First edition, 1931.)

Borst, A. (1993). *The Ordering of Time. From the Ancient Computus to the Modern Computer*, University of Chicago Press, Chicago.

Bos, H.J.M. et al. (eds.) (1980). *Studies on Christiaan Huygens*, Swets & Zeitlinger, Lisse.

Bos, H.J.M. (1980a). Christiaan Huygens—A biographical sketch. In H.J.M. Bos et al. (eds.), *Studies on Christiaan Huygens*, Swets & Zeitlinger, Lisse, pp.7–18.

Bos, H.J.M. (1980b). Huygens and mathematics. In H.J.M. Bos et al. (eds.), *Studies on Christiaan Huygens*, Swets & Zeitlinger, Lisse, pp. 126–146.

Bos, H.J.M. (1980c). Mathematics and Rational Mechanics. In G.S. Rousseau and R. Porter (eds.), *The Ferment of Knowledge: Studies in the Historiography of Eighteenth-Century Science*, Cambridge University Press, Cambridge, pp. 333–351.

Bos, H.J.M. (1986). Introduction. In R.J. Blackwell (trans.), *Christiaan Huygens' The Pendulum Clock*, Iowa State University Press, Ames, IA.

Boxer, C.R. (1965). *The Dutch Seaborne Empire, 1600–1800,* Hutchinson, London.

Boxer, C.R. (1969). *The Portuguese Seaborne Empire, 1415–1825,* Hutchinson, London.

Boxer, C.R. (1984). *From Lisbon to Goa, 1500–1750: Studies in Portuguese Maritime Enterprise*, Hutchinson, London.

Boxer, C.R. (1985). *Portuguese Conquest and Commerce in Southern Asia, 1500–1750*, Variorum Reprints, London.

Boxer, C.R.. (1988). *Dutch Merchants and Mariners in Asia, 1602–1795*, Hutchinson, London.

Boyer, C.B. (1968). *A History of Mathematics*, John Wiley, New York.

Boyle, R. (1772). *The Works of the Honourable Robert Boyle*, T. Birch (ed.), London.

Braverman, H. (1974). *Labour and Monopoly Capital*, Monthly Review Press, New York.

Brearley, H.C. (1919). *Time Telling Through the Ages*, Doubleday, Page & Co., New York.

Bricker, P. & Hughes, R.I.G. (eds.) (1990). *Philosophical Perspectives on Newtonian Science*, MIT Press, Cambridge MA.

Brock, W.H. (1989). History of science in British schools: Past, present and future. In M. Shortland and A. Warwick (eds.), *Teaching the History of Science*, Oxford, Basil Blackwell, pp. 30–41.

Brooke, J.H. (1991). *Science and Religion: Some Historical Perspectives*, Cambridge, Cambridge University Press.

Brooks, D.W. (1997). *Web-Teaching. A Guide to Designing Interactive Teaching for the World Wide Web*, Plenum Press, New York.

Brown, H.I. (1976). Galileo, the elements, and the tides. *Studies in History and Philosophy of Science* 7(4): 337–351.

Brown, H.I. (1977). *Perception, Theory and Commitment: The New Philosophy of Science*, University of Chicago Press, Chicago.

Brown, H.M. (1992). Vincenzo Galilei in Rome: His first book of lute music (1563) and its cultural context. In V. Coelho (ed.), *Music and Science in the Age of Galileo*, Kluwer Academic Publishers, Dordrecht, pp. 153–184.

Brown, J.R. (1994). *Smoke and Mirrors: How Science Reflects Reality*, Routledge, New York.

Brown, W. (1993). Classroom science goes into freefall, *New Scientist* December: 12–13.

Bruner, J.S. (1960). *The Process of Education*, Random House, New York.

Brush S.G. (1969). The role of history in the teaching of physics. *The Physics Teacher* 7(5): 271–280.

Brush, S.G. (1989). History of Science and Science Education. *Interchange* 20(2): 60–70.

Bruton, E. (1979). *The History of Clocks and Watches*, Obs, London.

Buchdahl, G. (1951). Science and logic: On Newton's second law of motion. *British Journal for Philosophy of Science* 2: 217–235.

Buchdahl, G. (1969). *Metaphysics and the Philosophy of Science*, Basil Blackwell, Oxford.

Buckley, M.J. (1971). *Motion and Motion's God*, Princeton University Press, Princeton.

Bunge, M. (1966). Mach's critique of Newtonian mechanics. *The American Journal of Physics* 34: 585–596. Reproduced in J. Blackmore (ed.), *Ernst Mach–A Deeper Look*, Kluwer Academic Publishers, Dordrecht, (1992), pp. 243-261.

Bunge, M. (1991). A critical examination of the new sociology of science: Part 1. *Philosophy of the Social Sciences* 21(4): 524–560.

Bunge, M. (1992). A critical examination of the new sociology of science: Part 2, *Philosophy of the Social Sciences* 22(1): 46–76.

Burgess, M. (1998). The scandalous neglect of Harrison's regulator science. In W.J.H. Andrewes (ed.), *The Quest for Longitude*, The Collection of Historical Scientific Instruments, Harvard University, Cambridge, MA, 2nd Edition, pp. 256–278.

Burke, J.G. (ed.) (1984). *The Uses of Science in the Age of Newton*, University of California Press, Berkeley, CA.

Burtt, E.A. (1932). *The Metaphysical Foundations of Modern Physical Science* (second edition), Routledge & Kegan Paul, London.

Butterfield, H. (1949). *The Origins of Modern Science 1300–1800*, G. Bell and Sons, London.

Butts, R.E. & Davis, J.W. (eds.) (1970). *The Methodological Heritage of Newton*, University of Toronto Press, Toronto.

Buzeika, A. (1993). A study of entry levels and progress in mathematics of primary teacher education students at Auckland College of Education. Thesis submitted in partial requirement for the Diploma of Mathematics Education, University of Auckland.

Bybee, R.W. (ed.) (1985). *Science, Technology, Society*, Yearbook of the National Science Teachers Association, NSTA, Washington.

Bybee, R.W. (1990). Teaching history and the nature of science in science courses: A rationale. *Science Education* 75(1):143–156.

Bybee, R.W., Ellis, J.D, Giese, J.R. & Parisi, L. (1992). *Teaching About the History and Nature of Science and Technology: A Curriculum Framework*, Colorado Springs, BSCS/SSEC.

Cajori, F. (1929). *A History of Physics*. Macmillan, New York (orig. 1899).

Callahan, J.F. (1948). *Four Views of Time in Ancient Philosophy*, Harvard University Press, Cambridge, MA.

Cambell, J. (ed.) (1957). *Man and Time: Papers from the Eranos Yearbooks*, Princeton University Press, Princeton.

Cantor, G. (1991). *Michael Faraday: Sandemanian and Scientist*, St. Martin's Press, New York.

Capek, M. (1973). Time. *Dictionary of the History of Ideas* iv: 389–398.

Capek, M. (ed.) (1976). *The Concepts of Space and Time: Their Structure and Development*, (*Boston Studies in the Philosophy of Science* Vol. 22), Reidel, Dordrecht.

Capra, F. (1975). *The Tao of Physics*, Shambhala Press, Berkeley.

Cardwell, D.S.L. (1972). *Technology, Science and History*, Heinemann, London.

Carr, M., Barker, M., Bell, B., Biddulph, F., Jones, A., Kirkwood, V., Pearson, J., & Symington, D. (1994). The constructivist paradigm and some implications for science content and pedagogy. In P. Fensham, R. Gunstone and R. White (eds.), *The Content of Science: A Constructivist Approach to its Teaching and Learning*, Falmer Press, London, pp. 147–160.

Carson, R.(1997). Science and the ideals of liberal education. *Science & Education* 6(3): 225–238.

Cartwright, N. (1983). *How the Laws of Physics Lie*, Clarendon Press, Oxford.

Cartwright, N. (1992). Natures and the experimental method. In J. Earman (ed.), *Inference, Explanation, and Other Frustrations: Essays in the Philosophy of Science*, University of California Press, Berkeley.

Cartwright, N. (1993). How we relate theory to observation. In P. Horwich (ed.), *World Changes*, MIT Press, Cambridge, MA, pp. 259–273.

Cassirer, E. (1943). Newton and Leibniz. *Philosophical Review* 52(4): 366–391.

Cawthorn, E.R. & Rowell, J.A. (1978). Epistemology and science education', *Studies in Science Education* 5: 31-59.

Center, Y., Wheldall, K.,and Freeman, L. (1992). Evaluating the effectiveness of reading recovery: A critique. *Educational Psychology* 12(3–4): 263–274.

Chalmers, A.F. (1976). *What Is This Thing Called Science?*, University of Queensland Press, St Lucia, Queensland. (Third edition, Hackett Publishing Company, Indiananapolis, 1999.)

Chalmers, A.F. (1993). The lack of excellency of Boyle's mechanical philosophy. *Studies in History and Philosophy of Science* 24: 541–564.

Chalmers, A.F. (1998). Retracing the ancient steps to atomic theory. *Science & Education* 7(1): 69–84.

Champagne, A.B, Klopfer, L.E., & Anderson, J. (1980). Factors influencing learning of classical mechanics. *American Journal of Physics* 48: 1074–1079.

Chandler, B. (1998). Longitude in the context of mathematics In W.J.H. Andrewes (ed.), *The Quest for Longitude*, The Collection of Historical Scientific Instruments, Harvard University, Cambridge, MA, 2nd Edition, pp. 34–42.

Chapin, S.L. (1994). Geodesy. In I. Grattan-Guinness (ed.), *Companion Encyclopedia of the History and Philosophy of the Mathematical Sciences*, Routledge, London, pp. 1089–1100.

Chapman, M. (1986). The structure of exchange: Piaget's sociological theory. *Human Development* 29: 181–194.

Chaunu, P. (1979). *European Expansion in the Later Middle Ages*, K. Bertram (trans.), Amsterdam.

Chomsky, N. (1959). Review of B.F. Skinner *Verbal Behavior. Language* 35: 26–58.

Christianson, G.E. (1996). *Isaac Newton and the Scientific Revolution*, Oxford University Press, New York.

Cipolla, C. (1967). *Clocks and Culture: 1300–1700*, Collins, London.

Clagett, M. (1959). *The Science of Mechanics in the Middle Ages*, University of Wisconsin Press, Madison, WI.

Clagett, M. (1995). *Ancient Egyptian Science. A Source Book, Vol. Two, Calendars, Clocks and Astronomy*, American Philosophical Society, Philadelphia.

Clancy, R. & Richardson, A. (1988). *So Came They South*, Shakespeare Head Press, Sydney.

Clarke, D.M. (1982). *Descartes' Philosophy of Science*, Manchester University Press, Manchester.

Clarke, M.L. (1974). *Paley: Evidences for the Man*, University of Toronto Press, Toronto.

Clavelin, M. (1974). *The Natural Philosophy of Galileo,* Harvard University Press, Cambridge, MA.

Clement, J. (1982). Students' preconceptions in introductory mechanics. *American Journal of Physics* 50: 66–71.

Clement, J. (1983). A conceptual model discussed by Galileo and intuitively used by physics students. In D. Genter & A.L. Stevens (eds.), *Mental Models*, Erlbaum, Hillsdale, NJ, pp. 325–339.

Coady, T. (ed.) (2000). *Why Universities Matter*, Allen & Unwin, Sydney.

Cochaud, G. (1989). The process skills of science. Unpublished paper, Australian Science Teachers Association Annual Conference.

Coelho, V. (ed.) (1992). *Music and Science in the Age of Galileo*, Kluwer Academic Publishers, Dordrecht.

Cohen, H.F. (1994). *The Scientific Revolution: A Historiographical Inquiry*, University of Chicago Press, Chicago.

Cohen, I.B. (1980). *The Newtonian Revolution*, Cambridge University Press, Cambridge.

Collier, B. & MacLachlan, J. (1998). *Charles Babbage and the Engines of Perfection*, Oxford University Press, New York.

Collingwood, R.G. (1945). *The Idea of Nature*, Oxford University Press, Oxford.

Collins, A. (1989). Assessing biology teachers: Understanding the nature of science and its influence on the practice of teaching. In D.E. Herget (ed.), *The History and Philosophy of Science in Science Teaching*, Florida State University, pp. 61–70.

Collins, A. (1995). National science education standards in the United States: A process and a product. *Studies in Science Education* 26: 7–37.

Collins, A. (1998). National science education standards: A political document, *Journal of Research in Science Teaching* 35(7): 711–727.

Collins, H.M. & Pinch, T. (1992). *The Golem: What Everyone Should Know about Science*, Cambridge University Press, Cambridge.

Comenius, J.A. (1657/1610). *The Great Didactic*, translated from the Latin, and edited by, M.W. Keatinge, Adam and Charles Black, London.

Conant, J.B. (1945). *General Education in a Free Society: Report of the Harvard Committee*, Harvard University Press, Cambridge, MA.

Conant, J.B. (1951). *Science and Common Sense*, Yale University Press, New Haven.

Conant, J.B. (ed.) (1957). *Harvard Case Histories in Experimental Science*, 2 vols., Harvard University Press, Cambridge (orig. 1948).

Conlin, M.F. (1999). The popular and scientific reception of the Foucault pendulum in the United States. *Isis* 90(2): 181–204.

Conroy, J. (1994). Effect on science A-levels of growth in balanced science GCSE. *Education in Science* April: 29.

Cook, A. (1998). *Edmond Halley: Charting the Heavens and the Seas*, Clarendon Press, Oxford.

Cooper, H.J. (1948). *Scientific Instruments Pt.II*, Hutchinson, London.

Cossman, G.W. (1989). A comparison of the image of science found in two future oriented guideline documents for science education. In D.E. Herget (ed.), *The History and Philosophy of Science in Science Teaching*, Florida State University, Tallahassee, FL., pp. 83–105.

Costabel, P. (1975). Mathematics and Galileo's inclined plane experiments. In M.L.R. Bonelli & W. Shea (eds.), *Reason, Experiment, and Mysticism*, Macmillian, London, pp. 177–187.

Cousins, F.W. (1969). *Sundials: A Simplified Approach by Means of the Equatorial Dial*, John Baker, London.

Cover, J.A. (1997). Non-basic time and reductive strategies: Leibniz's theory of time. *Studies in the History and Philosophy of Science* 28: 289–318.

Crombie, A.C. (1952). *Augustine to Galileo*, Heinemann, London.

Crombie, A.C. (1970). Premio Galileo 1968, *Physis* xii, 106–108. Reproduced in A.C. Crombie, *Science, Optics and Music in Medieval and Early Modern Thought*, The Hambledon Press, London, 1990, pp. 359–362.

Crombie, A.C. (1981). Philosophical presuppositions and the shifting interpretations of Galileo. In J. Hintikka et al. (eds.), *Theory Change, Ancient Axiomatics, and Galileo's Methodology*, Reidel, Boston, pp. 271–286. Reproduced in A.C. Crombie, *Science, Optics and Music in Medieval and Early Modern Thought*, The Hambledon Press, London, 1990, pp. 345–362.

Crombie, A.C. (1990). *Science, Optics, and Music in Medieval and Early Modern Thought*, The Hambledon Press, London.

Crombie, A.C. (1994). *Styles of Scientific Thinking in the European Tradition*, three volumes, Duckworth, London.

Cromer, A. (1993). *Uncommon Sense: The Heretical Nature of Science*, Oxford University Press, New York.

Cronin, V. (1961). *The Wise Man from the West*, R. Hart-Davis, London.

Crosby, A.W. (1997). *The Measure of Reality: Quantification and Western Society, 1250–1600*, Cambridge University Press, Cambridge.

Crowe, M.J. (1990). *Theories of the World from Antiquity to the Copernican Revolution*, Dover Publications, New York.

Crowley, F. (1998). *Degrees Galore: Australia's Academic Teller Machines*, Wild & Woolley, Sydney.

Dagher, Z.R. (1994). Does the use of analogies contribute to conceptual change? *Science Education* 78(6): 601–614.

Dahlin, B. (2001). The primacy of cognition or perception? A phenomenological critique of the theoretical basis of science education. *Science & Education* 10(6).

Dale, R. (1992). *Timekeeping*, Oxford University Press, Oxford.

Damerow, P., Freudenthal, G., McLaughlin, P., & Renn, J. (1992). *Exploring the Limits of Preclassical Mechanics. A Study of Conceptual Development in Early Modern Science: Free Fall and Compounded Motion in the Work of Descartes, Galileo, and Beeckman*, Springer, New York.

Dante, A. (1310/1967). *The Divine Comedy*, H. Longfellow (trans.), Dolphin Books, New York.

Darling-Hammond, L. & Hudson, L. (1990). Precollege science and mathematics teachers: Supply, demand and quality', *Review of Educational Research* 16: 223–264.

Daumas, M. (1972). *Scientific Instruments of the Seventeenth and Eighteenth Centuries and Their Makers*, B.T. Batsford, London.

Davies, A.C. (1978). The life and death of a scientific instrument: The marine chronometer, 1770–1920, *Annals of Science* 35: 509–525.

Davies, P. (1983). *God and the New Physics*, Simon and Schuster, New York.

Davies, P. (1992). *The Mind of God: Science and the Search for Ultimate Meaning*, Simon and Schuster, New York.

Davson-Galle, P. (1990). Philosophy of science done in the "philosophy for children" manner in lower-secondary schools. In D.E. Herget (ed.), *More The History and Philosophy of Science in Science Teaching*, Florida State University, Tallahassee, pp. 223–230.

Dawkins, K. & Glatthorn, A. (1998). Using historical case studies in biology to explore the nature of science: A professional development program for high school teachers. In W.F. McComas (ed.), *The Nature of Science in Science Education: Rationales and Strategies*, Kluwer Academic Publishers, Dordrecht, pp. 163–176.

Dear, P.R. (1988). *Mersenne and the Learning of the Schools*, Ithaca, Cornell University Press.

Dearden, R.F. (1967). Instruction and learning by discovery. In R.S. Peters (ed.), *The Concept of Education*, Routledge & Kegan Paul, London, pp.135–155.

Deason, G.B. (1986). Reformation theology and the mechanistic conception of nature. In D.C. Lindberg & R.L. Numbers (eds.), *God & Nature*, University of California Press, Berkeley, pp. 167–191.

Descartes, R. (1644/1983). *Principles of Philosophy*, V.R. & R.P. Miller (trans.), Reidel, Dordrecht.

Devitt, M. (1991). *Realism & Truth*, 2nd ed., Basil Blackwell, Oxford.

Devons, S. & Hartman, L. (1970). A history of physics laboratory. *Physics Today* 23: 44–49.

Dewey, J. (1910). Science as subject-matter and as method. *Science* 31: 121–127. Reproduced in *Science & Education* 1995, 4(4): 391–398.

Dewey, J. (1916). *Democracy and Education*, Macmillan Company, New York.

Di Sessa, A.A. (1982). Unlearning Aristotelian physics: A study of knowledge-based learning. *Cognitive Science* 6: 37–75.

Diffie, B.W. & Winius, G.D. (1977). *Foundations of the Portuguese Empire, 1415–1580*, University of Minnesota Press, Minneapolis.

Dijksterhuis, E.J. (1961/1986) *The Mechanization of the World Picture*, Princeton University Press, Princeton, NJ.

Dillenberger, J. (1961). *Protestant thought & natural science: A historical study,* Collins, London.

Dilworth, C. (1996). *The Metaphysics of Science. An Account of Modern Science in Terms of Principles, Laws and Theories*, Kluwer Academic Publishers, Dordrecht.

Dobson, R.D. (1985). Galileo Galilei and Christiaan Huygens. *Antiquarian Horology* 15: 261–270.

Dougherty, J.P. (1991). Abstraction and imagination in human understanding. In D.O. Dahlstrom (ed.), *Nature and Scientific Method*, Catholic University of America Press, Washington, D.C., pp. 51–62.

Drake, E.T. (1996). *Restless Genius: Robert Hooke and His Early Thoughts*, Oxford University Press, Oxford.

Drake, S. (1967). Mathematics, astronomy and physics in the work of Galileo. In C.S. Singleton (ed.), *Art, Science, and History in the Renaissance*, Johns Hopkins Press, Baltimore, pp. 305–330.

Drake, S. (1974). Mathematics and discovery in Galileo's physics. *Historia Mathematica* 1: 129–150.

Drake, S. (1976). The evolution of *De Motu. Physis* 14: 321–348.

Drake, S. (1978). *Galileo At Work*, University of Chicago Press, Chicago. (Reprinted Dover Publications, New York, 1996.)

Drake, S. (1980). *Galileo*, Oxford University Press, Oxford.

Drake, S. (1990). The laws of pendulum and fall. In his *Galileo: Pioneer Scientist*, University of Toronto Press, Toronto, pp. 9–31.

Drake, S. (1992). Music and philosophy in early modern science. In V. Coelho (ed.), *Music and Science in the Age of Galileo*, Kluwer Academic Publishers, Dordrecht, pp. 3–16.

Drake, S. & Drabkin, I.E. (eds.). (1969). *Mechanics in Sixteenth-Century Italy*, University of Wisconsin Press, Madison, WI.

Driver, R. & Bell, B. (1986). Students' thinking and the learning of science: A constructivist view. *School Science Review* 67: 443–456.

Driver, R., Asoko, H., Leach, J., Mortimer, E., & Scott, P. (1994). Constructing scientific knowledge in the classroom. *Educational Researcher* 23(7): 5–12.

Drover, C. (1954). Medieval monastic water clocks. *Antiquarian Horology* 5: 54–58.

Dugas, R. (1988). *A History of Mechanics*, Dover, New York. (orig. 1955).

Duhem, P. (1906/1954). *The Aim and Structure of Physical Theory*, trans. P.P. Wiener, Princeton University Press, Princeton.

Duhem, P. (1913). *Etudes sur Léonard de Vinci*, 3 vols., Paris.

Duhem, P. (1916/1991). *German Science*, trans. J. Lyon & S.L. Jaki, Open Court Publishers, La Salle, IL.

Duit, R. (1991). The role of analogy and metaphor in learning science. *Science Education* 75: 649–672.

Duit, R., Roth, W.-M., Komorek, M., & Wilbers, J. (1998). Conceptual change cum discourse analysis to understand cognition in a unit on chaotic systems: Towards an integrative perspective on learning in science. *International Journal of Science Education* 20(9): 1059–1073.

Duncan, D.E. (1998). *The Calendar: The 5,000 Years Struggle to Align the Clock and Heavens, and What Happened to the Missing Ten Days*, Fourth Estate Publishers, London.

Dunham, W. (1990). *Journey through Genius: The Great Theorems of Mathematics*, Penguin, London.

Duschl, R.A. (1985). Science education and philosophy of science: Twenty-five years of mutually exclusive development. *School Science and Mathematics* 87(7): 541–555.

Duschl, R.A. (1990). *Restructuring Science Education: The Importance of Theories and Their Development*, Teachers College Press, New York.

Eckstein, S.G. & Kozhevnikov, M. (1997). Parallelism in the development of children's ideas and the historical development of projectile motion Theories. *International Journal of Science Education* 19(9): 1057–1073.

Eddington, A. (1928/1978). *The Nature of the Physical World*, University of Michigan Press, Ann Arbor.

Edwardes, E.L. (1977). *The Story of the Pendulum Clock*, J. Sherratt, Altrincham.

Edwardes, E.L. (1980). The suspended foliot and new light on early pendulum clocks. *Antiquarian Horology* 12: 614–626.

Egan, K. (1986). *Teaching As Story Telling*, University of Chicago Press, Chicago.

Egan, K. (1997). *The Educated Mind: How Cognitive Tools Shape Our Understanding*, University of Chicago Press, Chicago.

Eger, M. (1987). Philosophy of science in teacher education. In J.D. Novak (ed.), *Misconceptions and Educational Strategies*, Cornell University, vol. I, pp. 163–176.

Egner, R.E. & Denonn, L.E. (eds.). (1961). *The Basic Writings of Bertrand Russell*, George Allen & Unwin, London.

Eijkelhof, H.M.C. & Kortland, K. (1988). Broadening the aims of physics education. In P. Fensham (ed.), *Development and Dilemmas in Science Education*, Falmer Press, London, pp. 282–305.

Einstein, A. & Infeld, L. (1938/1966). *The Evolution of Physics*, Simon and Schuster, New York.

Eisenhart, M., Finkel, E. & Marion, S.F. (1996). Creating the conditions for scientific literacy: A re-examination. *American Educational Research Journal* 33(2): 261–295.

Eisner, E.W. (1992). Curriculum ideologies. In P.W. Jackson (ed.), *Handbook of Research on Curriculum*, Macmillan, New York, pp. 302–326.

Elam, S. (ed.). (1964). *Education and the Structure of Knowledge*, Rand McNally, Chicago.

Elbers, G.W. & Duncan, P. (eds.). (1959). *The Scientific Revolution: Challenge and Promise*, Public Affairs Press, Washington D.C.

Elkana, Y. (ed.) (1974a). *The Interaction Between Science and Philosophy*, Humanities Press, Atlantic Highlands, NJ.

Elkana, Y. (1974b). *The Discovery of the Conservation of Energy*, Hutchinson Educational, London.

Ellis, B.D. (1965). The origin and nature of Newton's laws of motion. In R.G. Colodny (ed.), *Beyond the Edge of Certainty*, Englewood Cliffs, NJ., pp. 29–68.

Emerson, R.L. (1990). The organisation of science and its pursuit in early modern Europe. In R.C. Olby et al. (eds.), *Companion to the History of Modern Science*, Macmillan, New York, pp. 960–979.

Ennis, R.H. (1979). Research in philosophy of science bearing on science education. In P.D. Asquith & H.E. Kyburg (eds.), *Current Research in Philosophy of Science*, PSA, East Lansing, pp. 138–170.

Fantoli, A. (1994). *Galileo: For Copernicanism and for the Church*, (G.V. Coyne trans.), Vatican Observatory Publications, Vatican City. (Distributed by University of Notre Dame Press.)

Fauvel, J. & Gray, J. (eds.) (1987). *The History of Mathematics: A Reader*, Macmillan, London.

Fauvel, J., Flood, R., Shortland, M. & Wilson, R. (eds.) (1988). *Let Newton Be!*, Oxford University Press, Oxford.

Fensham, P.J., Gunstone, R., & White, R. (eds.) (1994). *The Content of Science: A Constructivist Approach to its Teaching and Learning*, Falmer Press, London.

Fensham, P.J. (1992), Science and technology. In P.W. Jackson (ed.), *Handbook of Research on Curriculum*, Macmillan, New York, pp. 789–829.

Fermi, L. & Bernadini, G. (1961). *Galileo and the Scientific Revolution*, Basic Books, New York.

Feyerabend, P.K. (1975). *Against Method*, New Left Books, London.

Findlay, J.N. (1968). Time: A treatment of some puzzles. In A.G.N. Flew (ed.), *Logic and Language, First Series*, Basil Blackwell, Oxford, pp. 37–54.

Finley, F., Allchin, A., Rhees, D. & Fifield, S. (eds.) (1995). *Proceedings of the Third International History, Philosophy and Science Teaching Conference*, University of Minnesota, Minneapolis.

Finocchiaro, M.A. (1980). *Galileo and the Art of Reasoning*, Dordrecht, Reidel.

Fisher, A. (1992). Why Johnny can't do science and math. *Popular Science* 241(3): 50–55.

Fleck, L. (1935/1979). *Genesis and Development of a Scientific Fact*, T.J.Trenn and R.K.Merton (eds.), University of Chicago Press, Chicago.

Fosnot, C.T. (ed.) (1996). *Constructivism: Theory, Perspectives, and Practice*, Teachers College Press, New York.

Fraassen, B.C. van (1970). *An Introduction to the Philosophy of Space and Time*, Random House, New York.

Fraisse, P. (1963). *The Psychology of Time*, Harper & Row, New York.

Fraisse, P. (1982). The adaptation of the child to time. In W.J. Friedman (ed.), *The Developmental Psychology of Time*, Academic Press, New York.

Franklin, A. (1986). *The Neglect of Experiment*, Cambridge University Press, Cambridge.

Franklin, J. (1999). Diagrammatic reasoning and modelling in the imagination: The secret weapons of the scientific revolution. In G. Freeland & A. Corones (eds.), *1543 And All That: Image and Word, Change and Contintuity in the Proto-Scientific Revolution*, Kluwer Academic Publishers, Dordrecht, pp. 53–115.

Fraser, J.T. (ed.) (1966). *The Voices of Time*, G. Braziller, New York.

Fraser, J.T. (ed.) (1981). *The Voices of Time: A Cooperative Survey of Man's Views of Time as Expressed by the Sciences and the Humanities*, 2nd edit., University of Massachusetts Press, Amherst.

Fraser, J.T. (1982). *The Genesis and Evolution of Time*, The Harvester Press, Brighton.

Fraser, J.T. (1987). *Time the Familiar Stranger*, Microsoft Press, Redmond, WA.

Fraser, J.T. & Lawrence, N., & Park (eds.) (1975). *The Study of Time*, Springer-Verlag, New York.

Fraser, J.T., Lawrence, N., & Park, D. (eds.) (1978). *The Study of Time III*, Springer-Verlag, New York.

Fraser, J.T., Lawrence, N., & Park, D. (eds.) (1981). *The Study of Time IV*, Springer-Verlag, New York.

Fredette, R. (1972). Galileo's *De Motu Antiquiora.. Physis* 14: 321–348.

Freier, G.D. (1965). *University Physics: Experiment and Theory*, Appleton-Century-Crofts, New York.

French, A.P. (1965). *Newtonian Mechanics*, W.W. Norton & Co., New York.

French, A.P. (1983). Pleasures and dangers of bringing history into physics teaching. In F. Bevilacqua & P.J. Kennedy (eds.), *Using History of Physics in Innovatory Physics Education*, Proceedings of an International Conference, Pavia, Italy, pp. 211–243.

French, A.P. (1989). Learning from the past; Looking to the future. *American Journal of Physics* 57: 587–592.

Freudenthal, G. (1988). Towards a social history of Newtonian mechanics. Boris Hessen and Henryk Grossmann revisited. In I. Hronszky, M. Feher, and & B. Dajka (eds), *Scientific Knowledge Socialized*, Kluwer Academic Publishers, Dordrecht, pp.193–211.

Friedman, W.J. (ed.) (1982). *The Developmental Psychology of Time*, Academic Press, New York.

Fuller, S. (2000). From Conant's Education Strategy to Kuhn's Research Strategy, *Science & Education* 9(1–2):21–37.

Galileo, G. (1586/1961). *La Bilancetta*. In L. Fermi and G. Bernardini, *Galileo and the Scientific Revolution*, Basic Books, New York, pp. 133–140.

Galileo, G. (1590/1960). *De Motu*. In I.E. Drabkin and S. Drake (eds), *Galileo Galilei On Motion and On Mechanics*, University of Wisconsin Press, Madison, WI., pp. 13–114.

Galileo, G. (1600/1960). *On Mechanics*. In I.E. Drabkin and S. Drake (eds), *Galileo Galilei On Motion and On Mechanics*, University of Wisconsin Press, Madison, WI., pp. 147–182.

Galileo, G. (1615/1957). Letter to Madame Christina of Lorraine, Grand Duchess of Tuscany concerning the use of Biblical quotations in matters of science. In S. Drake (ed.), *Discoveries and Opinions of Galileo*, Doubleday, New York, pp. 175–216.

Galileo, G. (1623/1957). *The Assayer*. In S. Drake (ed.), *Discoveries and Opinions of Galileo*, Doubleday, New York, pp. 229–280.

Galileo, G. (1633/1953). *Dialogue Concerning the Two Chief World Systems*, S. Drake (trans.), University of California Press, Berkeley. (second revised edition, 1967)

Galileo, G. (1638/1954). *Dialogues Concerning Two New Sciences*, trans. H. Crew & A. de Salvio, Dover Publications, New York (orig. 1914).

Galison, P. (1987). *How Experiments End*, University of Chicago Press, Chicago.

Gallagher, L.L. (1961). *China in the Sixteenth Century: The Journals of Matthew Ricci, 1583–1610*, Random House, New York.

Garber, D. (1986). Learning from the past: Reflections on the role of history in the philosophy of science. *Synthese* 67(1): 91–114.

Gaukroger, S. (1981). Aristotle on the function of sense perception. *Studies in History and Philosophy of Science* 12: 75–89.

Gauld, C.F. (1977). The role of history in the teaching of science. *Australian Science Teachers Journal* 23(3): 47–52.

Gauld, C.F. (1991). History of science, individual development and science teaching. *Research in Science Education* 21: 133–140.

Gauld, C.F. (1995). The Newtonian solution to the problem of impact in the 17th and 18th centuries and the teaching of Newton's third law today. In F. Finley et al. (eds.), *Proceedings of the Third International History, Philosophy, and Science Teaching Conference*, University of Minneapolis, MN., volume one, pp. 441–452.

Gauld, C.F. (1998a). Solutions to the problem of impact in the 17th and 18th Centuries and Teaching Newton's Third Law Today, *Science & Education* 7(1): 49–67.

Gauld, C.F. (1998b). Colliding pendulums, conservation of momentum and Newton's Third Law. *Australian Science Teachers Journal* 44(3): 37–38.

Geary, D.C. (1995). Reflections of evolution and culture in children's cognition: Implications for mathematical development and instruction, *American Psychologist* 50(1), 24–37.

Geymonat, L. (1957/1965). *Galileo Galilei: A Biography and Inquiry into His Philosophy of Science*, McGraw-Hill, New York.

Giere, R.N. (1988). *Explaining Science: A Cognitive Approach*, University of Chicago Press, Chicago.

Giere, R.N. (1994). The Cognitive structure of scientific theories. *Philosophy of Science* 64: 276–296.

Giere, R.N. (1999). *Science Without Laws*, University of Chicago Press, Chicago.

Gill, M.L. & Lennox, J.G. (eds.) (1994). *Self-Motion from Aristotle to Newton*, Princeton University Press, Princeton, NJ.

Gillispie, C.C. (1960). *The Edge of Objectivity*, Princeton University Press, Princeton.

Gimpel, J. (1992). *The Medieval Machine: The Industrial Revolution of the Middle Ages*, Pimplico, London.

Gingerich, O. (1998). Cranks and opportunists: "Nutty" solutions to the longitude problem. In W.J.H. Andrewes (ed.), *The Quest for Longitude*, The Collection of Historical Scientific Instruments, Harvard University, Cambridge, MA, 2nd Edition, pp. 134–148.

Gjertsen, D. (1986). *The Newton Handbook*, Routledge, London.

Gjertsen, D. (1989). *Science and Philosophy: Past and Present*, Penguin, Harmondsworth.

Glasersfeld, E. von (1989). Cognition, construction of knowledge and teaching. *Synthese* 80(1): 121–140.

Goff, J. le (1980). *Time, Work and Culture in the Middle Ages*, trans. Arthur Goldhammer, Chicago University Press, Chicago.

Good, R.G. & Demastes, S. (1995). The diminished role of nature in postmodern views of science and science education. In F. Finley et al. (eds.), *Proceedings of the Third International History, Philosophy, and Science Teaching Conference*, University of Minnesota, Minneapolis, pp. 480–487.

Good, R.G., Wandersee, J., and St. Julien, J. (1993). Cautionary notes on the appeal of the new "ism" (constructivism) in science education. In K. Tobin (ed.), *Constructivism in Science and Mathematics Education*, AAAS, Washington DC, pp. 71–90.

Good, R.G. & Shymansky, J. (2001). Nature-of-science literacy in *Benchmarks* and *Standards*: postmodern/relativist or modern/realist? *Science & Education* 10(6).

Goodman, D. & Russell, C.A. (eds.) (1991). *The Rise of Scientific Europe*, Hodder & Straughton, London.

Goodman, D. (1991). Iberian science: Navigation, empire and counter-reformation. In D. Goodman & C. A. Russell (eds.), *The Rise of Scientific Europe*, Hodder & Straughten, London, pp. 117–144.

Goudsmit, S.A. & Clariborne, R. (eds.) (1966). *Time*, Time-Life Books, Netherlands.

Gould, R. T. (1923). *The Marine Chronometer, Its History and Development*, J.D. Potter, London. Reprinted by The Holland Press, London, 1978.

Graham, L.R. (1993). *The Ghost of the Executed Engineer. Technology and the Fall of the Soviet Union*, Harvard University Press, Cambridge, MA.

Graham, P.A. (1998). Educational dilemmas for Americans. *Daedalus* 127(1): 225–236.

Gramsci, A. (1971). On education. In Q. Hoare & G. Nowell-Smith (eds.), *Selections from the Prison Notebooks of Antonio Gramsci*, International Publishers, New York, pp. 24–43.

Grant, E. (1964). Motion in the void and the principle of inertia in the Middle Ages. *Isis* 55: 265–292.

Grant, E. (1971). *Physical Science in the Middle Ages*, Cambridge University Press, Cambridge.

Grant, J. (1980). *The Book of Time*, Jacaranda, Milton.

Green, A. R. (1926). *Sundials*, Macmillan, New York.

Greenberg, J.L. (1995). *The Problem of the Earth's Shape from Newton to Clairaut: The Rise of Mathematical Science in Eighteenth-Century Paris and the Fall of 'Normal' Science*, Cambridge University Press, Cambridge.

Gross, P.R. & Levitt, N. (1994). *Higher Superstition: The Academic Left and Its Quarrels with Science*, Johns Hopkins University Press, Baltimore.

Gross, P.R., Levitt, N., & Lewis, M.W. (eds.). (1996). *The Flight from Science and Reason*, New York Academy of Sciences, New York, (distributed by Johns Hopkins University Press, Baltimore).

Grout, D.J. & Palisca, C.V. (1960). *A History of Western Music*, W.W. Norton, New York.

Grove, J.W. (1989). *In Defence of Science. Science, Technology and Politics in Modern Society*, University of Toronto Press, Toronto.

Guerlac, H. (1981). *Newton on the Continent*, Cornell University Press, Ithaca.

Haber, F.C. (1975). The cathedral clock and the cosmological clock metaphor. In J.T. Fraser & N. Lawrence (eds.), *The Study of Time II*, Springer-Verlag, Berlin, pp. 399–416.

Hacking, I. (1983). *Representing and Intervening*, Cambridge University Press, Cambridge.

Hacking, I. (1992). "Style" for historians and philosophers. *Studies in History and Philosophy of Science* 23(1): 1–20.

Hackmann, W.D. (1979). The relationship between concept and instrument design in eighteenth-century experimental science. *Annals of Science* 36: 205–224.

Hall, A.R. (1950). Robert Hooke and horology. *Notes and Records of the Royal Society of London* 8: 167–177.

Hall, A.R. (1957). Newton on the calculation of central forces. *Annals of Science* 13: 62–71.

Hall, A.R. (1963). *From Galileo to Newton: 1630–1720*, Harper & Row, New York.

Hall, A.R. (1965). Galileo and the science of motion. *British Journal for the History of Science* 2: 185–199.

Hall, A.R. (1970). On the historical singularity of the scientific revolution of the seventeenth century. In J.H. Elliot & H.G. Koenigsberger (eds.), *The Diversity of History: Essays in Honour of Sir Herbert Butterfield*, Routledge Kegan Paul, London, pp. 201–221.

Hall, A.R. (1978). Horology and criticism: Robert Hooke. *Studia Copernicana* 16: 261–281.

Hall, A.R. (1980). *Philosophers at War*, Cambridge University Press, Cambridge.

Hall, A.R. (1992). *Isaac Newton. Adventurer in Thought*, Blackwell Publishers, Oxford.

Hall, B.S. (1976). The new Leonardo. *Isis* 67: 463–475.

Hall, B.S. (1978). The scholastic pendulum. *Annals of Science* 35: 441–462.

Hall, R. (1951). The history of time. *Discovery* 12: 1–78.

Halloun, I. & Hestenes, D. (1985). The initial knowledge State of College pysics students. *American Journal of Physics* 53: 1043–1055.

Hamburg, D.A. (1992). *Today's Children: Creating a Future for a Generation in Crisis*, Random House, New York.

Hand, B., Treagust, D.F., & Vance, K. (1997). Student perceptions of the social constructivist classroom. *Science Education* 81(5): 561–575.

Hanson, N.R. (1958). *Patterns of Discovery*, Cambridge University Press, Cambridge.

Hanson, N.R. (1959). Broad and the Laws of Dynamics. In P.A. Schilpp (ed.), *The Philosophy of C.D. Broad*, Tudor Publishing Company, New York, pp. 281–312.

Hanson, N.R. (1965a). Aristotle (and others) on motion through air. *Review of Metaphysics* 19: 133–147.

Hanson, N.R. (1965b). Newton's first law: A philosopher's door into natural philosophy. In R.G. Colodny (ed.), *Beyond the Edge of Certainty*, Prentice Hall, Englewood-Cliffs, NJ, pp. 6–28.

Harding, S.G. (1986). *The Science Question in Feminism*, Cornell University Press, Ithaca.

Harding, S.G. (1991). *Whose Science? Whose Knowledge?*, Cornell University Press, Ithaca.

Harding, S.G. (1994). After the neutrality ideal: Science, politics, and "strong objectivity". In M.C. Jacob (ed.), *The Politics of Western Science*, Humanities Press, NJ., pp. 81–101.

Harley, J.B. & Woodward, D. (eds.) (1987). *The History of Cartography*, vol. 1, University of Chicago Press, Chicago.

Harman, P.M. (1982). *Energy, Force and Matter: The Conceptual Development of Nineteenth-Century Physics,* Cambridge University Press, Cambridge.

Harner, L. (1982). Talking about the past and future. In W.J. Friedman (ed.), *The Developmental Psychology of Time*, Academic Press, New York, pp. 141–170.

Harper, W. & Smith, G.E. (1995). Newton's new way of inquiry. In J. Leplin (ed.) *The Creation of Ideas in Physics: Studies for a Methodology of Theory Construction*, Kluwer Academic Publishers, Dordrecht, pp. 113–166.

Harré, R. (1964). *Matter and Method*, Macmillan & Co., London.

Hart, H.H. (1950). *Sea Route to the Indies*, W. Hodge, London.

Hastings, P. (1993). A look at the grasshopper escapement. *Horological Journal* 136(2): 48–53.

Hatfield, G. (1990). Metaphysics and the new science. In D.C. Lindberg & R.S. Westman (eds.) *Reappraisals of the Scientific Revolution*, Cambridge University Press, Cambridge, pp. 93–166.

Hawking, S.W. (1988). *A Brief History of Time*, Bantam Books, London.

Heath, T. (1913). *Aristarchus of Samos: The Ancient Copernicus*, Clarendon Press, Oxford. (Reprinted, Dover 1981.)

Heath-Stubbs, J. & Salman, P. (1984). *Poems of Science*, Penguin, Harmondsworth.

Hegel, G.W.F. (1810/1910). *The Phenomenology of Mind*, (J.B. Baillie, trans., ed.), Harper and Row, New York.

Hegel, G.W.F. (1842/1970). *Hegel's Philosophy of Nature*, (M.J. Petry, trans., ed.), George Allen & Unwin, London.

Heilbron, J.L. (1986). *The Dilemmas of an Upright Man: Max Planck as Spokesman for German Science*, University of California Press, Berkeley.

Heilbron, J.L. (1989). The politics of the meter stick. *American Journal of Physics* 57: 988–992.

Helden, A. van: 1998, Longitude and the Satellites of Jupiter. In W.J.H. Andrewes (ed.), *The Quest for Longitude*, The Collection of Historical Scientific Instruments, Harvard University, Cambridge, MA, 2nd Edition, pp. 86–100.

Helvétius, C. (1758/1810). *Essays on the Mind*, translated from the French, Albion, London.

Hemmendinger, D. (1984). Galileo and the phenomena: On making the evidence visible. In R.S. Cohen & M.W. Wartofsky (eds.), *Physical Sciences and the History of Physics*, Reidel, Dordrecht, pp. 115–143.

Hempel, C.G. (1945). Studies in the logic of confirmation. *Mind* 54: 1–26; 97–121. Reprinted in his *Aspects of Scientific Explanation*, Macmillan, New York, 1965, pp. 3–51.

Hempel, C.G. (1965). *Aspects of Scientific Explanation*, Macmillan, New York.

Hempel, C.G. & Oppenheim, P. (1948). Studies in the logic of explanation. *Philosophy of Science* 15: 135–175. Reprinted in his *Aspects of Scientific Explanation*, Macmillan, New York, 1965, pp. 245–290.

Herget, D.E. (ed.) (1989). *The History and Philosophy of Science in Science Teaching*, Florida State University, Tallahassee FL.

Herget, D.E. (ed.) (1990). *The History and Philosophy of Science in Science Teaching*, Florida State University, Tallahassee FL.

Herivel, J. (1965). *The Background to Newton's 'Principia'*, Clarendon Press, Oxford.

Hesse, M.B. (1961). *Forces and Fields: The Concept of Action at a Distance in the History of Physics*, Thomas Nelson & Sons, London.

Hesse, M.B. (1963). *Models and Analogies in Science*, London.

Hessen, B.M. (1931). The social and economic roots of Newton's *Principia'*. In *Science at the Crossroads*, Kniga, London. Reprinted in G. Basalla (ed.) *The Rise of Modern Science: External or Internal Factors?*, D.C. Heath & Co., New York, 1968, pp. 31–38.

Hewson, P.W. (1985). Epistemological commitments in the learning of science: Examples from dynamics. *European Journal of Science Education* 7(2): 163–172.

Hiebert, E.N. (1962). *Historical Roots of the Principle of Conservation of Energy*, University of Wisconsin Press, Madison.

Higgins, A. (1989). Medieval Notions of the Structure of Time. *Journal of Medieval and Renaissance Studies* 19: 227–250.

Hill, D. (ed. & trans.) (1976). *Archimedes "On the Construction of Water-Clocks,"* Turner & Devereux, Paris.

Hill, D. (1993). *Islamic Science and Engineering*, Edinburgh University Press, Edinburgh.

Hill, D, & al-Hassan, A. (1986). *Islamic Technology: An Illustrated History*, Cambridge University Press, London.

Hills, S. (ed.) (1992). *The History and Philosophy of Science in Science Education*, two volumes, Queen's University, Kingston.

Hirsch, E.D. (1996). *The Schools We Need & Why We Don't Have Them*, Doubleday, New York.

Hirst, P.H. (1974). Liberal education and the nature of knowledge. In his *Knowledge and the Curriculum*, Routledge & Kegan Paul, London, pp. 16–29.

Hirst, P.H. (1991). Educational aims: Their nature and content. *Philosophy of Education*, 1991: 40–53.

Hobbes, T. (1651/1962). *Leviathan*, J. Plamenatz (ed.), Collins, London.

Hodson, D. (1986). Philosophy of science and the science curriculum. *Journal of Philosophy of Education* 20: 241–251. Reprinted in M.R. Matthews (ed.), *History, Philosophy and Science Teaching: Selected Readings*, OISE Press, Toronto, 1991, pp. 19–32.

Hodson, D. (1988). Toward a philosophically more valid science curriculum. *Science Education* 72: 19–40.

Hodson, D. (1992). Assessment of practical work: Some considerations in philosophy of science. *Science & Education* 1(2): 115–144.

Hogben, L. (1936). *Mathematics for the Millions*, George, Allen & Unwin, London.

Hogben, L. (1940). *Science for the Citizen*, 2nd edition, George, Allen & Unwin, London. (First edition, 1938.)

Hogben, L. (1973). *Maps, Mirrors and Mechanics*, Heinemann, London.

Holmes, R. (1664). A narrative concerning the success of pendulum-watches at sea for the longitudes. *Philosophical Transactions* 1: 13–15.

Holton, G. et al. (1967). Symposium on the Project Physics Course. *The Physics Teacher* 5(5): 196–231.

Holton, G. (1952). *Introduction to Concepts and Theories in Physical Science*, Princeton University Press, Princeton. Second edition (revised with S.G. Brush) 1985.

Holton, G. (1978). On the educational philosophy of the project physics course. In his *The Scientific Imagination: Case Studies*, Cambridge University Press, Cambridge, pp. 284-298.

Holton, G. (1985). *Introduction to Concepts and Theories in Physical Science*, second edition, revised with Stephen G. Brush, Addison-Wesley, New York. (First edition, 1973)

Holton, G. (1986). Metaphors in science and education. In his *The Advancement of Science and Its Burdens*, Cambridge University Press, Cambridge, pp. 229–252.

Holton, G. (1993). *Science and Anti-Science*, Harvard University Press, Cambridge, MA.

Holton, G. (1995). How can science courses use the history of science? In his *Einstein, History and Other Passions*, American Institute of Physics, Woodbury, NY, pp. 257–264.

Holton, G. (1996). Science education and the sense of self. In P.R. Gross, N. Levitt & M.W. Lewis (eds.), *The Flight from Science and Reason*, New York Academy of Science, New York, pp. 551–560.

Holton, G. & Roller, D.H.D. (1958). *Foundations of Modern Physical Science*, Addison-Wesley, Reading.

Hooke, R. (1705/1969). *The Posthumous Works of Robert Hooke*, R. Waller (ed.), London. Reprinted 1969 with Introduction by R.S. Westfall, Johnson Reprint Corporation.

Hoskin, M. (ed.) (1997). *The Cambridge Illustrated History of Astronomy*, Cambridge University Press, Cambridge.

Hough, R. (1994). *Captain James Cook: A Biography*, Hodder & Stoughton, London.

Howe, A.C. & Stubbs, H.S. (1997). Empowering science teachers: A model for professional development. *Journal of Science Teacher Education* 8(3): 167–182.

Howse, D. (1980). *Greenwich Time and the Discovery of Longitude*, Oxford University Press, Oxford.

Howse, D. (1985). 1884 and longitude zero. *Vistas in Astronomy*, P. Beer, A.J. Meadows, & A.E. Roy (eds.), 11–22.

Howse, D. (1998). The lunar-distance method of measuring longitude. In W.J.H. Andrewes (ed.), *The Quest for Longitude*, The Collection of Historical Scientific Instruments, Harvard University, Cambridge, MA, 2nd Edition, pp. 150-161.

Huff, T.E. (1993). *The Rise of Early Modern Science*, Cambridge University Press, Cambridge.

Hughes, P. (1993). All part of life's rich tapestry. *New Scientist*, December: 47–48.

Hume, D. (1777/1902). *Enquiries Concerning the Human Understanding and Concerning the Principles of Morals*, (L.A. Selby-Bigge, ed.), Clarendon Press, Oxford.

Hume, D. (1779/1963). *Dialogues Concerning Natural Religion*. In R. Wollheim (ed.), *Hume on Religion*, Collins, London, pp. 99–204.

Humphreys, W.C. (1967). Galileo, falling bodies and inclined planes: An attempt at reconstructing Galileo's discovery of the law of squares. *British Journal for the History of Science* 3(11): 225–244.

Hunt, I.E. & Suchting, W.A. (1969). Force and "Natural Motion." *Philosophy of Science* 36: 233–251.

Hunter, M. & Schaffer, S. (eds.) (1989). *Robert Hooke: New Studies*, Boydell Press, Woodbridge, England.

Hurd, P.D. (1961). *Biological Education in American Secondary Schools 1890-1960*, American Institute of Biological Science, Washington, DC.

Hurd, P.D. (1977). Reflections and the quest for perspective. In R.L. Steiner (ed.), *Science Education: Past or Prologue: 1978 AETS Yearbook*, Association for the Education of Science Teachers, Columbus, OH., pp. 106–117.

Hurd, P.D. (1998). Scientific Literacy: New Minds for a Changing World', *Science Education* 82(3): 407–416.

Huxley, A. (1947). *Science, Liberty and Peace*, Chatto & Windus, London.

Huxley, T.H. (1868). A liberal education; and where to find it. In his *Science & Education*, Appleton, New York, 1897 (orig. 1885). Reprinted with Introduction by C. Winick, Citadel Press, New York, 1964, pp. 72–100.

Huygens, C. (1669). Instructions concerning the use of pendulum-watches for finding the longitude at sea. *Philosophical Transactions* 4(47), 937–953.

Huygens, C. (1673/1986). *Horologium Oscillatorium. The Pendulum Clock or Geometrical Demonstrations Concerning the Motion of Pendula as Applied to Clocks*, (R.J. Blackwell, trans.), Iowa State University Press, Ames.

Huygens, C. (1690/1945). *Treatise on Light*, translated and edited by S.P. Thompson, University of Chicago Press, Chicago.

Hyman, A. (1984). *Charles Babbage: Pioneer of the Computer*, Oxford University Press, Oxford.

Inhelder, B. & Piaget, J. (1958). *The Growth of Logical Thinking*, Basic Books, New York. (French original, 1955).

Irzik, G. (1995). Popper's Epistemology and World Three'. In I. Kuçuradi & R.S. Cohen (eds.), *The Concept of Knowledge: The Ankara Seminar*, Kluwer Academic Publishers, Dordrecht, pp. 83–95.

Iverson, S. & Tunmer, W.E. (1992). Phonological processing skills and the reading recovery program. *Journal of Educational Psychology* 85: 112–126.

Jacob, M.C. (1976). *The Newtonians and the English Revolution 1689–1720*, The Harvester Press, Hassocks, Sussex.

Jacob, M.C. (1986). Christianity and the Newtonian Worldview. In D.C. Lindberg & R.L. Numbers (eds.), *God & Nature: Historical Essays on the Encounter between Christianity and Science*, University of California Press, Berkeley, pp. 238–255.

Jaki, S.L. (1974). *Science & Creation*, Scottish Academic Press, Edinburgh.

James, J. (1993). *The Music of the Spheres: Music, Science and the Natural Order of the Universe*, Abacus, London.

Jammer, M. (1954). *Concepts of Space*, Harvard University Press, Harvard.

Jammer, M. (1957). *Concepts of Force: A Study in the Foundations of Dynamics*, Harvard University Press, Cambridge, MA.

Jammer, M. (1961). *Concepts of Mass in Classical and Modern Physics,* Harvard University Press, Cambridge, MA.

Janich, P. (1985). *Protophysics of Time: Constructivie Foundation and History of Time Measurement*, Reidel, Boston.

Jardine, L. (1999). *Ingenious Pursuits: Building the Scientific Revolution*, Little, Brown and Company, 1999.

Jeans, J. (1943/1981). *Physics and Philosophy,* Dover Publications, New York.

Jenkins, E.W. (1990). History of science in schools: Retrospect and prospect in the UK. *International Journal of Science Education* 12(3): 274–281. Reprinted in M.R. Matthews (ed.) *History, Philosophy and Science Teaching: Selected Readings*, OISE Press, Toronto, 1991, pp.33–42.

Jenkins, E.W. (1994). HPS and school science education: Remediation or reconstruction?. *International Journal of Science Education* 16(6): 613–624.

Jenkins, E.W. (1996). The "Nature of Science" as a curriculum Component. *Journal of Curriculum Studies* 28: 137–150.

Jespersen, J. & Fitz-Randolph, J. (1982). *From Sundials to Atomic Clocks*, Dover Publications, New York. (Originally, National Bureau of Standards, Washington, 1977.)

Jones, L.V. (1988). School achievement trends in mathematics and science, and what can be done to improve them. *Review of Research in Education* 15: 307–341.

Jones, W.P. (1966). *The Rhetoric of Science: A Study of Scientific Ideas and Imagery in Eighteenth-Century English Poetry*, Routledge & Kegan Paul, London.

Jonkers, A.R.T. (1996). Finding longitude at sea: Early attempts in Dutch navigation. *De Zeventiende Eeuw* 12(1): 186–197.

Jourdain, P.E. (1913). Robert Hooke as a Precursor of Newton. *Monist*, 23: 353–384.

Jungwirth, E. (1987). Avoidance of logical fallacies: A neglected aspect of science-education and science-teacher education. *Research in Science & Technological Education* 5(1): 43–58.

Kant, I. (1787/1933). *Critique of Pure Reason*, 2nd edit., N.K. Smith (trans.), Macmillan, London.

Kant, I. (1968). *Selected Precritical Writings and Correspondence with Beck*, (G.K. Kenford & D.E. Walford, trans.), Manchester University Press, Manchester.

Kassler, J.C. (1982). Music as model in early science. *History of Science* XX: 103–131.

Kelly, G.J. (1997). Research traditions in comparative context: A philosophical challenge to radical constructivism. *Science Education* 81(3): 355–375.

Kenealy, P. (1989). Telling a coherent Story: A role for the history and philosophy of science in a physical science course. In D.E. Herget (ed.), *The History and Philosophy of Science in Science Teaching*, Florida State University, Tallahassee, pp. 209–220.

Kimball, B. (1986). *A History of the Idea of Liberal Education*, Teachers College Press, New York.

King, A.L. (1998). "John Harrison, Clockmaker at Barrow; Near Barton upon Humber; Lincolnshire": The Wooden Clocks, 1713–1730'. In W.J.H. Andrewes (ed.), *The Quest for Longitude*, The Collection of Historical Scientific Instruments, Harvard University, Cambridge, MA, 2nd Edition, pp. 168–187.

Kipnis, N. (1992). *Rediscovering Optics*, BENA Press, Minneapolis MN.

Kipnis, N. (1996). The "historical-investigative" approach to teaching science. *Science & Education* 5(3): 277–292.

Kitchener, R.F. (1985a). Genetic epistemology, history of science and genetic psychology. *Synthese* 65(1): 3–32.

Kitchener, R.F. (1985b). A Bibliography of Philosophical Work on Piaget. *Synthese* 65(1): 139–151.

Kitchener, R.F. (1986). *Piaget's Theory of Knowledge: Genetic Epistemology and Scientific Reason*, Yale University Press, New Haven.

Kitcher, P. (1988). The child as parent of the scientist. *Mind and Language* 3(3): 217–228.

Kline, H.A. (1988). *The Science of Measurement: A Historical Survey*, Dover Publications, New York.

Kline, M. (1959). *Mathematics and the Physical World*, Dover Publications, New York.

Kline, M. (1972). *Mathematical Thought from Ancient to Modern Times*, 3 vols., Oxford University Press, Oxford.

Klopfer, L.E. (1969). *Case Histories and Science Education*, Wadsworth Publishing Company, San Francisco.

Klopfer, L.E. & Champagne, A.B. (1990). Ghosts of crisis past. *Science Education* 74(2): 133–154.

Knight, D. (1986). *The Age of Science*, Basil Blackwell, Oxford.

Knox, R. (1980). The moving earth in space. In C. Wilson (ed.), *The Book of Time*, Westbridge Books, North Pomfret, Vermont, pp. 45–83.

Koertge, N. (1969). Towards an integration of content and method in the science curriculum. *Curriculum Theory Network* 4: 26–43. Reprinted in *Science & Education* 5(4): 391–402, (1996).

Koertge, N. (1977). Galileo and the problem of accidents. *Journal of the History of Ideas* 38: 389–409.

Koertge, N. (1981). Methodology, ideology and feminist critiques of science. In P. D. Asquith & R. N. Giere (eds.), *Proceedings of the Philosophy of Science Association 1980*, Edwards Bros, Ann Arbor, pp. 346–359.

Koertge, N. (1996). Feminist epistemology: Stalking an un-dead horse. In P.R. Gross, N. Levitt & M.W. Lewis (eds.), *The Flight from Science and Reason*, New York Academy of Sciences, New York, pp. 413–419.

Koestler, A. (1964). The Act of Creation, Hutchinson, London.

Koyré, A. (1939/1978). *Galileo Studies*, (J. Mepham, trans.), Harvester Press, Hassocks, Sussex.

Koyré, A. (1943a). Galileo and the scientific revolution of the seventeenth century, *Philosophical Review* 52: 333–348. Reprinted in his *Metaphysics and Measurement*, 1968, pp. 1–15.

Koyré, A. (1943b). Galileo and Plato. *Journal of the History of Ideas* 4: 400–428. Reprinted in his *Metaphysics and Measurement,*1968, pp. 16–43.

Koyré, A. (1953). An experiment in measurement. *Proceedings of the American Philosophical Society* 7: 222–237. Reproduced in his *Metaphysics and Measurement*, 1968, pp. 89–117.

Koyré, A. (1957). *From the Closed World to the Infinite Universe*, The Johns Hopkins University Press, Baltimore.

Koyré, A. (1960). Galileo's treatise "de motu gravium": The use and abuse of imaginary experiment. *Revue d'Histoire des Sciences* 13: 197–245. Reprinted in his *Metaphysics and Measurement*, 1968, pp. 44–88.

Koyré, A. (1968). *Metaphysics and Measurement*, Harvard University Press, Cambridge.

Kozol, J. (1991). *Savage Inequalities*, Harper, New York.

Krips, H. (1980). Aristotle on the infallibility of normal observation. *Studies in the History and Philosophy of Science* 12: 79–86.

Kuhn, T.S. (1957). *The Copernican Revolution*, Random House, New York.

Kuhn, T.S. (1959). The essential tension: Tradition and innovation in scientific research. *The Third University of Utah Research Conference on the Identification of Scientific Talent*, University of Utah Press, Salt Lake City. Reprinted in his *The Essential Tension*, University of Chicago Press, Chicago, pp. 225–239.

Kuhn, T.S. (1970). *The Structure of Scientific Revolutions*, (2nd edition), University of Chicago Press, Chicago. (First edition 1962.)

Kula, W. (1986). *Measures and Man*, Princeton University Press, Princeton NJ.

Kumar, D.D. & Berlin, D.F. (1993). Science-technology-society policy implementation in the USA: A literature review. *The Review of Education* 15: 73–83.

Kutschmann, W. (1986). Scientific instruments and the senses: Towards an anthropological histori-ography of the natural sciences. *International Studies in Philosophy of Science* 1(1): 106–123.

Labaree, D.F. (1997a). Public goods, private goods: The American struggle over educational goals. *American Educational Research Journal* 34(1): 39–81.

Labaree, D.F. (1997b). *How to Succeed in School Without Really Learning: The Credentials Race in American Education*, Yale University Press, New Haven, CT.

Lafuente, A. & Sellé, M.A. (1985). The problem of longitude at sea in the 18th Century in Spain. *Vistas in Astronomy*, P. Beer, A.J. Meadows, and A.E. Roy (eds.), 28: 243–252.

Lakatos, I. (1970). Falsification and the methodology of scientific research programmes. In I. Lakatos & A. Musgrave (eds.) *Criticism and the Growth of Knowledge*, Cambridge University Press, Cambridge, pp. 91–196.

Lakatos, I. (1971). History of science and its rational reconstructions. In R.C. Buck & R.S. Cohen (eds.), *Boston Studies in the Philosophy of Science* 8: pp. 91–135.

Lakatos, I. (1978). History of science and its rational reconstructions. In J. Worrall & G. Currie (eds.), *The Methodology of Scientific Research Programmes: Volume I*, Cambridge University Press, Cambridge, pp. 102–138.

Landes, D.S. (1983). *Revolution in Time. Clocks and the Making of the Modern World*, Harvard University Press, Cambridge, MA.

Landes, D.S. (1998). Finding the point at sea. In W.J.H. Andrewes (ed.), *The Quest for Longitude*, The Collection of Historical Scientific Instruments, Harvard University, Cambridge, MA, 2nd Edition, pp. 20–30.

Lang, H. S. (1992). *Aristotle's Physics and Its Medieval Varieties*, State University of New York Press, Albany, NY.

Lankshear, C. (1998). Meanings of literacy in contempory educational reform Proposals. *Educational Theory* 48(3): 351–372.

Lattery, M.J. (2001). Thought experiments in physics education: a simple and practical example. *Science & Education* 10(6).

Laudan, L. (1968). The *Vis Viva* controversy. A post-mortem, *Isis* LIX: 131–143.

Laudan, L. (1981a). The clock metaphor and hypotheses: The impact of Descartes on English meth-odological thought. In his *Science and Hypothesis*, Reidel, Dordrecht, pp. 27–58.

Laudan, L. (1981b). The pseudo-science of science? *Philosophy of Social Science* 11: 173–198.

Laudan, L. (1990). *Science and Relativism*, University of Chicago Press, Chicago.

Law, J. (1987). On the social explanation of technical change: The case of the Portuguese maritime expansion. *Technology and Culture* 28:227–252.

Lawson, A. (ed.) (1993). The role of analogy in science and science teaching. A special issue of *Journal of Research in Science Teaching* 30(10).

Laymon, R. (1978). Newton's *Experimentum Crucis* and the Logic of Idealization and Theory refu-tation. *Studies in the History and Philosophy of Science* 9: 51–77.

Laymon, R. (1982). Scientific realism and the hierarchical counterfactual path from data to theory. In P.D Asquith & T. Nickles (ed.), *PSA 1982*, pp. 107–121.

Laymon, R. (1985). Idealizations and the testing of theories by experimentation. In P. Achinstein & O. Hannaway (eds.) *Observation, Experiment, and Hypothesis in Modern Physical Science*, MIT Press, Cambridge, MA, pp. 147–173.

Layton, A.D. & Powers, S.R. (1949). *New Directions in Science Teaching*, McGraw-Hill, New York.

Leach, J. (1997). Students' understanding of the nature of science. In G. Welford, J. Osborne & P. Scott (eds.), *Research in Science Education in Europe: Current Issues and Themes,* London, Falmer Press, pp 269–282.

Leatherdale, W.H. (1974). *The Role of Analogy, Model and Metaphor in Science*, Oxford University Press, Oxford.

Lederman, N.G. (1992). Students' and teachers' conceptions of the nature of science: A review of the research. *Journal of Research in Science Teaching* 29(4): 331–359.

Lederman, N.G., Kuerbis, P.J., Loving, C.C., Ramey-Gassert, L., Roychoudhury, A., & Spector, B.S. (1997). Professional knowledge standards for science teacher educators. *Journal of Science Teacher Education* 8(4): 233–240.

Leibniz, G.W. (1686/1989). *Discourse on Metaphysics.* In *Philosophical Essays*, R. Ariew & D. Garber (trans.), Hackett Publishing Company, Indianapolis, pp. 35–68.

Leibniz, G.W. (1981). *New Essays on Human Understanding*, (P. Remnant & J. Bennett, trans. and eds.), Cambridge University Press, Cambridge

Lemann, N. (1997). The reading wars. *The Atlantic Monthly* 280(5): pp. 128–134.

Lennon, T.M. (1989). Physical and metaphysical atomism: 1666–1682. In J.R. Brown & J. Mittelstrass (eds.), *An Intimate Relation: Studies in the History and Philosophy of Science'*, Kluwer Academic Publishers, Dordrecht, pp. 81–96.

Lennox, J.G. (1986). Aristotle, Galileo, and the "mixed sciences." In W.A. Wallace (ed.), *Reinterpreting Galileo*, Catholic Univerity of America Press, Washington, pp. 29–51.

Lenzen, V.F. (1938). Procedures of empirical science. In O. Neurath, R. Carnap, & C.W. Morris (eds.), *International Encyclopedia of Unified Science*, Vol. 1, Pt.1, University of Chicago Press, Chicago, pp. 279–339.

Leopold, J.H. (1993). Christiaan Huygens, the royal society and horology. *Antiquarian Horology* 21(1): 37–42.

Leopold, J.H. (1996). The longitude timekeepers of Christiaan Huygens. In W.J.H. Andrewes (ed.) *The Quest for Longitude: The Proceedings of the Longitude Symposium, Harvard University, Cambridge, Massachusetts, November 4–6, 1993*, Collection of Historical Scientific Instruments, Harvard University, Cambridge, MA., pp. 102–114.

Leopold, J.H. (1998). The longitude timekeepers of Christiaan Huygens. In W.J.H. Andrewes (ed.) *The Quest for Longitude: The Proceedings of the Longitude Symposium, Harvard University, Cambridge, Massachusetts, November 4–6, 1993*, Collection of Historical Scientific Instruments, Harvard University, Cambridge, MA., pp. 102–114.

Levin, I. (1977). The development of time concepts in young children: Reasoning about duration. *Child Development* 48: 435–444.

Levin, I. & Simons, H. (1986). The nature of children's and adult's concepts of time, speed, and distance and their sequence in development: Analysis via circular motion. In I. Levin (ed.), *Stage and Structure: Reopening the Debate*, Abex Publishing Corporation, Norwood, NJ, pp. 77–105.

Levitt, N. (1998). *Prometheus Bedeviled: Science and the Contradictions of Contempory Culture*, Rugters University Press, Piscataway, NJ.

Lewin, K. (1931). The conflict between Aristotelian and Galilean modes of thought in contemporary psychology. *Journal of General Psychology* 5: 141–177. Reproduced in his *A Dynamic Theory of Personality*, McGraw Hill, New York, 1935.

Lindberg, D.C. & Westman, R.S. (eds.) (1990). *Reappraisals of the Scientific Revolution*, Cambridge University Press, Cambridge.

Lipman, M. & Sharp, A.M. (eds.) (1978). *Growing Up with Philosophy*, Temple University Press, Philadelphia.

Lipman, M. (1991). *Thinking in Education*, Cambridge University Press, Cambridge.

Lippincott, K. (ed.) (1999). *The Story of Time*, Merrell Holberton, London.

Lloyd, H.A. (1957). Mechanical timekeepers. In C.J. Singer, E.J. Holmyard, A.R. Hall & T.I. Williams (eds.), *History of Technology* (5 vols.) 3: 648–675.

Lloyd, H.A. (1958). *Some Outstanding Clocks Over Seven Hundred Years, 1250–1950*, Leonard Hill, London.

Lloyd, H.A. (1970). *Old Clocks*, Dover Publications, New York.

Locke, J. (1690/1924). *An Essay Concerning Human Understanding*, (A.S. Pringle-Pattison, ed.), Clarendon Press, Oxford.

Lonergan, B. (1993). *Collected Works of Bernard Lonergan: Topics in Education*, (R.M. Doran & F.E. Crowe, eds.), University of Toronto Press, Toronto.

Loving, C.C. (1991). The scientific theory profile: A philosophy of science model for science teachers. *Journal of Research in Science Teaching* 28(9): 823–838.

Lux, D.S. (1991). Societies, circles, academies, and organizations: A historiographic essay on seventeenth-century science. In P. Baker & R. Ariew (eds.), *Revolution and Continuity: Essays in the History and Philosophy of Early Modern Science*, Catholic University of America Press, Washington, DC, pp. 23–44.

Macey, S.L. (1980). *Clocks and Cosmos: Time in Western Life and Thought*, Garland Publishing Company, Hamden, CT.

Macey, S.L. (1991). *Time: A Bibliographical Guide*, Garland Publishing Company, Hamden, CT.

Macey, S.L. (ed.) (1994). *Encyclopedia of Time*, Garland Publishing, New York.

Mach, E. (1872/1911). *The History and Root of the Principle of the Conservation of Energy*, Open Court Publishing Company, Chicago.

Mach, E. (1883/1960). *The Science of Mechanics*, Open Court Publishing Company, LaSalle Il.

Mach, E. (1886/1986). On instruction in the classics and the sciences. In his *Popular Scientific Lectures*, Open Court Publishing Company, La Salle, pp. 338–374.

Mach, E. (1894/1986). On the principle of comparison in physics. In his *Popular Scientific Lectures*, Open Court Publishing Company, La Salle, pp. 236–258.

Mach, E. (1896/1976). On thought experiments. In his *Knowledge and Error*, Reidel, Dordrecht, pp. 134–147.

Mach, E. (1898/1986). *Popular Scientific Lectures*, 5th edition, Open Court Publishing Company, La Salle.

Machamer, P. (1978). Aristotle on natural place and natural motion. *Isis* 69: 377–387.

Machamer, P. & Woody, A. (1994). The balance as a model for understanding the motion of bodies: Galileo and classroom physics. *Science & Education* 3(3): 215–244.

Mackenzie, J. (1998). Forms of knowledge and forms of discussion. *Educational Philosophy and Theory* 30(1): 27–49.

MacLachlan, J. (1973). A test of an "imaginery experiment" of Galileo's. *Isis* 64: 374–379.

MacLachlan, J. (1976). Galileo's experiments with pendulums: Real and imaginary. *Annals of Science* 33: 173–185.

MacLachlan, J. (1997). *Galileo Galilei: First Physicist*, Oxford University Press, New York.

Maddison, F. (1969). *Medieval Scientific Instruments and the Development of Navigational Instruments in the XVth and XVIth Centuries,* Coimbra.

Maddison, F. (1991). Tradition and innovation: Columbus' first voyage and Portuguese navigation in the fifteenth century. In J. A. Levenson (ed.), *Circa 1492: Art in the Age of Exploration*, Yale University Press, New Haven.

Mahoney, M.S. (1980). Christiaan Huygens: The measurement of time and of longitude at sea In H.J.M. Bos et al. (eds.), *Studies on Christiaan Huygens*, Swets & Zeitlinger, Lisse, pp. 234–270.

Mahoney, M.S. (1990). Infintesimals and transcendent relations: The mathematics of motion in the late seventeenth century. In D.C. Lindberg & R.S. Westman (eds.), *Reappraisals of the Scientific Revolution*, Cambridge University Press, Cambridge, pp. 461–491.

Maier, A. (1982). *On the Threshold of Exact Science: Collected Essays*, University of Pennsylvania Press, Philadelphia.

Maltin, M. (1993). Some notes on the medieval clock in Salisbury cathedral. *Antiquarian Horology* 20(5): 438–442.

Mann, C.R. (1912). *The Teaching of Physics for Purposes of General Education*, Macmillan, New York.

Manuel, D.E. (1981). Reflections on the role of history and philosophy of science in school science education. *School Science Review* 62(221): 769–771.

Martel, E. (1991). How valid are the Portland baseline essays? *Educational Leadership* Dec/Jan: 20–23.

Martin, M. (1972). *Concepts of Science Education*, Scott, Foresman & Co., New York (reprint, University Press of America, 1985)

Martin, M. (1974). The relevance of philosophy of science for science education. *Boston Studies in Philosophy of Sciences* 32: 293–300.

Martins, R. de A. (1993). Huygens's reaction to Newton's gravitational theory. In J.V. Field & F.A.J.L. James (ed.), *Renaissance and Revolution: Humanists, Scholars, Craftsmen and Natural Philosophers in Early Modern Europe*, Cambridge University Press, Cambridge, pp. 203–214.

Marx, G. (1994). *The Voice of the Martians*, Roland Eötvös Physical Society, Budapest.

Mathew, K.M. (1988). *History of the Portuguese Navigation in India (1497–1600)*, Mittal Publications, Delhi.

Matthews, M.R. (1980). *A Marxist Theory of Schooling: A Study in Epistemology and Education*, Harvester Press, Brighton.

Matthews, M.R. (1987a). Galileo's pendulum and the objects of science. In B. & D. Arnstine (eds.), *Philosophy of Education*, Philosophy of Education Society, Normal, IL, pp. 309–319.

Matthews, M.R. (1987b). Experiment as the objectification of theory: Galileo's revolution. In J. Novak (ed.), *Misconceptions and Educational Strategies in Science and Mathematics*, Cornell University, Ithaca, pp. 289–299.

Matthews, M.R. (1989a). Galileo and pendulum motion: A case for history and philosophy in the science classroom. *Research in Science Education* 19: 187–197.

Matthews, M.R. (ed.). (1989b). *The Scientific Background to Modern Philosophy*, Hackett Publishing Company, Indianapolis.

Matthews, M.R. (1990). Ernst Mach and contemporary science education reforms. *International Journal of Science Education* 12(3): 317–325.

Matthews, M.R. (ed.) (1991). *History, Philosophy and Science Teaching: Selected Readings*, OISE Press, Toronto.

Matthews, M.R. (1992a). Old wine in new bottles: A problem with constructivist epistemology. In H. Alexander (ed.), *Philosophy of Education 1992*, Proceedings of the Forty-Eighth Annual Meeting of the Philosophy of Education Society, Philosophy of Education Society, Normal, IL., pp. 303–311.

Matthews, M.R. (1992b). Constructivism and the empiricist legacy. In M.K. Pearsall (ed.), *Scope, Sequence and Coordination of Secondary School Science: Relevant Research*, National Science Teachers Association, Washington, D.C., pp. 183–196.

Matthews, M.R. (1992c). History, philosophy and science teaching: The present rapprochement. *Science & Education* 1(1): 11–48.

Matthews, M.R. (1993). Constructivism and science education: Some epistemological problems. *Journal of Science Education and Technology* 2(1): 359–370.

Matthews, M.R. (1994a). *Science Teaching: The Role of History and Philosophy of Science*, Routledge, New York.

Matthews, M.R. (1994b). Discontent with constructivism. *Studies in Science Education* 24: 165–172

Matthews, M.R. (1995a). *Challenging New Zealand Science Education*, Dunmore Press, Palmerston North.

Matthews, M.R. (1995b). Lessons from the past: Philosophy of science and US science teaching in the 1960s— The work of J.T. Robinson. In F. Finley et al. (eds.), *Proceedings of the Third International History, Philosophy and Science Teaching Conference*, University of Minnesota, Minneapolis, Volume 2, pp. 703–716.

Matthews, M.R. (1997a). James T. Robinson's account of philosophy of science and science teaching: Some lessons for today from the 1960s. *Science Education* 81(3): 295–316.

Matthews, M.R. (1997b). Israel Scheffler on the role of history and philosophy of science in science teacher education. *Studies in Philosophy and Education* 16(1–2): 159–173.

Matthews, M.R. (ed.) (1998). *Constructivism in Science Education: A Philosophical Examination*, Kluwer Academic Publishers, Dordrecht.

Matthews, M.R. (2000). Constructivism in science and mathematics education. In D.C. Phillips (ed.) *National Society for the Study of Education 100th Yearbook*.

Mattila, J.O. (1991). The Foucault pendulum as a teaching aid. *Physics Education* 26: 120–123.

Maurice, K. & Mayr, O. (eds.) (1980). *Clockwork Universe. German Clocks and Automata 1550–1650*, Neale Watson Academic Publications, New York.

Maurice, K. (1980). Jost Burgi, or on innovation. In K. Maurice & O. Mayr (eds.), *Clockwork Universe. German Clocks and Automata 1550–650*, Neale Watson Academic Publications, New York, 1980, pp. 87–102.

Maxwell, J.C. (1877). *Matter and Motion*, Dover Publications, New York.

May, W.E. & Holder, L. (1973). *A History of Marine Navigation*, Norton, New York.

May, W.E. (1973). *A History of Marine Navigation*, Norton, New York.

May, W.E. (1976). How the chronometer went to sea. *Antiquarian Horology* March: 638–663.

Mayr, O. (1986). *Authority, Liberty and Automatic Machinery in Early Modern Europe*, The Johns Hopkins University Press, London.

Mays, W. (1982). Piaget's sociological theory. In S. & C. Mogdill (eds.), *Jean Piaget: Consensus and Controversy*, Holt, Rinehart & Winston, London, pp. 31–50.

McCloskey, M. & Kargon, R. (1988). The meaning and use of historical models in the study of intuitive physics. In S. Strauss (ed.), *Ontology, Phylogeny and Historical Development*, Ablex Publishing Corporation, Norwood, NJ, pp. 49–67.

McCloskey, M. (1983a). Intuitive physics. *Scientific American* 248: 114–122.

McCloskey, M. (1983b). Naive theories of motion. In D. Gentner & A.L. Stevens (eds.), *Mental Models*, Lawerence Erlbaum, Hillsdale, NJ., pp. 299–324.

McComas, W.F. (ed.) (1998). *The Nature of Science in Science Teacher Education: A Handbook for Practitioners*, Kluwer Academic Publishers, Dordrecht.

McCourt, F. (1996). *Angela's Ashes: A Memoir of a Childhood*, Harper Collins, London.

McDermott, L.C. (1984). Research on conceptual understanding in mechanics. *Physics Today* 37: 24–32.

McFadden, C.P. (1989). Redefining the school curriculum. In D.E. Herget (ed.), *The History and Philosophy of Science in Science Teaching*, Florida State University, Tallahassee, pp. 259–270.

McGuire, J.E. (1994). Natural motion and its causes: Newton on the "vis insita" of bodies. In M.L. Gill and J.G. Lennox (eds.) *Self-Motion: From Aristotle to Newton*, Princeton University Press, Princeton NJ, pp. 305–329.

McKeon, R. (1941). *The Basic Works of Aristotle*, Random House, New York.

McKeon, R. (1994). *On Knowing. The Natural Sciences*, compiled by D.B Owen, edited by D.B. Owen & Z.K. McKeon, University of Chicago Press, Chicago.

McMullin, E. (ed.) (1963a). *The Concept of Matter in Greek and Medieval Philosophy*, University of Notre Dame Press, Notre Dame.

McMullin, E. (ed.) (1963b). *The Concept of Matter in Modern Philosophy*, University of Notre Dame Press, Notre Dame.

McMullin, E. (1970). The history and philosophy of science: A taxonomy. *Minnesota Studies in the Philosophy of Science* 5: 12–67.

McMullin, E. (1975). History and philosophy of science: A marriage of convenience? *Boston Studies in the Philosophy of Science* 32: 515–531.

McMullin, E. (1978). The conception of science in Galileo's work. In R.E. Butts & J.C. Pitt (eds.), *New Perspectives on Galileo*, Reidel Publishing Company, Dordrecht, pp. 209–258.

McMullin, E. (1985). Galilean Idealization. *Studies in the History and Philosophy of Science* 16: 347–373.

McMullin, E. (1990). Conceptions of science in the scientific revolution. In D.C. Lindberg & R.S. Westman (eds.), *Reappraisals of the Scientific Revolution*, Cambridge University Press, Cambridge, pp. 27–92.

McReynolds, P. (1980). The clock metaphor in psychology. In T. Nickles (ed.), *Scientific Discovery: Case Studies*, Reidel Publishing Company, Dordrecht, pp. 97–112.

McTighe, T.P. (1967). Galileo's Platonism: A reconsideration. In E. McMullin (ed.), *Galileo Man of Science*, Basic Books, New York, pp. 365–388.

Meichtry, Y.J. (1993). The impact of science curricula on student views about the nature of science. *Journal of Research in Science Teaching* 30(5), 429–444.

Melsen, A.G. van (1952). *From Atomos to Atom*, Duquesne University Press, Pittsburgh.

Mendelssohn, K. (1977). *Science and Western Domination*, Readers Union, Newton Abbot.

Menut, A.D. & Denomy A.J. (eds) (1968). *Nicole Oresme "Le Livre du ciel et du monde."* University of Wisconsin Press, Madison.

Merton, R.K. (1942). The normative structure of science. In his *The Sociology of Science: Theoretical and Empirical Investigations* (N.W. Storer, ed.), University of Chicago Press, Chicago, 1973, pp. 267–280.

Mertz, D.W. (1982). The concept of structure in Galileo: Its role in the methods of proportionality and *Ex Suppositione* as applied to the tides. *Studies in the History and Philosophy of Science* 9: 229–242.

Mesnage, P. (1969). The building of clocks. In M. Daumas (ed.), Maurice Daumas (trans.), *A History of Technology and Invention Through the Ages*, Crown, New York, vol. 2.

Messerly, J.G. (1996). Psychogenesis and the history of science: Piaget and the problem of scientific change. *Modern Schoolman* 73, 295–307.

Mestre, J.P. (1991). Learning and instruction in pre-college physical science. *Physics Today* September: 56–62.

Mettrie, J.O. la (1748/1912). *Man a Machine*, translated from the French, Open Court Publishing Company, Chicago.

Ministry of Research, Science and Technology (MORST) (1991). *Survey of Attitudes to, and Understanding of, Science and Technology in New Zealand*, Publication No.4, MORST, Wellington.

Milham, W.I. (1945). *Time & Timekeepers. Including the History, Construction, Care, and Accuracy of Clocks and Watches*, Macmillan, London.

Millar, R. & Osborne, J. (1998). *Beyond 2000: Science Education for the Future*, School of Education, King's College, London.

Miller, J.D. (1983). Scientific literacy: A conceptual and empirical review. *Daedalus* 112(2): 29–47.

Miller, J.D. (1987). Scientific literacy in the United States. In E. David & M. O'Connor (eds.) *Communicating Science to the Public*, John Wiley, London.

Miller, J.D. (1992). *The Public Understanding of Science and Technology in the United States, 1990*, National Science Foundation, Washington D.C.

Mills, J.A. (1998). *Control: A History of Behavioral Psychology*, New York University Press, New York.

Mills, J.F. (1983). *Encyclopedia of Antique Scientific Instruments*, Aurum Press, London.

Ministry of Research, Science and Technology (MORST) (1991). *Survey of Attitudes to, and Understanding of, Science and Technology in New Zealand*, Publication No. 4, MORST, Wellington.

Mintzes, J.J., Wandersee, J.H., & Novak, J.D. (eds.) (1998). *Teaching Science for Understanding. A Human Constructivist View*, Academic Press, San Diego.

Mischel, T. (ed.) (1971). *Cognitive Development and Epistemology*, Academic Press, New York.

Misgeld, W., Ohly, K.P., Rühaak, H.,& Wiemann, H. (eds.) (1994). *Historisch-genetisches Lernen in den Naturwissenschaften*, Deutscher Studien Verlag, Weinheim.

Misgeld, W., Ohly, K.P., & Strobl, G. (2000). The historical-genetical approach to science teaching at the Oberstufen-Kolleg. *Science & Education* 9(4).

Mitchell, S. (1993). Mach's mechanics and absolute space and time. *Studies in the History and Philosophy of Science* 24(4): 565–583.

Mittelstrass, J. (1972). The Galilean revolution: The historical fate of a methodological insight. *Studies in the History and Philosophy of Science* 2: 297–328.

Mittelstrass, J. (1977). Changing concepts of the *A Priori*. In R. Butts & J. Hintikka (eds.), *Historical and Philosophical Dimensions of Logic, Methodology and Philosophy of Science*, Kluwer Academic Publishers, Dordrecht, pp. 113–128.

Modrak, D.K.W. (1987). *Aristotle: The Power of Perception*, University of Chicago Press, Chicago.

Monk, M. & Osborne, J. (1997). Placing the history and philosophy of science on the curriculum: A model for the development of pedagogy. *Science Education* 81(4): 405–424.

Monte, G. del (1581/1969). *Mechaniche*. In S. Drake and I.E. Drabkin (ed.), *Mechanics in Sixteenth-Century Italy*, University of Wisconsin Press, Madison, WI, pp. 241–329.

Montellano, B.R.O. de (1996). Afrocentric pseudoscience: The miseducation of African Americans. In P.R. Gross, N. Levitt, & M.W. Lewis (eds.), *The Flight from Science and Reason*, New York Academy of Sciences, New York, pp. 561–573.

Moody, E.A. (1951). Galileo and avempace: The dynamics of the leaning tower experiment. *Journal of the History of Ideas* 12: 163–193, 375–422. Reprinted in his *Studies in Medieval Philosophy, Science and Logic*, University of California Press, Berkeley, 1975, pp. 203–286.

Moody, E.A. (1966). Galileo and his precursors. In C.L. Golino (ed.), *Galileo Reappraised*, University of California Press, Berkeley, CA. Reprinted in his *Studies in Medieval Philosophy, Science and Logic*, University of California Press, Berkeley, 1975, pp. 393–408.

Moody, E.A. (1975). *Studies in Medieval Philosophy, Science and Logic*, University of California Press, Berkeley.

Morgan, C. (1980). From sundial to atomic clock. In C. Wilson (ed.), *The Book of Time*, Westbridge Books, North Pomfret, Vermont, pp. 84–128.

Morison, S.E. (1942). *Admiral of the Ocean Sea: A life of Christopher Columbus*, Little Brown, Boston.

Morison, S.E. (1974). *The European Discovery of America: The Southern Voyages, 1492–1616*, Oxford University Press, Oxford.

Moss, J.D. (1986). The rhetoric of proof in Galileo's writings on the Copernican system. In W.A. Wallace (ed.), *Reinterpreting Galileo*, Catholic University of America Press, Washington, pp. 179–204.

Moss, J.D. (1993). *Novelties in the Heavens: Rhetoric and Science in the Copernican Controversy*, University of Chicago Press, Chicago.

Mumford, L. (1934). *Technics and Civilization*, Harcourt Brace Jovanovich, New York.

Murdoch, J.E. & Sylla, E.D. (1978). The science of motion. In D.C. Lindberg (ed.), *Science in the Middle Ages*, University of Chicago Press, Chicago, pp. 206–264.

Murschel, A. & Andrewes, W.J.H. (1998). Translations of the earliest documents describing the principal methods used to find longitude at sea. In W.J.H. Andrewes (ed.), *The Quest for Longitude*, The Collection of Historical Scientific Instruments, Harvard University, Cambridge, MA, 2nd Edition, pp. 376–392.

Musgrave, A. (1974). The objectivism of Popper's epistemology. In P.A. Schilpp (ed.), *The Philosophy of Karl Popper*, Open Court Publishing Co., La Salle, IL., pp. 560–596.

Musgrave, A. (1999). How to do without inductive logic. *Science & Education* 8(4): 395–412.

Nagel, E. (1961). *The Structure of Science*, Routledge & Kegan Paul, London.

Naylor, R.H. (1974a). Galileo's simple pendulum. *Physis* 16: 23–46.

Naylor, R.H. (1974b). The evolution of an experiment: Guidobaldo del Monte and Galileo's *Discourse* demonstration of the parabolic trajectory. *Physis* 16: 323–348.

Naylor, R.H. (1974c). Galileo and the problem of free fall. *British Journal for the History of Science* 7: 105–134.

Naylor, R.H. (1976a). Galileo: Real experiment and didactic experiment. *Isis* 67(238): 398–419.

Naylor, R.H. (1976b). Galileo: Search for the parabolic trajectory. *Annals of Science* 33: 153–174.

Naylor, R.H. (1977). Galileo's theory of motion: Processes of conceptual change in the period 1604–10. *Annals of Science* 34: 365–392.

Naylor, R.H. (1980). The role of experiment in Galileo's early work on the law of fall. *Annals of Science* 37: 363–378.

Naylor, R.H. (1989). Galileo's experimental discourse. In D. Gooding et al. (eds.), *The Uses of Experiment*, Cambridge University Press, Cambridge, 117–134.

National Curriculum Council (NCC) (1988). *Science in the National Curriculum*, NCC, York.

National Curriculum Council(NCC) (1991). *Science for Ages 5 to 16*, DES, London.

National Commission on Excellence in Education (NCEE) (1983). *A Nation At Risk: The Imperative for Education Reform*, U.S. Department of Education, Washington D.C.

National Research Council (NRC) (1994). *National Science Education Standards: Draft*, National Academy Press, Washington.

National Science Board (NSB) (1996). *National Indicators in Science and Mathematics Education*, National Science Foundation, Washington.

National Science Resources Center (NSRC) (1994). *Measuring Time*, Carolina Biological Supply Company, Burlington, NC.

National Science Teachers Association (NSTA) (1982). *Science-Technology-Society: Science Education for the 1980's*, NSTA, Washington.

National Science Teachers Association (NSTA) (1992). *Scope, Sequence and Coordination: Volume One, The Content Core*, NSTA, Washington.

Needham, J. and Ling, W. (1954–65). *Science and Civilisation in China*, vols. 1–4, Cambridge University Press, Cambridge.

Needham, J., Ling, W., & Price, D.J. de Solla (1960). *Heavenly Clockwork*, Cambridge University Press, Cambridge.

Nelson, L.H. & J. (eds.) (1996). *Feminism, Science, and the Philosophy of Science*, Kluwer Academic Publishers, Dordrecht.

Nersessian, N.J. (1989). Conceptual Change in Science and in Science Education. *Synthese* 80(1): 163–184. In M.R. Matthews (ed.), *History, Philosophy and Science Teaching: Selected Readings*, OISE Press, Toronto.

Nersessian, N.J. (1992). Constructing and instructing: The role of "abstraction techniques" in creating and learning physics. In R.A. Duschl & R.J. Hamilton (eds.), *Philosophy of Science, Cognitive Psychology, and Educational Theory and Practice*, State University of New York Press, Albany, NY, pp. 48–68.

Nersessian, N.J. & Resnick, L. (1989). Comparing historical and intuitive explanations of motion: Does "Naive Physics" have a structure? *Proceedings of the Cognitive Science Society* 11: 412–420.

Neugebauer, O. (1969). *The Exact Sciences in Antiquity*, 2nd edition, Dover, New York.

Newitt, M. (1986). Prince Henry and the origins of Portuguese expansion. In M. Newitt (ed.), *The First Portuguese Colonial Empire*, Exeter University Press, Exeter.

Newton, I. (1729/1934). *Mathematical Principles of Mathematical Philosophy*, (A. Motte, trans., revised F. Cajori), University of California Press, Berkeley.

Newton, I. (1730/1979). *Opticks*, 4th edition, I.B. Cohen and D.H.D. Roller (eds.), Dover, New York. First edition, 1704.

Nicholson, T. (1993). Our illiteracy. *North and South,* November:38–39.

Nicholson, T. (1994). The best in the world? *Metro*, May: Letters Page.

Nicholson, T. (1997). Closing the gap on reading failure: Social class, phonemic awareness and learning to read. In B.A. Blachman (ed.), *Foundations of Reading Acquisition,* Lawrence Erlbaum Mahwah, NJ.

Nicholson, T. (1998). Social Class and Reading Achievement. In G.B. Thompson & T. Nicholson (eds.), *Learning to Read: Theory and Research,* Teachers College Press, New York.

Nicholson, T. (2000). *Reading the Writing on the Wall: Debates, Challenges and Opportunities in the Teaching of Reading*, Dunmore Press, Palmerston North, New Zealand.

Nielsen H. & Thomsen, P. (1990). History and philosophy of science in the Danish curriculum. *International Journal of Science Education* 12(4): 308–316.

Nola, R. (1995). Objectivism and constructivism in knowledge, science and science education. In F. Finley et al. (eds.), *Proceedings of the Third International History, Philosophy, and Science Teaching Conference*, University of Minnesota, Minneapolis, Vol. 2, pp. 834–847.

Nola, R. (1997). Constructivism in science and in science education: A critique. *Science & Education* 6(1–2): 5—83.

North, J.D. (1975). Monasticism and the first mechanical clocks. In J.T. Fraser & N. Lawrence (eds.), *The Study of Time*, New York.

Novak, J.D. (1977). *A Theory of Education*, Cornell University Press, Ithaca. Paperback edition, 1986.

Novak, J.D. (1983). Overview. In H. Helm & J.D. Novak (eds.) *Proceedings of the International Seminar on Misconceptions in Science and Mathematics*, Cornell University, Ithaca, pp. 1–4.

Nowak, L. (1980). *The Structure of Idealization*, Reidel, Dordrecht.

Nussbaum, J. (1983). Classroom conceptual change: The lesson to be learned from the history of science. In H. Helm & J.D. Novak (eds.), *Misconceptions in Science & Mathematics*, Department of Education, Cornell University, pp. 272–281.

Nussbaum, J. (1998). History and philosophy of science and the preparation for constructivist teaching: The case of particle theory. In J.J. Mintzes, J.H. Wandersee & J.D.Novak (eds.), *Teaching Science for Understanding. A Human Constructivist View*, Academic Press, San Diego, pp. 165–194.

O'Loughlin, M. (1992). Rethinking science education: Beyond Piagetian constructivism toward a sociocultural model of teaching and learning. *Journal of Research in Science Teaching* 29(8): 791–820.

O'Neil, W.M. (1975). *Time and the Calendars*, Sydney University Press, Sydney.

Oldenburg, H. (1676). Review of *Longitude Found*. *Philosophical Transactions* 11: 774.

Olmsted, J.W. (1942). The scientific expedition of Jean Richer to Cayenne (1672–1673). *Isis* 34: 117–128.

Olson, R. (1990). Historical reflections on feminist critiques of science: The scientific background to modern feminism. *History of Science* 28: 125–145.

Osborne, J. (1996). Beyond constructivis. *Science Education* 80(1): 53–82.

Osborne, R.J. & Freyberg, P. (1985). *Learning in Science: The Implications of Children's Science*, Heinemann, London.

Owen, J.D. (1995). *Why Our Kids Don't Study: An Economist's Perspective*, Johns Hopkins University Press, Baltimore, MD.

Pais, A. (1994). *Einstein Lived Here*, Clarendon Press, Oxford.

Paley, W. (1802/1964). *Natural Theology; or Evidence for the Existence and Attributes of the Deity Collected from the Appearances of Nature*, Bobbs-Merrill, Indianapolis.

Palisca, C.V. (1961). Scientific empiricism in musical thought. In H.H. Rhys (ed.), *Seventeenth Century Science and the Arts*, Princeton University Press, Princeton, pp. 91–137.

Pap, A. (1945). *The A Priori in Physical Theory*, King's Crown Press, New York.

Parker, J. (ed.) (1965). *Merchants and Scholars: Essays in the History of Exploration and Trade*, University of Minnesota Press, Minneapolis, MN.

Passmore, J. (1978). *Science and Its Critics*, Rutgers University Press, Rutgers NJ.

Patterson, L.D. (1952). Pendulums of Wren and Hooke. *Osiris* 10: 277–321.

Pearsall, M.K. (ed.) (1992). *Scope, Sequence and Coordination of Secondary School Science: Relevant Research*, National Science Teachers Association, Washington.

Pedersen, O. (1994). *The Book of Nature*, University of Notre Dame Press, Notre Dame.

Pentz, M.J. (ed.) (1970). *Science: Its Origins, Scales and Limitations*, Open University Press, Bletchley, Bucks.

Perl, M.R. (1969). Physics and metaphysics in Newton, Leibniz and Clarke. *Journal of the History of Ideas* 30: 507–526.

Peters, R.S. (1966). *Ethics and Education*, George Allen and Unwin, London.

Petry, M.J. (1993).Classifying the motion: Hegel on the pendulum. In M.J. Petry (ed.), *Hegel and Newtonism*, Kluwer Academic Publishers, Dordrecht, pp. 291–316.

Pfundt, H. & Duit, R. (1994). *Bibliography of Students' Alternative Frameworks & Science Education*, 4th Edit., Institute for Science Education, University of Kiel.

Phillippes, H. (1657a). *The Advancement of the Art of Navigation*, London.

Phillippes, H. (1657b). *The Geometrical Seaman*, London.

Phillips, D.C. (1995). The good, the bad and the ugly: The many faces of constructivism. *Educational Researcher* 24(7): 5–12.

Phillips, D.C. (1997). Coming to terms with radical social constructivisms. *Science & Education* 6(1–2),:85–104.

Physical Science Study Committee (PSSC) (1960). *Physics*, D.C.Heath & Co., Boston.

Piaget, J. & Garcia, R. (1989). *Psychogenesis and the History of Science*, Columbia University Press, New York.

Piaget, J. (1929). *The Child's Conception of the World*, Routledge & Kegan Paul, London.

Piaget, J. (1946/1970). *The Child's Conception of Time*, Routledge & Kegan Paul, London.

Piaget, J. (1972). *Psychology and Epistemology: Towards a Theory of Knowledge*, Penguin, Harmondsworth.

Pinnick, C. (1994). Feminist epistemology: Implications for philosophy of science. *Philosophy of Science* 61: 646–657.

Pitt, J.C. (1989). Apologia pro simplicio: Galileo and the limits of knowledge. In J.R. Brown & J. Mittelstrass (eds.), *An Intimate Relation: Studies in the History and Philosophy of Science*, Kluwer Academic Publishers, Dordrecht, pp. 1–22.

Pogo, A. (1935). Gemma Frisius, his method of determining differences of longitude by transporting time-pieces (1530) and his treatise on triangulation (1533). *Isis* 22: 469–485.

Poincaré, H. (1905/1952). *Science and Hypothesis*, Dover Publications, New York.

Popper, K.R. (1934/1959). *The Logic of Scientific Discovery*, Hutchinson, London.

Popper, K.R. (1945). *The Open Society and Its Enemies*, Routledge & Kegan Paul, London.

Popper, K.R. (1963). *Conjectures and Refutations*, Routledge & Kegan Paul, London.

Popper, K.R. (1972). *Objective Knowledge*, Clarendon Press, Oxford.

Postman, N. (1985). *Amusing Ourselves to Death*, Methuen, New York.

Pratt, V. (1987). *Thinking Machines. The Evolution of Artificial Intelligence*, Basil Blackwell, Oxford.

Price, D.J. de S. (1956). The prehistory of the clock. *Discovery* April: 153–157.

Price, D.J. de S. (1961). Celestial clockwork in Greece and China. In *Science Since Babylon*, Yale University Press, New Haven, pp. 23–44.

Price, D.J. de S. (1964). Automata and the origins of mechanism and mechanistic philosophy. *Technology and Culture* 5(1): 9–23.

Price, D.J. de S. (1967). Piecing together an ancient puzzle: The tower of the winds. *National Geographic*, April: 587–596.

Price, D.J. de S. (1975). Clockwork before the clock and timekeepers before timekeeping. In J.T. Fraser & N. Lawrence (eds.), *The Study of Time II*, Springer-Verlag, Berlin, pp. 367–380.

Price, H. (1996). *Time's Arrow and Archimedes' Point*, Oxford University Press, Oxford.

Priestley, F.E.L. (1970). The Clarke-Leibniz controversy. In R.E. Butts & J.W. Davis (eds.), *The Methodological Heritage of Newton*, University of Toronto Press, Toronto, pp. 34–56.

Proverbio, E. (1985). The contribution of the mechanical clock to the improvement of navigation. *Vistas in Astronomy* 28, P. Beer, A.J. Meadows, & A.E. Roy (eds.), 95–104.

Ptolemy, C. (1991). *The Geography*, trans. & ed. E.L. Stevenson, Dover, New York.

Pukies, J. (1979). *Das Verstehen der Naturwissenschaften*, Georg Westermann Verlag, Braunschweig

Pumfrey, S., Rossi, P.L., & Slawinski, M. (eds.): 1991, *Science, Culture and Popular Belief in Renaissance Europe*, Manchester University Press, Manchester.

Putnam, H. (1978). *Meaning and the Moral Sciences*, Routledge & Kegan Paul, London.

Putnam, H. (1981). The impact of science on modern conceptions of rationality. *Synthese* 46: 359–382.

Pyenson, L. (1993). The ideology of Western rationality: History of science and the European civilizing mission. *Science & Education* 2(4): 329–344.

Pyle, A. (1997). *Atomism and Its Critics,* Thoemmes Press, Bristol.

Quill, H. (1966). *John Harrison: The Man Who Found Longitude*, New York.

Quine, W.V.O. (1951). Two dogmas of empiricism. *Philosophical Review*. Reprinted in his *From a Logical Point of View*, Harper & Row, New York, 1953, pp. 20–46.

Quine, W.V.O. (1953)., *From a Logical Point of View*, Harper & Row, New York.

Quinones, R.J. (1972). *The Renaissance Discovery of Time*, Harvard University Press, Cambridge.

Raju, C.K. (1994), *Time: Towards a Consistent Theory*, Kluwer Academic Publishers, Dordrecht.

Randall, A.G. (1998). The timekeeper that won the longitude prize. In W.J.H. Andrews (ed.), *The Quest for Longitude,* The Collection of Historical Scientific Instruments, Harvard University, Cambridge, MA, 2nd Edition, pp. 235–254.

Randles, W.G.L. (1985). Portuguese and Spanish attempts to measure longitude in the 16th Century. *Vistas in Astronomy* 28: P. Beer, A.J. Meadows, & A.E. Roy (eds.), 235–242.

Raven, C.E. (1953). *Natural Religion and Christian Theology*, Cambridge University Press, Cambridge.

Reichenbach, H. (1927/1958). *Philosophy of Space and Time*, Dover Publications, New York.

Reichenbach, H. (1951). *The Rise of Scientific Philosophy*, University of California Press, Berkeley.

Reif, F. & Allen, S. (1992). Cognition for interpreting scientific concepts: A study of zcceleration. *Cognition and Instruction* 9(1): 1–44.

Reif, F. (1995). Understanding and teaching important scientific thought processes. *Journal of Science Education and Technology* 4(4): 261–282.

Renn, J., Damerow, P., Rieger, S., & Camerota, M. (1998). *Hunting the White Elephant: When and How did Galileo Discover the Law of Fall*, Max Planck Institute for the History of Science, Preprint 97, Berlin.

Rhys, H.H. (ed.) (1961). *Seventeenth Century Science and the Arts*, Princeton University Press, Princeton.

Richards, E.G. (1998). *Mapping Time: The Calendar and its History,* Oxford University Press, Oxford.

Richards, J.R. (1996). Why feminist epistemology isn't. In P.R. Gross, N. Levitt & M.W. Lewis (eds.), *The Flight from Science and Reason*, New York Academy of Sciences, New York, pp. 385–412.

Riess, F. (1995). Teaching science and the history of science by redoing historical experiments. In F. Finley et al. (eds.) *Proceedings of the Third International History, Philosophy and Science Teaching Conference*, University of Minnesota, Minneapolis, vol. 2, pp. 958–966.

Riess, F. & Nielsen, T. (eds.) (2000). *History and Philosophy in German Science Education.* Special issue of *Science & Education* 9(4).

Roberts, D. (1992). *The Collector's Guide to Clocks*, Apple Press, London.

Roberts, D.A. (1982). Developing the concept of "curriculum emphases" in science education. *Science Education* 66: 243–260.

Roberts, G.K. (1991). Scientific academies across Europe. In D. Goodman & C.A. Russell (eds.), *The Rise of Scientific Europe, 1500–1800*, Hodder & Stoughton, Sevenoaks, Kent, pp. 227–252.

Robertson, J.D. (1931). *The Evolution of Clockwork*, Cassell, London.

Robin, N. & Ohlsson, S. (1989). Impetus then and now: A detailed comparison between Jean Buridan and a single contemporary subject. In D.E. Heget (ed.), *The History and Philosophy of Science in Science Teaching*, Florida State University, pp. 292–305.

Robinson, J.T. (1965). Science teaching and the nature of science. *Journal of Research in Science Teaching* 3: 37–50.

Robinson, J.T. (1968). *The Nature of Science and Science Teaching*, Wadsworth, Belmont CA.

Robinson, J.T. (1969). Philosophy of science: Implications for teacher education." *Journal of Research in Science Teaching* 6(1): 99–104.

Rodriguez, A.J. (1997). The dangerous discourse of invisibility: A critique of the National Research Council's National Science Education Standards. *Journal of Research in Science Teaching* 34(1): 19–37.

Rohr, R.J. (1965). *Sundials: History, Theory and Practice*, Toronto.

Ronan, C.A. (1974). *Galileo*, G.P. Putnam's Sons, New York.

Ronneberg, C.E. (1967). Laboratory projects in physical science for general students. *Science Education* 51(2): 152–161.

Rooke, L., Moray, R., & Hooke, R. (1667). Directions for observations and experiments to be made by masters of ships, pilots, and other fit persons on their sea-voyages. *Philosophical Transactions* 2: 433–448.

Rooke, L. (1665). Directions for sea-men bound for far voyages. *Philosopical Transactions* 1: 140–143.

Rorty, R. (1963). Matter and event. In E. McMullin (ed.), *The Concept of Matter in Modern Philosophy*, University of Notre Dame Press, Notre Dame, IN., pp. 221–248.

Rose, P.L. (1980). Monte Guidobaldo, Marchese del. In C.C. Gillispie (ed.), *Dictionary of Scientific Biography*, Scribners, New York, IX: 487–489.

Rosenberg, N. & Birdzell, L. (1990). Science, technology and the western miracle. *Scientific American* 263(5): 18–25.

Ross, A. (ed.) (1996). *Science Wars*, Duke University Press, Durham.

Rossum, G. D-V. (1996). *History of the Hour: Clocks and Modern Temporal Orders*, Chicago University Press, Chicago.

Roth, W.-M. (1993). Construction sites: Science labs and classrooms. In K. Tobin (ed.) *The Practice of Constructivism in Science Education*, AAAS Press, Washington, D..C, pp. 145–170

Roth, W.-M., Tobin, K., & Shaw, K. (1997). Cascades of inscriptions and the re-presentation of nature: How numbers, graphs, and money come to re-present a rolling ball. *International Journal of Science Education* 19 (9): 1075–1091.

Royal Ministry of Church, Education and Research (RMCER). (1994). *Core Curriculum for Primary, Secondary, and Adult Education in Norway*, RMCER, Oslo, Norway.

Rubba, P. et al. (1996). A new scoring procedure for the *Views on Science–Technology–Society* Instrument. *International Journal of Science Education* 18(4): 387–400.

Ruby, J.E. (1986). The origins of scientific "Law." *Journal of the History of Ideas* XLVII (3): 341–360.

Rühaak, H. (1994). Das Historisch-Genetische Konzept am Oberstufen-Kolleg. Eine Bestandsaufnahme. In W. Misgeld et al. (eds.), *Historisch-genetisches Lernen in den Naturwissenschaften*, Deutscher Studien Verlag, Weinheim, pp. 11–34.

Ruse, M. (ed.) (1988). *But Is It Science? The Philosophical Question in the Creation/Evolution Controversy*, Prometheus Books, Albany, NY.

Ruskai, M.B. (1996). Are "feminist perspectives" in mathematics and science feminist?" In P.R. Gross, N. Levitt & M.W. Lewis (eds.), *The Flight from Science and Reason*, New York Academy of Sciences, New York, pp. 437–442.

Russell, P.E. (1984). *Prince Henry the Navigator: The Rise and Fall of a Culture Hero*, Clarendon Press, Oxford.

Russell, T. (1993). An Alternative Conception: Representing Representations. In P.J. Black & A.M. Lucas (eds.), *Children's Informal Ideas in Science*, Routledge, London, pp. 62–84.

Russell, T.L. (1981). What history of science, how much and why?' *Science Education* 65:51–64.

Russo, F. (1987). Galileo and the theology of his time. In P. Poupard (ed.), *Galileo Galilei: Towards a Resolution of 350 Years of Debate — 1633–1983*, Duquesne University Press, Pittsburgh.

Rutherford, F.J. & Ahlgren, A. (1990). *Science for All Americans*, Oxford University Press, New York.

Rutherford, F.J., Holton, G. & Watson, F.G. (eds.) (1970). *The Project Physics Course: Text*, Holt, Rinehart, & Winston, New York.

Rutherford, F.J. (1972). A humanistic approach to science teaching. *National Association of Secondary School Principals Bulletin* 56(361): 53-63.

Rutherford, F.J. (2001). Fostering the history of science in American science education: the role of Project 2061. *Science & Education* 10(6).

Sachs, J. (1995). *Aristotle's "Physics." A Guided Study*, Rutgers University Press, New Brunswick, NJ.

Sadler, D.H. (1968). *Man is not Lost. A Record of Two Hundred Years of Astronomical Navigation with the Nautical Almanac, 1767–1967*, London.

Sagan, C. (1997). *The Demon-Haunted World: Science as a Candle in the Dark*, Headline Book Publishing, London.

Sarlemijn, A. (1993). Pendulums in Newtonian mechanics. In M.J. Petry (ed.), *Hegel and Newtonism*, Kluwer Academic Publishers, Dordrecht, pp.267–290.

Scarr, A.J.T. (1967). *Metrology and Precision Engineering*, McGraw-Hill, London.

Science Council of Canada (SCC). (1984). *Science for Every Student: Educating Canadians for Tomorrow's World*, Report 36, SCC, Ottawa.

Schecker, H. (1992). The paradigmatic change in mechanics: Implications of historical processes on physics education. *Science & Education* 1(1): 71–76.

Scheffler, I. (1970). Philosophy and the curriculum. In his *Reason and Teaching*, London, Routledge, 1973, pp. 31–44. Reprinted in *Science & Education* 1(4): 385–394.

Scheffler, I. (1973). *Reason and Teaching*, Bobbs- Merrill, Indianapolis.

Schlegel, R. (1968). *Time and the Physical World*, Dover Publications, New York.

Schliesser, E. & Smith, G.E. (1996). Huygens's 1688 Report to the Directors of the Dutch East India company on the measurement of longitude at sea and its implications for the non-uniformity of gravity. *De Zeventiende Eeuw* 12(1): 198–214.

Schmidt, W.H., McKnight, C.C., & Raizen, S.A. (1997). *A Splinted Vision: An Investigation of U.S. Science and Mathematics Education*, Kluwer Academic Publishers, Dordrecht.

Schneider, C.G. & Shoenberg, R. (1998). Contemporary understandings of liberal education. *Liberal Education*, Spring, 32–37.

Schon, D.A. (1963). *Displacement of Concepts*, Tavistock Publications, London.

Schramm, M. (1963). Aristotelianism: Basis and obstacle to scientific progress in the Middle Ages. *History of Science* 2: 91–113.

Schwab, J.J. (1945). The nature of scientific knowledge as related to liberal education. *Journal of General Education* 3: 245–266. Reproduced in I. Westbury & N.J. Wilkof (eds.) *Joseph J. Schwab: Science, Curriclum, and Liberal Education*, University of Chicago Press, Chicago, 1978, pp. 68–104.

Schwab, J.J. (1958). The teaching of science as inquiry. *Bulletin of Atomic Scientists* 14: 374–379. Reprinted in J.J. Schwab & P.F. Brandwein (eds.), *The Teaching of Science*, Harvard University Press, Cambridge MA.

Schwab, J.J. (1962). The concept of the structure of a discipline. *The Educational Record* 43: 197–205.

Schwab, J.J. (1964). Structure of the disciplines: Meaning and significances. In G.W. Ford & L. Pugno (eds.), *The Structure of Knowledge and the Curriculum*, Rand McNally & Co., Chicago.

Schweitzer, A. (1910). *The Quest for the Historical Jesus*, Adam and Charles Black, London.

Scott, P., Asoko, H., Driver, R., & Emberton, J. (1994). Working from children's ideas: Planning and teaching a chemistry topic from a constructivist perspective. In P. Fensham, R. Gunstone & R. White (eds.) *The Content of Science: A Constructivist Approach to its Teaching and Learning*, Falmer Press, London, pp. 201–220.

Scott, W.L. (1960). The significance of "Hard Bodies" in the history of scientific thought. *Isis* 50: 199–210.

Scriven, M. (1961). The key property of physical laws – Inaccuracy. In H. Feigl & G. Maxwell (eds.), *Current Issues in the Philosophy of Science*, Holt, Rinehart & Winston, New York, pp. 91–101.

Segre, M. (1980). The role of experiment in Galileo's physics. *Archive for History of Exact Sciences* 23: 227–252.

Segre, M. (1991). *In the Wake of Galileo*, Rutgers University Press, New Brunswick, NJ.

Sequeira, M. & Leite, L. (1991). Alternative conceptions and history of science in physics teacher education. *Science Education* 75(1): 45–56.

Settle, T.B. (1961). An experiment in the history of science. *Science* 133: 19–23.

Settle, T.B. (1983). Galileo and early experiment. In R. Aris et al. (eds.), *Springs of Scientific Creativity*, University of Minnesota Press, Minneapolis, MN, pp. 3–20.

Shamos, M.H. (ed.) (1959). *Great Experiments in Physics. First Hand Accounts from Galileo to Einstein*, Holt, Rinehart & Winston, New York.

Shamos, M.H. (1995). *The Myth of Scientific Literacy*, Rutgers University Press, New Brunswick.

Shapere, D. (1974). *Galileo: A Philosophical Study*, University of Chicago Press, Chicago.

Shapere, D. (1977). What can the theory of knowledge learn from the history of knowledge? *The Monist* LX(4): 488–508. Reproduced in his *Reason and the Search for Knowledge*, Reidel, Dordrecht, The Netherlands, pp. 182–202.

Shapin, S. (1994). *A Social History of Truth*, Chicago University Press, Chicago.

Sharratt, M. (1994). *Galileo. Decisive Innovator*, Cambridge University Press, Cambridge.

Shea, W.R. (1970). Galileo's claim to fame: The proof that the earth moves from the evidence of the tides. *British Journal for the History of Science* 5: 111–127.

Shiland, T.W. (1998). The atheoretical nature of the national science education standards. *Science Education* 82(5): 615–617.

Shulman, L.S. (1986). Those who understand: Knowledge growth in teaching. *Educational Researcher* 15(2): 4-14.

Siegel, H. (1997). Science education: Multicultural and universal. *Interchange* 28 (2–3): 97–108.

Siegler, R., & Richards, D.D. (1979). Development of time, speed and distance concepts. *Developmental Psychology* 15: 288–296.

Simon, B. (1981). Why no pedagogy in England? In B. Simon & W. Taylor (eds.), *Education in the Eighties: The Central Issues*, Batsford Academic & Educational, London, pp. 124–145.

Singer, C.J., Price, D.J., & Taylor, E.G.R. (1957). Cartography, survey and navigation to 1400. In C.J. Singer et al. (eds.), *A History of Technology, vol. 3*, Oxford, pp. 501–529.

Sivan, N. (1984). Why the scientific revolution did not take place in China—or Didn't It?. In E. Mendelshohn (ed.), *Transformation and Tradition in the Sciences*, Cambridge University Press, Cambridge.

Slakey, T.J. (1961). Aristotle on sense perception. *Philosophical Review* 70: 470–484.

Sleeswyk, A. and Hulden, B. (1990). The three waterclocks described by Vitruvius. *History and Technology* 8: 25–50.

Slezak, P. (1994a). Sociology of science and science education: Part I: *Science & Education* 3(3): 265–294.

Slezak, P. (1994b). Sociology of science and science education: Part II: *Science & Education* 3(4): 329–356.

Slezak, P. (1996). Review of Paul Davies, *The Mind of God. Science & Education* 5(2): 201–212.

Smith, D.E. (1923). *History of Mathematics*, Ginn & Co., Boston.

Smith, L.D. (1986). *Behaviorism and Logical Positivism*, Stanford University Press, Stanford.

Smith, V.E. (1965). *Science and Philosophy*, Bruce Publishing Company, Milwaukee.

Sneed, J.D. (1979). *The Logical Structure of Mathematical Physics*, 2nd Edition, Reidel, Dordrecht.

Sobel, D. (1995). *Longitude: The True Story of a Lone Genius Who Solved the Greatest Scientific Problem of His Time*, Walker & Co., New York.

Solomon, J. (1983). *Science in a Social Context (SISCON)*, Basil Blackwell, Oxford.

Solomon, J. (1994). The rise and fall of constructivism. *Studies in Science Education* 23: 1–19.

Solomon, J. (1991). Teaching about the nature of science in the British National Curriculum. *Science Education* 75(1): 95–104.

Solomon, J. & Aikenhead, G.S. (eds.) (1994). *STS Education: International Perspectives on Reform*, Teachers College Press, New York.

Sorabji, R. (ed.) (1987). *Philoponus and the Rejection of Aristotelian Science*, Cornell University Press, Ithaca.

Sorabji, R. (1988). *Matter, Space and Motion. Theories in Antiquity and Their Sequel*, Cornell University Press, Ithaca.

Spence, J.D. (1984). *The Memory Palace of Matteo Ricci*, Penguin, New York.

Sprat, T. (1667). *The History of the Royal Society of London*, London.

Stayer, M.S. (ed.) (1988). *Newton's Dream*, McGill-Queen's University Press, Kingston.

Stedman, L.C. (1987). The political economy of recent educational reform reports. *Educational Theory* 37(1): 69–76.

Stein, H. (1990). On Locke, "the Great Huygenius, and the imcomparable Mr. Newton". In P. Bricker & R.I.G. Hughes (eds.), *Philosophical Perspectives on Newtonian Science*, MIT Press, Cambridge, pp. 17–47.

Stimson, A. (1998). The longitude problem: The navigator's story. In W.J.H. Andrewes (ed.), *The Quest for Longitude*, The Collection of Historical Scientific Instruments, Harvard University, Cambridge, MA, 2nd Edition, pp. 72–84.

Stinner, A. & Williams, H. (1993). Conceptual change, history, and science stories. *Interchange* 24(1–2): 87–104.

Stokes, G.G. (1851). On the motion of pendulums. *Transactions of the Cambridge Philosophical Society* IX. Republished in his *Mathematical and Physical Papers*, pps. 1–141.

Stove, D.C. (1991). *The Plato Cult and Other Philosophical Follies*, Basil Blackwell, Oxford.

Stuewer, R. (1994). History and science. In W. Misgeld et al. (eds.), *Historisch-genetisches Lernen in den Naturwissenschaften*, Deutscher Studien Verlag, pp. 41–68.

Sturmy, S. (1683). *The Art of Dialling*, London.

Suchting, W.A. (1986a). Marx and "the problem of knowledge." In his *Marx and Philosophy*, Macmillan, London, pp. 1–52.

Suchting, W.A. (1986b). *Marx and Philosophy*, Macmillan, London.

Suchting, W.A. (1992). Constructivism deconstructed. *Science & Education* 1(3): 223–254.

Suchting, W.A. (1995). The nature of scientific thought. *Science & Education* 4(1): 1–22.

Suchting, W.A. (1997). Reflections on Peter Slezak and the "Sociology of Scientific Knowledge'", *Science & Education* 6(1-2): 151-195.

Suchting, W.A. forthcoming, The historical origin of the concept of scientific law. *Science & Education*.

Summers, M.K. (1982). Philosophy of science in the science teacher education curriculum. *European Journal of Science Education* 4: 19–28.

Sund, R.B. & Trowbridge, L.W. (eds.) (1967). *Teaching Science by Inquiry*, Charles Merrill, Columbus.

Swartz, R.J. (ed.) (1965). *Perceiving, Sensing and Knowing*, Doubleday & Co., New York.

Swift, J.N. (1988). The tyranny of terminology: Biology. *The Science Teachers Bulletin* 60(2): 24–26.

Szamosi, G. (1990). Polyphonic music and classical physics: The origin of Newtonian time. *History of Science* 28: 175–191.

Tait, H. (1990). *Clocks and Watches*, British Museum Publications, London.

Taylor, E.G.R. (1956). *The Haven-Finding Art: A History of Navigation from Odysseus to Captain Cook*, Hollis & Carter, London.

Taylor, E.G.R. (1966). *The Mathematical Practitioners of Hanoverian England 1714–1840*, Cambridge.

Taylor, E.G.R. (1971). *The Haven-Finding Art*, Elsevier, New York.

Taylor, E.G.R. & Richey, M.W. (1962). *The Geometrical Seaman: A Book of Early Nautical Instruments*, Hollis & Carter, for the Institute of Navigation, London.

Taylor, L.W. (1941). *Physics, the Pioneer Science*, Houghton and Mifflin, Boston. Reprinted Dover, New York, 1959.

Taylor, P.C. (1998). Constructivism: Value added. In B.J. Fraser & K.G. Tobin (eds.) *International Handbook of Science Education*, Kluwer Academic Publishers, Dordrecht, pp. 1111–1123.

Telfer, E. (1975). Autonomy as an Educational Ideal II. In S.C. Brown (ed.), *Philosophers Discuss Education*, Macmillan, London, pp.19–35.

Thayer, H.S. & Randall, J.H. (eds.) (1953). *Newton's Philosophy of Nature: Selections from His Writings*, Hafner, New York.

Thompson, E.P. (1967). Time, work-discipline and industrial capitalism. *Past and Present* 38: 56–97.

Thomson, J.J. (ed.) (1918). *Natural Science in Education*, HMSO, London. (Known as the *Thomson Report*.)

Thrower, N.J.W. (ed.) (1984). *Sir Francis Drake and the Famous Voyage, 1577–1580. Essays Commemorating the Quadricentennial of Drake's Circumnavigation of the Earth*, University of California Press, Berkeley, CA.

Tobin, K. (ed.) (1993). *The Practice of Constructivism in Science and Mathematics Education*, AAAS Press, Washington DC.

Tobin, K. (1998). Sociocultural perspectives on the teaching and learning of science. In M. Larochelle, N. Bednarz & J. Garrision (eds.), *Constructivism and Education*, Cambridge University Press, pp. 195–212.

Tobin, K. & McRobbie, C.J. (1997). Beliefs about the nature of science and the enacted science curriculum. *Science & Education* 6(4): 355–371.

Todhunter, I. (1873/1962). *A History of the Mathematical Theories of Attraction and of the Shape of the Earth*, Dover Publications, New York.

Toulmin, S.E. & Goodfield, J. (1965)., *The Discovery of Time*, Harper & Row, New York.

Trefil, J.S. (1978). *Physics as a Liberal Art*, Pergamon Press, Oxford.

Trefil, J.S. (1996). Scientific literacy. In P.R. Gross, N. Levitt & M.W. Lewis (eds.), *The Flight from Science and Reason*, New York Academy of Sciences, New York, pp. 543–550.

Trusted, J. (1991). *Physics and Metaphysics: Theories of Space and Time*, Routledge, London.

Turner, A.J. (1985). *Water-clocks, Sand-glasses, Fire-clocks*, The Time Museum, Rockford IL.

Turner, A.J. (1985a). France, Britain and the Resolution of the Longitude Problem Problem in the 18th Century. *Vistas in Astronomy* 28 nos.1–2: 315–319.

Turner, A.J. (1989). Sundials: History and classification. *History of Science* 27: 303–318.

Turner, A.J. (1993). *Of Time and Measurement: Studies in the History of Horology and Fine Technology*, Variorum, Aldershot, UK.

Turner, A.J. (1998). In the Wake of the Act, but Mainly Before. In W.J.H. Andrewes (ed.), *The Quest for Longitude*, The Collection of Historical Scientific Instruments, Harvard University, Cambridge, MA, 2nd Edition, pp. 116–132.

United States Department of Education. (1991). *America 2000. An Educational Strategy*, U.S. Department of Education, Washington, D.C.

United States Department of Education. (1995). *Adult Literacy in America*, US Department of Education, Washington, DC.

Various Authors: 1992, *Education: Putting the Record Straight*, Network Educational Press, Stafford, UK.

Viennot, L. (1979). Spontaneous reasoning in elementary dynamics. *European Journal of Science Education* 1: 205–221.

Vitruvius (1960). *The Ten Books on Architecture*. M. Morgan (trans.) Dover, New York.

Vosnaidou, S. 7 Ortony, A. (eds.) (1989). *Similarity and Analogical Reasoning*, Cambridge University Press, Cambridge.

Vygotsky, L.S. (1934/1962). *Thought and Language*, MIT Press, Cambridge.

Vygotsky, L.S. (1978). *Mind in Society: The Development of Higher Psychological Processes*, Harvard University Press, Cambridge.

Wagenschein, M. (1962). *Die Padagogische Dimension der Physik*, Westermann, Braunschweig.

Wagner, P.A. 7 Lucas, C.J. (1977). Philosophic inquiry and the logic of elementary school science education. *Science Education* 61(4): 549–558.

Wallace, W.A. (1977). *Galileo's Early Notebooks: The Physical Questions*, University of Notre Dame Press, Notre Dame, IN.

Wallace, W.A. (1981a). Galileo and reasoning *ex suppositione*. In W.A. Wallace *Prelude to Galileo*, Reidel, Dordrecht, pp. 129–159.

Wallace, W.A. (1981b). *Prelude to Galileo: Essays on Medieval and Sixteenth-Century Sources of Galileo's Thought*, Reidel Publishing Company, Dordrecht.

Wallace, W.A. (1984). *Galileo and His Sources*, Princeton University Press, Princeton.

Wandersee, J.H. (1985). Can the history of science help science educators anticipate students' misconceptions? *Journal of Research in Science Teaching* 23(7): 581–597.

Wandersee, J.H. (1990). On the value and use of the history of science in teaching today's science: Constructing historical vignettes. In D.E. Herget (ed.), *More History and Philosophy of Science in Science Teaching*, Florida State University, Tallahassee, pp. 278–283.

Wandersee, J.H. (1992). The historicality of cognition: Implications for science education research: *Journal of Research in Science Teaching* 4(4): 423–434.

Wandersee, J.H. & Roach, L.M. (1998). Interactive historical vignettes'. In J.J. Mintzes, J.H. Wandersee, & J.D.Novak (eds.), *Teaching Science for Understanding. A Human Constructivist View*, Academic Press, San Diego, pp. 281–306.

Wartofsky, M. (1968). Metaphysics as a heuristic for science. In R.S. Cohen and M.W. Wartofsky (eds.), *Boston Studies in the Philosophy of Science* 3: 123–172. Republished in his *Models*, Reidel, Dordrecht, 1979, pp. 40–89.

Wartofsky, M.W. (1976). The relation between philosophy of science and history of science. In R.S. Cohen, P.K. Feyerabend, & M.W. Wartofsky (eds.), *Essays in Memory of Imre Lakatos*, Reidel, Dordrecht, pp. 717–738. (*Boston Studies in the Philosophy of Science* 39.) Republished in his *Models*, Reidel, Dordrecht, 1979, pp. 119–139.

Wartofsky, M.W. (1979a). The model muddle: Proposals for an immodest realism. In his *Models: Representation and Scientific Understanding*, Reidel, Dordrecht, pp.1–11.

Wartofsky, M.W. (1979b). *Models: Representation and Scientific Understanding*, Reidel, Dordrecht.

Waters, D.W. (1958). *The Art of Navigation in England in Elizabethan and Early Stuart Times*, Hollis & Carter, London.

Waters, D.W. (1967). Science and the techniques of navigation in the renaissance. In C. Singleton (ed.), *Art, Science and History in the Renaissance*, Johns Hopkins University Press, Baltimore.

Waters, D.W. (1968). *The Rutters of the Sea*, Yale University Press, New Haven, CT.

Waters, D.W. (1983). Nautical astronomy and the problem of longitude. In J.G. Burke (ed.), *The Uses of Science in the Age of Newton*, University of California Press, Berkeley, pp. 143–170.

Waugh, A.E. (1973). *Sundials: Their Theory and Construction*, Dover, New York.

Weber, R.L., White, M.W., & Manning, K.V. (1959). *College Physics*, 3rd edition, McGraw-Hill, New York.

Weisheipl, J.A. (1959). *The Development of Physical Theory in the Middle Ages*, Sheed & Ward, New York.

Weisheipl, J.A. (1965). The Principle *Omne quod movetur ab alio movetur* in medieval physics. *Isis* 56: 26–45. Also in his *Nature and Motion in the Middle Ages*, Catholic University of America Press, Washington, D.C. 1985, pp. 75–98.

Weisheipl, J.A. (1985). *Nature and Motion in the Middle Ages*, Catholic University of America Press, Washington D.C.

Welch, K.F. (1972). *Time Measurement: An Introductory History*, David & Charles, Newton Abbot.

Welch, W.W. & Walberg, H.W.A. (1972). A national experiment in curriculum evaluation. *American Educational Research Journal* 9: 373–383.

Welch, W.W. (1973). Review of the research and evaluation program of Harvard project physics. *Journal of Research in Science Teaching* 10: 365–378.

Wertsch, J.V. (1985). *Vygotsky and the Social Formation of Mind*, Harvard University Press, Cambridge.

Westaway, F.W. (1919). *Scientific Method: Its Philosophy and Practise*, London.

Westaway, F.W. (1929). *Science Teaching*, Blackie and Son, London.

Westbury, I. & Wilkof, N.J. (eds.) (1978). *Joseph J. Schwab: Science, Curriculum, and Liberal Education*, University of Chicago Press, Chicago.

Westfall, R.S. (1966). The problem of force in Galileo's physics. In C.L. Golino (ed.), *Galileo Reappraised*, University of California Press, Berkeley, pp. 67–95.

Westfall, R.S. (1971). *Force in Newton's Physics: The Science of Dynamics in the Seventeenth Century*, Macdonald, London.

Westfall, R.S. (1980). *Never at Rest: A Biography of Isaac Newton*, Cambridge University Press, Cambridge.

Westfall, R.S. (1985). Science and patronage: Galileo and the telescope. *Isis* 76: 11–30.

Westfall, R.S. (1986). Newton and the acceleration of gravity. *Archive for History of Exact Sciences* 35: 255–272.

Westfall, R.S. (1988). Newton and the scientific revolution. In M.S. Stayer, *Newton's Dream*, McGill-Queen's University Press, Kingston, pp. 4–18.

Westfall, R.S. (1990). Making a world of precision: Newton and the construction of a quantitative physics. In F. Durham & R.D. Purrington (eds.), *Some Truer Method. Reflections on the Heritage of Newton*, Columbia University Press, New York, pp. 59–87.

Westfall, R.S. (1993). *The Life of Isaac Newton*, Cambridge University Press, Cambridge.

Wheldall, K. (1999). Literacy levels still falling—You wouldn't read about it. *Sydney Morning Herald*, 4th October, p. 17.

Whitaker, R.J. (1983). Aristotle is not dead: Student understanding of trajectory motion. *American Journal Physics* 51(4): 352–357.

White, L. (1962). *Medieval Technology and Social Change*, Oxford University Press, Oxford.

White, L. (1966). Pumps and pendula: Galileo and technology. In C.L. Golino (ed.), *Galileo Reappraised*, University of California Press, Berkeley, pp. 96–110.

White, R.T. & Gunstone, R.F. (1989). Metalearning and conceptual change. *International Journal of Science Education* 11: 577–586.

Whitehead, A.N. (1925). *Science and the Modern World*, Macmillan, New York.

Whitehead, A.N. (1947). Technical education and its relation to science and literature. In his *The Aims of Education and Other Essays*, Williams & Norgate, London, pp. 43–60.

Whitrow, G.J. (1971). The laws of motion. *British Journal for the History of Science* 5(19): 217–234.

Whitrow, G.J. (1972). *What is Time?*, Thames and Hudson, London.

Whitrow, G.J. (1980). *The Natural Philosophy of Time*, Oxford University Press, Oxford.

Whitrow, G.J. (1988). *Time in History*, Oxford University Press, Oxford.

Wilkening, F. (1982). Children's knowledge about time, distance, and velocity. In W.J. Friedman (ed.), *The Developmental Psychology of Time*, Academic Press, New York.

Williams, J.E.D. (1992). *From Sails to Satellites: The Origins and Development of Navigational Science,* Oxford University Press, Oxford.

Wilson, C. (ed.) (1980). *The Book of Time*, Westbridge Books, North Pomfret, VT.

Wilson, E.O. (1978). *On Human Nature*, Bantam Books, New York.

Wisan, W.L. (1974). The new science of motion: A study of Galileo's *De Motu Locali. Archive for History of exact Sciences* 13: 103–306.

Wisan, W.L. (1978). Galileo's scientific method: A reexamination. In R.E. Butts & J.C. Pitt (eds.), *New Perspectives on Galileo*, Reidel Publishing Company, Dordrecht, pp. 1–58.

Wisan, W.L. (1981). Galileo and the emergence of a new scientific style. In J. Hintikka, D. Gruender, & E. Agazzi (eds.), *Theory Change, Ancient Axiomatics, and Galileo's Methodology*, Vol. I, Reidel, Dordrecht, pp. 311–339.

Wise, M.N. (ed.) (1995). *The Values of Precision*, Princeton University Press, Princeton.

Wittgenstein, L. (1958). *Philosophical Investigations*, Basil Blackwell, Oxford.

Wolf, A. (1950). *A History of Science, Technology, and Philosophy in the 16th and 17th Centuries*, George Allen & Unwin, London.

Wolfe, C.J.E. (ed.) (1889). *Mémoires sur le pendule*, 2 vols., Gauthier-Villars, Paris.

Wolff, M. (1987). Philoponus and the rise of preclassical dynamics. In R. Sorabji (ed.), *Philoponus and the Rejection of Aristotelian Science*, Cornell University Press, Ithaca, pp. 84–120.

Wolpert, L. (1992). *The Unnatural Nature of Science*, Faber & Faber, London.

Woolnough, B.E. (ed.) (1991). *Practical Science*, Open University Press, Milton Keynes.

Wright, M. (1989). Robert Hooke's longitude timekeeper. In M. Hunter & S. Schaffer (eds.), *Robert Hooke: New Studies,* D.S. Brewer, Woodbridge, England, pp. 63–118.

Yager, R.E. (ed.) (1993). *The Science, Technology, Society Movement*, National Science Teachers Association, Washington D.C.

Yeany, R.H. (1991). A unifying theme in science education? *NARST News* 33(2): 1–3.

Yoder, J.G. (1988). *Unrolling Time: Christiaan Huygens and the Mathematization of Nature*, Cambridge University Press, Cambridge.

Yoder, J.G. (1991). Christian Huygens Great Treasure. *Tractrix* 3: 1–13.

Zeidler, D.L. (1997). The central role of fallacious thinking in science education. *Science Education* 81(4): 483–496.

Zerubavel, E. (1982). The standardization of time: A sociohistorical perspective. *American Journal of Sociology* 88(1): 1–23.

Zilsel, E. (1942). The genesis of the concept of scientific law. *The Philosophical Review* 51: 245–267.

Zubrowski, B. (1988). *Clocks: Building and Experimenting with Timepieces*, Morrow Junior Books, New York.

Zukav, G. (1979). *The Dancing Wu Li Masters*, Fontana, London.

Credits

Grateful acknowledgement is made to the following individuals, companies, and institutions for their permission to reproduce these diagrams, drawings, and portraits.

Figure 3. Eratosthenes' observations at Syene and Alexandria (Hogben, 1940, p. 83), George, Allen & Unwin, London.

Figure 4. The calculation of Eratosthenes (Hogben, 1940, p. 83), George Allen & Unwin, London.

Figure 5. Latitude ϕ (Williams, 1992, p. 11), Oxford University Press, Oxford.

Figure 6. Longitude λ (Williams, 1992, p. 11), Oxford University Press, Oxford.

Figure 7. Latitude by observation of pole star (Waters, 1958, p. 49), Hollis & Carter, London.

Figure 8. Latitude by observation of sun (Waters, 1958, p. 51), Hollis & Carter, London.

Figure 9. Latitude by noon sight. (a) Summer in northern hemisphere; (b) winter in southern hemisphere (Waters, 1958, p. 58), Hollis & Carter, London.

Figure 10. Portuguese exploration of the African coast (Williams, 1992, p.7), Oxford University Press, Oxford.

Figure 11. Using Jacob's Staff (Mills, 1983, p. 137), Aurum Press, London.

Figure 12. Quadrant, with pinholes for sighting the pole star and a plumb-line which crossed a graduated scale giving the star's altitude (Goodman and Russell, 1991, p. 119), Hodder & Straughton, London.

Figure 13. Portuguese caravel (Mendelssohn, 1977, p. 27), Thames and Hudson, London.

Figure 14. 1508 Portuguese map of Africa (Mendelsshohn, 1977, p. 36), Thames and Hudson, London.

Figure 17. 1681 Survey of France (Paris as Prime Meridian), (Howse, 1980, p. 17), The British Library.

Figure 18. Title page of Frisius' Cosmographicae (Pogo, 1935, p. 474), U.S. History of Science Society.

Figure 19. Without time longitude is indeterminate (Williams, 1992, p. 85), Oxford University Press, Oxford.

Figure 20. Time and longitude determination (Hogben, 1936, p. 169), George, Allen & Unwin, London.

Figure 21. Hogarth's Longitude Lunatic, (Howse 1980, p. 55), Oxford University Press, Oxford.

Figure 22. Annual movement of noon shadow in Cairo (lat. 30°) (Hogben, 1936, p. 41), George, Allen & Unwin, London.

Figure 23. Board with vertical rod for finding meridan and true solar noon (Milham, 1945, p. 32), Macmillan, London.

Figure 24. Modern sundial (Milham, 1945, p. 33), Macmillan, London.

Figure 25. Relation of sundial to sun (northern hemisphere) (Milham, 1945, p. 34), Macmillan, London.

Figure 26. Latitude by noon altitude at the equinox (Williams, 1992, p. 10), Oxford University Press, Oxford.

Figure 27. 10th Century Moorish sundial, latitude L degrees (Hogben, 1936, p. 379), George, Allen & Unwin, London.

Figure 28. Ctesibius's water clock, 250 BC (Bond, 1948, p. 59), Collier & Son, New York.

Figure 29. An ancient clepsydra (Milham, 1945, p. 49), Macmillan, London.

Figure 30. Foliot controlling power from a falling weight (Bell and Bell, 1963, p. 57), Viking Press, New York.

Figure 31. The Strasbourg cathedral clock (woodcut, 1574) (Hoskin, 1997, p. 138), Cambridge University Press, Cambridge.

Figure 32. de Vick's clock (Brearley, 1919, p. 81), Doubleday, Page & Co., New York.

Figure 33. Controlling mechanism of de Vick's clock (Milham, 1945, p. 85), Macmillan, London.

Figure 34. German city in 1497, showing Tower Clock (Rossum, 1996, p. 149), Chicago University Press, Chicago.

Figure 35. Pictorial reconstruction of Su Sung's heavenly clockwork (Needham, Ling, and Solla Price, 1962, p. i), Cambridge University Press, Cambridge.

Figure 36. Su Sung's escapement mechanism (Jespersen and Fitz-Randolph, 1982, p. 26), Dover Publications, New York.

Figure 37. Leonardo da Vinci's sketches for a mechanical escapement with foliot (Edwardes, 1977, p. 227), J. Sherratt, Altrincham.

Figure 38. Leonardo's pendulum drawing (Bedini and Reti, 1974, p. 262), McGraw-Hill Book Company, New York.

Figure 39. Bar pulsilogium (Wolf, 1950, p. 433), George Allen & Unwin, London.

Figure 40. Circular pulsilogium (Wolf, 1950, p. 433), George Allen & Unwin, London.

Figure 41. Galileo's original drawing for pendulum clock (Bell and Bell, 1963, p. 58), Viking Press, New York.

Figure 42. Galileo's escapement mechanism (Bell and Bell, 1963, p. 58), Viking Press, New York.

Figure 43. Galileo's 1590 inclined plane construction (Galileo, 1590/1960, p. 63), University of Wisconsin Press, Madison.

Figure 44. Galileo's 1590 composite construction (Galileo, 1590/1960, p. 65), University of Wisconsin Press, Madison.

Figure 45. Galileo's 1600 construction (Galileo, 1600/1961, p. 316), University of Wisconsin Press, Madison.

Figure 46. Guidobaldo del Monte (Ronan, 1974, p. 77), G.P. Putnam's Sons, New York.

Figure 47. Ball in rim (Drake, 1978, p. 71), University of Chicago Press, Chicago.

Figure 48. Chords and planes construction (Drake, 1978, p. 71), University of Chicago Press, Chicago.

Figure 49. Ball in hoop (Galileo, 1633/1953, p. 451), University of California Press, Berkeley.

Figure 50. Galileo's brachistochrone proof (Galileo, 1638/1954, p. 237), Dover Publications, New York

Figure 51. Galileo's brachistochrone proof (cont.) (Galileo, 1638/1954, p. 237), Dover Publications, New York

Figure 52. Mersenne's $T \propto \sqrt{l}$ proof (Crombie, 1994, p. 869), Duckworth, London.

Figure 53. Mersenne's gravitational constant experiment (Westfall, 1990, p. 71), Columbia University Press, New York.

Figure 54. Christiann Huygens (Gould, 1923, p. 26), J.D. Potter, London.

Figure 55. Cycloid curve generated by point on moving circle (Huygens, 1673/1986, p. 50), Iowa State University Press, Ames.

Figure 56. Cycloid curve with defining feature being parallelism of tangent at B to chord on generating circle (Yoder, 1988, p. 60), Cambridge University Press, Cambridge.

Figure 57. Huygens' 1659 geometric derivation of cycloidal osochronism (Yoder, 1988, p. 51), Cambridge University Press, Cambridge.

Figure 58. Huygens' isochrony proof (Huygens, 1673/1986, pp. 69), Iowa State University Press, Ames.

Figure 59. Contemporary demonstration of cycloidal isochrony (Dugas, 1988, p. 186), Dover, New York.

Figure 60. Cycloid as brachistochrone (Gjertsen, 1986, p. 146), Penguin, Harmondsworth.

Figure 61. Huygens' pendulum clock (Huygens, 1673/1986, p. 14), Iowa State University Press, Ames.

Figure 62. Huygens cycloidal suspension (Hogben, 1940, p. 280), George Allen & Unwin, London.

Figure 85. Collision on the vertical plane between unequal mass and amplitude pendulums (French, 1965, p. 440), W.W. Norton & Co., New York.

Figure 86. Newton's collision experiment (Newton, 1729/1934, p. 22), University of California Press, Berkeley.

Figure 87. Leibniz's thought experiment (Leibniz, 1686/1989), Hackett Publishing Company, Indianapolis.

Figure 88. Cavendish's pendulum illustration (Elkana, 1974b, p. 45), Hutchinson Educational, London.

Figure 89. Cradle pendulum (Freier, 1965, p. 45), Appleton-Century-Crofts, New York.

Figure 90. Robin's 1742 ballistic pendulum (Taylor, 1941, p. 205), Houghton Mifflin Co., Boston.

Figure 91. Foucault's intuition (PSSC, 1960, p. 340), D.C. Heath & Co., Boston.

Figure 93. Moving pendulum (Mathews, 1994a), Routledge, New York.

Figure 94. Acceleration of Bob (Mathews, 1994a). Routledge, New York.

Figure 95. Apparent annual movement of the sun (northern hemisphere) (Holton, 1985, p. 5), Princeton University Press, Princeton.

Figure 96. Variation in length of solar day (Jespersen and Fitz-Randolph, 1982, p. 61), Dover Publications, New York.

Figure 97. Difference between sidereal and solar days (Jespersen and Fitz-Randolph, 1982, p. 61), Dover Publications, New York.

Figure 98. Precession of the Equinoxes (O'Neil 1975, p. 21), University of Sydney Press, Sydney.

Figure 99. A simple pendulum (Weber, White, and Manning 1959, p. 137), McGraw-Hill, New York.

Figure 100. Pendulum and potential energy (Maxwell, 1877, p. 97), Dover Publications, New York.

Figure 101. Pendulum and inclined plane (Hogben, 1940, p. 286), George, Allen & Unwin, London.

Figure 102. Modern construction of Galileo's chord law (Pentz, 1970, p. 50), Open University Press, Bletchley, Bucks.

Figure 104. Junior school escapement (NSRC, 1994, p. 132), Carolina Biological Supply Company, Burlington, NC.

Figure 105. Clockface added to escapement (NSRC, 1994, p. 142), Carolina Biological Supply Company, Burlington, NC.

AUTHOR INDEX

425

SUBJECT INDEX